Generalized Additive Models for Location, Scale and Shape

An emerging field in statistics, distributional regression facilitates the modeling of the complete conditional distribution, rather than just the mean. This book introduces generalized additive models for location, scale and shape (GAMLSS) – one of the most important classes of distributional regression. Taking a broad perspective, the authors consider penalized likelihood inference, Bayesian inference, and boosting as potential ways of estimating models, and illustrate their usage in complex applications.

Written by the international team who developed GAMLSS, the text's focus on practical questions and problems sets it apart. Case studies demonstrate how researchers in statistics and other data-rich disciplines can use the model in their work, exploring examples ranging from fetal ultrasounds to social media performance metrics. The R code and datasets for the case studies are available on the book's companion website, allowing for replication and further study.

MIKIS D. STASINOPOULOS is Professor of Statistics at the School of Computing and Mathematical Sciences, University of Greenwich. He is, together with Professor Bob Rigby, coauthor of the original Royal Statistical Society article on GAMLSS. He has also coauthored three books on distributional regression, and in particular the theoretical and computational aspects of the GAMLSS framework.

THOMAS KNEIB is a Professor of Statistics at the University of Göttingen, where he is the Spokesperson of the interdisciplinary Centre for Statistics and Vice-Spokesperson of the Campus Institute Data Science. His main research interests include semiparametric regression, spatial statistics, and distributional regression.

NADJA KLEIN is Emmy Noether Research Group Leader in Statistics and Data Science and Professor for Uncertainty Quantification and Statistical Learning at TU Dortmund University and the Research Center Trustworthy Data Science and Security. Nadja is member of the Junge Akademie and associate editor for *Biometrics*, *JABES*, and *Dependence Modeling*. Her research interests include Bayesian methods, statistical and machine learning, and spatial statistics.

ANDREAS MAYR is a Professor at the Institute for Medical Biometry, Informatics, and Epidemiology at the University of Bonn. He has authored more than 100 research articles both in statistics as well as medical research and is currently Editor of the *Statistical Modelling Journal*, Associate Editor of the *International Journal of Biostatistics*, and Editorial Board Member of the *International Journal of Eating Disorders*.

GILLIAN Z. HELLER is Professor of Biostatistics at the NHMRC Clinical Trials Centre, University of Sydney. She has coauthored four books in the regression modeling area, the first directed towards actuarial applications of the generalized linear model, and the remaining three focusing on distributional regression, in particular the GAMLSS framework.

CAMBRIDGE SERIES IN STATISTICAL AND PROBABILISTIC MATHEMATICS

This series of high-quality upper-division textbooks and expository monographs covers all aspects of stochastic applicable mathematics. The topics range from pure and applied statistics to probability theory, operations research, optimization, and mathematical programming. The books contain clear presentations of new developments in the field and also of the state of the art in classical methods. While emphasizing rigorous treatment of theoretical methods, the books also contain applications and discussions of new techniques made possible by advances in computational practice.

A complete list of books in the series can be found at www.cambridge.org/statistics. Recent titles include the following:

Generalized Additive Models for Location, Scale and Shape

A Distributional Regression Approach, with Applications

Mikis D. Stasinopoulos
University of Greenwich

Thomas Kneib
University of Göttingen

Nadja Klein
TU Dortmund University

Andreas Mayr
University of Bonn

Gillian Z. Heller
University of Sydney

CAMBRIDGE
UNIVERSITY PRESS

Shaftesbury Road, Cambridge CB2 8EA, United Kingdom

One Liberty Plaza, 20th Floor, New York, NY 10006, USA

477 Williamstown Road, Port Melbourne, VIC 3207, Australia

314–321, 3rd Floor, Plot 3, Splendor Forum, Jasola District Centre, New Delhi – 110025, India

103 Penang Road, #05–06/07, Visioncrest Commercial, Singapore 238467

Cambridge University Press is part of Cambridge University Press & Assessment,
a department of the University of Cambridge.

We share the University's mission to contribute to society through the pursuit of
education, learning and research at the highest international levels of excellence.

www.cambridge.org
Information on this title: www.cambridge.org/9781009410069

DOI: 10.1017/9781009410076

First published 2024
First hardback edition 2024

A catalogue record for this publication is available from the British Library

Library of Congress Cataloging-in-Publication data
Names: Stasinopoulos, Mikis D., author. | Kneib, Thomas, author. | Klein, Nadja, 1987- author. |
Mayr, Andreas, 1983- author. | Heller, Gillian Z., author.
Title: Generalized additive models for location, scale and shape : a distributional regression approach,
with applications / Mikis D. Stasinopoulos, Thomas Kneib, Nadja Klein, Andreas Mayr,
Gillian Z. Heller.
Description: First edition. | Cambridge ; New York, NY : Cambridge University Press, 2024. | Series:
Cambridge series in statistical and probabilistic mathematics 56 | Includes bibliographical
references and index.
Identifiers: LCCN 2023035047 (print) | LCCN 2023035048 (ebook) | ISBN 9781009410069 (hardback)
| ISBN 9781009410076 (epub)
Subjects: LCSH: Regression analysis–Mathematical models. | Theory of distributions (Functional
analysis) Classification: LCC QA278.2 .S674 2024 (print) | LCC QA278.2 (ebook) |
DDC 519.5/36–dc23/eng/20231016
LC record available at https://lccn.loc.gov/2023035047
LC ebook record available at https://lccn.loc.gov/2023035048

ISBN   978-1-009-41006-9   Hardback

During the course of writing this book, we lost two people close to us.

Our dear friend and colleague
Professor Brian D. Marx (1960–2021)

and

María Belén Avila (1983–2020)

We dedicate this book to their memories.

Contents

Preface

What this Book is About

This book is devoted to a special class of regression models that takes a *distributional* perspective on regression modeling. Instead of focusing on the expectation of the response variable, as most classical regression approaches such as linear models, generalized linear models, and generalized additive models do, we deal with regression models that relate more general features of the response distribution to covariates. This yields characterizations of, for example, location, scale, and shape of the response distribution conditional on covariate information. Since the models also involve flexible forms of regression modeling based on a variety of covariate types, including nonlinear effects of continuous covariates, spatial effects, and random effects, the model class is called *generalized additive models for location, scale and shape (GAMLSS)*.

More precisely, GAMLSS builds upon the classical framework of generalized linear models but

- relaxes the assumption that the response distribution belongs to the exponential family such that any parametric distribution can be assumed for the response, and

- specifies separate regression predictors for any parameter of this distribution, in particular parameters determining scale and shape features.

The assumption of a parametric response distribution implies the immediate availability of the likelihood such that inferential procedures based on the likelihood can be used for inference. In this book, we will consider (penalized) maximum likelihood inference, Bayesian inference, and functional gradient descent boosting as specific algorithmic approaches for implementing inference in GAMLSS.

The main contribution of the GAMLSS framework is that it challenges the classical notion of modeling only the mean of the distribution of the response variable. Rather, all parameters (including location, scale, and shape) of that distribution may be modeled. This allows the analyst to approach a much wider range of problems, including heterogeneity in variance, positive or negative skewness, platykurtic or leptokurtic response distributions, heavy tails and extremes, overdispersion and underdispersion in count data, excess or shortage of certain parts of the support (e.g. zero-inflated count data), and multivariate responses. Importantly, the aspects *beyond the mean* are not treated as a nuisance but rather can be central to the analysis of interest. There are other approaches to distributional regression such as quantile regression or conditional transformation models that will not be considered here, but we provide a brief comparison in Section 1.6.

Types of Applications and Questions that Can be Approached with GAMLSS

GAMLSS can be useful in any area that relies on regression modeling for the generation of insights from observed data. This includes *exploratory* as well as *confirmatory* analyses (depending on the research question, data collection strategy, and exact model specification). Moreover, the goals of the analysis may range from *interpreting* the resulting model coefficients to pure *prediction-oriented* applications. For the latter, GAMLSS have the specific advantage of providing covariate-dependent predictive distributions rather than only a point prediction.

We illustrate the applicability of GAMLSS in various kinds of research questions in case studies in Part III. More precisely, we consider the following applications:

- fetal ultrasound (Chapter 8) where we are considering the prediction of birthweight from a number of covariates derived from ultrasound measurements to illustrate basic components and steps of GAMLSS analyses;

- speech intelligibility testing (Chapter 9) where a mixed discrete–bounded continuous response is considered in combination with random effects to adjust for a repeated measurements design;

- social media post performance (Chapter 10) to illustrate the consideration of overdispersed count data regression involving cyclic P-splines and a comparison of different inferential approaches for prediction;

- childhood undernutrition in India (Chapter 11) as a case study on the Bayesian approach to GAMLSS and its advantages including the extension to bivariate responses and the consideration of spatial effects;

- federal election outcomes in Germany (Chapter 12) as another case of Bayesian inference, dealing with multivariate fractional responses as well as nonlinear and spatial effects;

- riboflavin production (Chapter 13) as a case study on high-dimensional regression with variable selection via boosting.

Goal of this Book

With this book, we provide a comprehensive introduction to the concepts of GAMLSS that serves applied scientists interested in conducting data analyses with GAMLSS and statisticians interested in the statistical foundations and background of GAMLSS. Rather than focusing on one specific way of conducting statistical inference, we aim at a unified treatment of different inferential approaches, including applications high-lighting their specific advantages and disadvantages. For this, we draw on the ample experience of the authors in developing and applying penalized likelihood approaches, Bayesian inference, and functional gradient descent boosting. While the book includes a solid introduction to the foundations of the statistical inference approaches, our emphasis is always on the practical applications rather than a theoretical dogma. This is supported by devoting a considerable part of the book to hands-on examples in terms of complex applications that illustrate the dos and don'ts of GAMLSS modeling.

Readership of the Book and Assumed Background Knowledge

This book is written for:

- practitioners and applied researchers interested in understanding modern regression approaches and applying them to their own datasets,

- applied statisticians aiming at a better understanding of the relevance and application of GAMLSS as well as the different inferential approaches for GAMLSS,

- data analysts who are interested in models emphasizing interpretability over pure prediction power, and

- students in statistics and data science who wish to go beyond basic forms of regression modeling.

We assume that readers are familiar with the basic concepts of standard regression analysis (including the corresponding matrix algebra) as presented in, for example, Fahrmeir et al. (2021). All computations underlying the examples presented in the book have been conducted in the **R** environment for statistical computing. **R** commands are not provided in this text, but are available in the online supplementary material of the book published on `https://gamlssbook.bitbucket.io`.

Structure of the Book and Additional Resources

The book consists of three distinct parts.

- In Part I (Chapters 1 to 3), we introduce the basic concepts and history of GAMLSS and review some of the relevant ingredients for setting up a GAMLSS model, that is, different types of response distributions, and the various regression effects that can appear in the predictor specifications.

- Part II (Chapters 4 to 7) explains the theoretical background of the different inferential procedures that can be used to estimate GAMLSS models. This part comprises dedicated chapters on penalized likelihood inference, Bayesian inference, and functional gradient descent boosting.

- In Part III (Chapters 8 to 13), we present several practical examples and case studies to illustrate the different inferential approaches as well as various aspects relevant in GAMLSS modeling.

Depending on the reader's background and interest, there are different ways of reading the book. While Chapter 1 serves as a good starting point for any reader, it is not necessary to work through the rest of the book sequentially. Rather, it is possible to pick the inferential approach that appears most relevant, first, or to start with reading some of the case studies.

This book is the third in a series of texts on GAMLSS. The first, *Flexible Regression and Smoothing: Using GAMLSS in R* (Stasinopoulos et al., 2017), concerns the implementation of GAMLSS in the suite of **R** packages **gamlss**, **gamlss.dist**, **gamlss.add**, etc. The second, *Distributions for Modeling Location, Scale, and Shape: Using GAMLSS in R* (Rigby et al., 2019), describes the more than 100 continuous, discrete, and mixed distributions available in the package **gamlss.dist**, and also shows how more distributions can be generated by transformation, truncation, censoring, and zero-inflation. In contrast, this book is less associated with a specific group of **R** packages but serves as a general introduction to GAMLSS. The first two books are therefore not assumed knowledge, but they may nonetheless be useful as resources for distributions and details on the usage of the penalized likelihood **R** packages from the **gamlss** suite.

Additional resources for this book have been collected on the website `https://gamlssbook.bitbucket.io`. These comprise code and datasets for the case studies presented in the book, further supporting material and a list of errata.

Acknowledgments

We are indebted to Professor Bob Rigby, who was unable to participate in the authorship of this book, for his substantial intellectual contribution and support. We thank Stephan Klasen (1966–2020) for joint research on childhood malnutrition and distributional modeling thereof that provided the foundation for the application in Chapter 11; Stefan Lang and Nikolaus Umlauf, with whom we have collaborated on Bayesian inference for GAMLSS and corresponding statistical software; Marah-Lisanne Thormann for assistance with graphics in Chapters 3 and 6; Benjamin Hofner, Nora Fenske, and Matthias Schmid for joint research on boosting GAMLSS; Janek Thomas and Tobias Hepp for further enhancements; Jost von Petersdorff-Campen and Maarten Jung for assistance with code and graphics in Chapters 11 and 12; Annika Strömer for her help with the genetic application in Chapter 13; Tobias Wistuba for his help with graphics in Chapter 7; Stanislaus Stadlmann for his help with graphics in Chapter 10; Wenli Hu and Brett Swanson for joint research on models for speech intelligibility that provided the basis for Chapter 9; Fernanda de Bastiani for her contribution to the bucket plots; and the production team at Cambridge University Press for their great support throughout the production process of this book. Gillian Heller and Nadja Klein thank their previous employers, Macquarie University and Humboldt-Universität zu Berlin, respectively, for their support during the book-writing period. Finally, many of our productive collaborations were initiated and fostered at the annual meetings of the Statistical Modelling Society, the International Workshop on Statistical Modelling. To the IWSM community: Thank you for the very supportive research environment, and many useful discussions.

Notation and Terminology

Packages and code	
GAMLSS	the statistical model
gamlss	the **R** package
`gamlss()`	the **R** function
General mathematical notation	
$X^\top$	vector or matrix transpose
$f'(x)$	derivative of a function f
$\mathbb{1}\,(y = x)$	indicator function
$\mathbf{1}_n$	$(1, \ldots, 1)^\top$
$\boldsymbol{I}_n$	$n \times n$ identity matrix
Distributions	
y	a univariate response variable, or a single realization of the response variable
$\boldsymbol{y}$	the vector of observed values of the response variable y, namely $\boldsymbol{y} = (y_1, y_2, \ldots, y_n)^\top$
$\boldsymbol{x}_i$	covariates (features)
n	sample size
$\mathbb{P}(\cdot)$	probability
$f(y)$	probability function; probability mass function for discrete random variables and probability density function for continuous random variables; occasionally, for clarity, $f_y(y)$
$F(y)$	cumulative distribution function (cdf)
y_p	$F^{-1}(p)$, inverse cdf (quantile function)
$\mathcal{D}(\cdot)$	generic distribution
$\mathcal{E}(\cdot)$	exponential family distribution
$\mathcal{S}$	support of y
$\mathcal{N}(\mu, \sigma^2)$	(univariate) normal distribution with mean μ and variance σ^2
$\phi(\cdot)$	probability density function of $\mathcal{N}(0, 1)$ (the standard normal)
$\Phi(\cdot)$	cumulative distribution function of the standard normal
$\mathcal{N}_k(\boldsymbol{\mu}, \Sigma)$	k-dimensional normal distribution with mean vector $\boldsymbol{\mu}$ and variance–covariance matrix Σ
$\mathcal{U}(a, b)$	uniform distribution on the interval (a, b)

Intervals

$\mathbb{R}$	real line
$\mathbb{R}_+$	positive real line
$\mathbb{R}_{(0,1)}$	unit interval $(0,1)$
$\mathbb{N}$	nonnegative integers $(0,1,2,\ldots)$
$\mathbb{N}_+$	positive integers $(1,2,\ldots)$

Distribution parameters

K	number of distribution parameters
θ_k	kth distribution parameter, $k = 1,\ldots,K$
$\boldsymbol{\theta}$	vector of distribution parameters $(\theta_1,\ldots,\theta_K)^\top$ in GAMLSS
θ_1	first distribution parameter, sometimes denoted as μ
θ_2	second distribution parameter, sometimes denoted as σ
θ_3	third distribution parameter, sometimes denoted as ν
θ_4	fourth distribution parameter, sometimes denoted as τ
λ	a hyperparameter
$\boldsymbol{\lambda}$	vector of all hyperparameters in the model
$\boldsymbol{\vartheta}$	vector of all parameters in the model

Systematic part of the GAMLSS model

$\boldsymbol{X}_k$	$n \times p_k$ fixed effects design matrix for θ_k
p_k	number of columns in the design matrix $\boldsymbol{X}_k$ for the kth parameter
J_k	total number of smoothers for θ_k
q_{kj}	dimension of the random effect vector $\boldsymbol{\gamma}_{kj}$
$\boldsymbol{x}_{kj}$	jth explanatory variable vector for θ_k
$\boldsymbol{\beta}_k$	vector of fixed effect coefficients of length J_K
$\boldsymbol{z}_{kj}$	jth random effects explanatory variable vector
$\boldsymbol{Z}_{kj}$	$n \times q_{kj}$ random effect design matrix
$\boldsymbol{\alpha}_{kj}$	jth random effect coefficient vector of length q_{kj}
$\boldsymbol{B}_{jk}$	$n \times L_{jk}$ generic design matrix for θ_k
$\boldsymbol{\gamma}_k$	vector of basis coefficients for this effect
$\boldsymbol{\eta}_k$	predictor for θ_k
$g_k(\cdot)$	link function for θ_k
$s_{kj}(\boldsymbol{x}_{kj})$	jth nonparametric or nonlinear function in $\boldsymbol{\eta}_k$
$\boldsymbol{W}$	$n \times n$ diagonal matrix of weights
$\boldsymbol{w}$	n-dimensional vector of weights (the diagonal elements of $\boldsymbol{W}$)
$\boldsymbol{K}_{kj}$	jth smoothing or penalty matrix for θ_k

Likelihood and information criteria

L	likelihood function
ℓ	log-likelihood function
$\mathcal{I}$	Fisher's expected information matrix
$\mathcal{H}$	observed information matrix
df	total effective degrees of freedom used in the model
κ	penalty for each degree of freedom used in the model
GDEV	global deviance = minus twice the fitted log-likelihood

AIC	Akaike information criterion $= \text{GDEV} + 2 \times \text{df}$
BIC	Bayesian information criterion $= \text{GDEV} + \log(n) \times \text{df}$
GAIC	generalized AIC $= \text{GDEV} + \kappa \times \text{df}$

Residuals

$\boldsymbol{u}$	vector of (randomized) quantile residuals
$\boldsymbol{r}$	vector of normalized (randomized) quantile residuals
$\boldsymbol{\varepsilon}$	vector of (partial) residuals

Moment measures and functions

$\mathbb{E}(y)$	expected value (or mean) of random variable y
$\mathbb{V}(y)$	variance of random variable y
$\text{Cov}(\boldsymbol{y})$	variance–covariance matrix of the random vector $\boldsymbol{y}$
μ_y	location parameter (usually mean)
$\tilde{\mu}_k$	$\mathbb{E}(y^k)$, kth population moment about zero
μ_k	$\mathbb{E}\{[y - \mathbb{E}(y)]^k\}$, kth population central moment
γ_1	moment skewness $= \mu_3/\mu_2^{1.5}$
β_2	moment kurtosis $= \mu_4/\mu_2^2$
γ_2	moment excess kurtosis $= \beta_2 - 3$

Boosting notation

$\rho(y_i, \eta(x_i))$	loss function evaluated at observation i
$h(\cdot)$	base-learner
$h_{jk}^{[m]}(\cdot)$	base-learner at boosting iteration m for explanatory variable j in parameter θ_k

Part I

Introduction and Basics

1

Distributional Regression Models

This chapter serves as a general introduction to the types of regression models discussed in this book and has the following four major aims.

- Provide the context for the GAMLSS way of distributional regression modeling.

- Give a brief history of the development of GAMLSS and its relation to other regression approaches.

- Introduce a first example illustrating some advantages and potential of GAMLSS analyses.

- Compare GAMLSS with, and delineate from, competing distributional regression approaches.

We start with setting the scene in Section 1.1, where we describe how our presentation fits into the data analysis circle for working with research data. Afterwards, we introduce basic terminology for and ingredients of statistical regression models (Sections 1.2 and 1.3) and briefly summarize the historical developments that led to the introduction of GAMLSS (Section 1.4). We then introduce the structure of GAMLSS in a more general and formal way (Section 1.5) and close the chapter with a discussion of alternative distributional regression approaches (Section 1.6).

1.1 The Data Analysis Circle

The statistical analysis of research data can be seen as a circular process consisting of the following phases (see Figure 1.1):

(i) Planning the data collection: In the first phase, the process of collecting the data has to be decided upon. The exact strategy strongly depends on the research question of interest as well as the purpose of the analysis, and may involve aspects such as sample size calculation, defining the relevant study population, choosing an experimental design, defining the sampling process, investigating sources for already available data, etc. In addition, this step may feature developing a fixed data analysis plan (for confirmatory analysis, e.g. in clinical studies), pre-registration of the trial, and other aspects that determine steps taken later on in the analysis of the data.

3

Figure 1.1 The data analysis circle.

(ii) Collecting the data: The second phase implements data collection, which may range from generating data in experimental settings, to collecting observational data or scraping data from existing sources.

(iii) Processing the data: To turn the raw data collected in the second phase into a dataset to be analyzed, it is of utmost importance to pre-process the data. This includes diverse aspects such as checking for inconsistencies, treating missing values, implementing transformations, calculating indices, graphical visualizations, etc.

(iv) Analyzing the data: The fourth phase implements the actual analysis. Throughout this book, this analysis will be based on statistical modeling, but of course very different approaches may be taken depending on the purpose of the analysis. Graphical tools are of particular relevance here to check the validity of model decisions and assumptions.

(v) Interpreting the results: After having analyzed the data, interpreting the results with respect to the original research question is crucial. This includes deriving required quantities from the raw model results, communication with subject matter scientists, and also publishing the results in appropriate formats, communicating the results to the general public, or deriving policy advice. Again, graphical displays feature prominently in this task.

(vi) Archiving the data and code: To ensure reproducibility and to make data that have been collected with considerable effort accessible to other researchers, it has become common practice in many areas to archive research data in repositories (possibly including their publication and attaching a persistent identifier to the data) and to publish the code utilized for their analysis. Of course there may be limitations, for example owing to data confidentiality, proprietary rights, etc., but publishing at least a basic set of information including analysis code is nowadays considered good scientific practice.

(vii) Reusing the data and results: To close the data analysis circle, data collected for a specific purpose can often be reused for further analyses in different settings or for different research questions. Similarly, the results achieved in one analysis often form the basis for future research. In both cases, we considerably benefit from archived data and code as discussed in the previous step.

The focus in our book is on phases (iv) and (v) of the data analysis circle, that is, we are concerned with statistical modeling as a specific instance of data analysis and the thorough interpretation of the results obtained from the statistical model. However, the other aspects are of course also very relevant when performing data analyses, even though we do not discuss them in detail in this book. In particular, for a data analysis to be useful, the following properties are required from the data.

• The data should accurately represent the *population* under study. The population consists of the subjects we would like to investigate and the sample (that is, the data) should be representative of it. For example, this can be ensured by taking a random sample from the population.

• The data should be collected with *integrity* so there is no intentional bias.

• *Extreme values* in the data should be genuine values and not human or machine errors.

• *Missing values* should be properly treated rather than ignored, see, for example, van Buuren (2018).

• *Data contamination* should be small, if it exists at all. By data contamination we mean that part of the data is unintentionally corrupt. This phenomenon is usually treated by "robust" statistical analysis; for example, Aeberhard et al. (2021) propose robust methods for GAMLSS.

The decisions made in any of the steps of the data analysis circle depend of course very much on the purpose of the analysis. In particular, data are analyzed with very different aims, which often can be classified as follows.

• Confirmatory, when a specific hypothesis about an effect of interest is studied. This often goes along with the goal of establishing a causal link between observables, which requires either specific choices concerning the data collection process (e.g. an experimental set-up with random treatment assignment) or additional assumptions on observational data (e.g. using instrumental variables or graphical models to establish causal effects);

• Exploratory, when the goal is to derive additional (or new) knowledge about an empirical phenomenon of interest. In this case, the main focus is on identifying relevant associations rather than establishing causal relations;

• Prediction-oriented, when the analyst is interested in determining a model that not only describes the observed data well, but is also able to predict new observations. In this case, one may set interpretability aside in favour of better prediction, as is commonly done in machine learning methodology such as deep neural networks.

Most of our applications fall into the category of exploratory data analysis, but we will also address applications with an emphasis on prediction. In data mining, the models are typically exploratory and are used to find interesting patterns in the data. Medical statistics and econometrics often use confirmatory models, where the estimated coefficients of the model and their interpretation play an important role. It is here where likelihood-based and Bayesian ways of fitting the model are prominent. In machine learning (including classical boosting) and artificial intelligence, the focus is often on finding models that sacrifice interpretability for the sake of optimized predictive ability. Statistical boosting approaches (Chapter 7) represent a mixture between classical statistical modeling and machine learning, allowing the fitting of an interpretable model with competitive prediction accuracy.

1.2 Statistical Models

This book follows the general scientific principle that "all models are wrong but some are useful" (Box, 1979) in the sense that no statistical model will usually be complex enough to fully describe reality. However, statistical models can provide a reasonable approximation to reality, subject to a certain level of abstraction. Statistical models consist of a *structural component* that relies on a mathematical description of how certain input variables (the covariates in a regression model) determine properties of the distribution of a response variable. In contrast to deterministic models, statistical models are equipped with a *stochastic component* that represents the deviation between the real data generating process and the approximation by the model, as well as truly random aspects of the data generation such as measurement errors or uncertainty stemming from random sampling from a population. In the linear model, the structural component (the regression predictor of the model) and the stochastic component (the error term) are nicely separated, whereas this is no longer the case for GAMLSS.

The process of statistical modeling includes the steps of making reasonable assumptions concerning the data generating process, fitting the model to the observed data, checking the validity of the model, and interpreting the results. As emphasized previously, all statistical models are derived through simplifying assumptions. If the assumptions are correct, it is more likely that the conclusions from the model will also be useful.

General requirements for choosing a good model include the model's ability to

- answer the right scientific question,

- highlight important features of the data while ignoring the less relevant,

- provide a good trade-off between fidelity to the data and complexity such that we neither overfit the data (implying restrictions on the generalizability beyond the observed data for predictions) nor underfit the data (implying a potential bias in the conclusions).

In essence, a good statistical model should be able to "allow the data to tell its story."

Exploratory statistical modeling often relies on the idea of trying different models to the data and choosing the most appropriate, while accepting the principle that there could be more than one appropriate and useful model, or occasionally none, that adequately fits the data of interest.

1.3 Regression Models

Regression models are one specific instance of statistical models that consist of a *response* variable y (also denoted as the outcome, dependent variable, or the target variable), a number of *explanatory* variables x (also called predictors, covariates, independent variables, terms), and assumptions as to how the explanatory variables affect the response.

The set of possible values the response variable can assume (i.e. its *support*) is crucial in developing an appropriate model. The most important differentiation is between continuous and discrete responses, but additional differentiations are possible, for example relating to nonnegative responses, responses with skewed distributions, responses with bounded support, mixed discrete–continuous responses, nominal and ordinal discrete responses, etc. We will also consider models for multivariate responses. As we will see in Section 2.6, the support of the response can be the first criterion when choosing an appropriate distribution and we discuss a number of distribution classes in Chapter 2.

For the explanatory variables, we distinguish

- continuous covariates assuming values on the real line or an interval subset of the real line. For example, age and height both assume values on the positive real line. There are occasions when a continuous explanatory variable needs to be transformed. Skew distributed values, unusually large or small values, or the scaling of the explanatory variable are some of the reasons to transform continuous variables;

- spatial covariates representing either continuous coordinate information or discrete spatial information in terms of the assignment to a set of pre-specified regions;

- factors, namely categorical variables which can be *unordered*, for example, eye color where the *levels* of the factor do not have a specific order; or *ordered*, for example, disease level where the levels "severe", "moderate", "mild", "none" do have a specific order. Factors are also used as grouping or clustering variables, for example the identification variable for individuals in longitudinal data.

It is important to remember that not all of the available covariates may be needed to explain the behavior of the response variable. A central part of *statistical modeling* is to determine which of the covariates are indeed important, and in what form(s) they should be included.

1.4 From Linear Models to GAMLSS

In this section, we provide a brief history of the development of regression methodology from the linear model to GAMLSS. Our goal is not to provide a complete literature review or a complete enumeration of all regression approaches, but to motivate the important generalizations of and differences to the most well-known regression set-ups.

1.4.1 The Linear Model

Historically, the most popular regression model is the *linear model*, where the response variable y_i is related to a set of covariates $x_{i1}, \ldots, x_{ip}$ as

$$y_i = \beta_0 + \beta_1 x_{i1} + \cdots + \beta_p x_{ip} + \epsilon_i, \quad \text{for } i = 1, \ldots, n, \qquad (1.1)$$

where ϵ_i are the *errors* or *disturbances* that quantify the deviations between the structural part of the model $\beta_0 + \beta_1 x_{i1} + \cdots + \beta_p x_{ip}$ and the observed responses y_i. The very basic assumption about the error terms ϵ_i is that they are i.i.d. with zero mean and constant variance. The additive composition of the error terms and the structural component yields a separable model where additional assumptions on the errors entail easy interpretation and estimation of the model.

Note that in this book we use lower-case letters to denote the responses and covariates, irrespective of whether they are random variables or their realizations. Only in cases where it cannot be easily deduced from the context, will we make explicit notational distinction between random variables and realizations.

An extra assumption for the error terms, which is particularly helpful for uncertainty assessment and hypothesis testing, is that they are i.i.d. realizations from a normal distribution with zero mean and constant variance, that is, $\epsilon_i \overset{\text{ind}}{\sim} \mathcal{N}(0, \sigma^2)$. The assumption of zero means implies that the structural part of the model determines the (conditional) expectation of the responses such that

$$\mathbb{E}(y_i | x_{i1}, \ldots, x_{ip}) = \beta_0 + \beta_1 x_{i1} + \cdots + \beta_p x_{ip},$$

which also yields the famous ceteris paribus (everything else fixed) interpretation of the regression coefficients: When comparing two observations that differ in one unit in covariate x_j but have the same values for all other covariates, we expect a difference of β_j in the response.

More generally, normally distributed error terms imply

$$y_i | x_{i1}, \ldots, x_{ip} \overset{\text{ind}}{\sim} \mathcal{N}(\beta_0 + \beta_1 x_{i1} + \cdots + \beta_p x_{ip}, \sigma^2),$$

that is, the responses themselves follow a normal distribution such that inferences for the regression coefficients can be drawn based on the implied normal likelihood. This includes not only point estimates based on maximum likelihood theory, but also the assessment of uncertainties via standard errors, confidence intervals, or statistical tests.

In matrix notation, the linear model is expressed as

$$\boldsymbol{y} = \boldsymbol{X}\boldsymbol{\beta} + \boldsymbol{\epsilon}, \tag{1.2}$$

where $\boldsymbol{y}$ and $\boldsymbol{\epsilon}$ are n-dimensional vectors of the response variable and error terms, respectively, $\boldsymbol{X}$ is an $(n \times (p+1))$-dimensional design matrix containing the explanatory variables as columns (including a column of ones relating to the intercept β_0), and $\boldsymbol{\beta}$ is the $(p+1)$-dimensional vector of regression coefficients, which shall be estimated from the data. For the vector of error terms, we then obtain the n-dimensional multivariate normal distribution:

$$\boldsymbol{\epsilon} \sim \mathcal{N}_n(\boldsymbol{0}, \sigma^2 \boldsymbol{I}_n),$$

which in particular implies $\mathbb{E}(\boldsymbol{\epsilon}) = \boldsymbol{0}$ and $\mathrm{Cov}(\boldsymbol{\epsilon}) = \sigma^2 \boldsymbol{I}_n$. Similarly, for the responses, we find

$$\boldsymbol{y}|\boldsymbol{X} \sim \mathcal{N}_n(\boldsymbol{X}\boldsymbol{\beta}, \sigma^2 \boldsymbol{I}_n) \tag{1.3}$$

and therefore $\mathbb{E}(\boldsymbol{y}|\boldsymbol{X}) = \boldsymbol{X}\boldsymbol{\beta}$ and $\mathrm{Cov}(\boldsymbol{y}|\boldsymbol{X}) = \sigma^2 \boldsymbol{I}_n$. The corresponding likelihood for σ^2 and $\boldsymbol{\beta}$ is given by

$$L(\boldsymbol{\beta}, \sigma^2) = \left(\frac{1}{\sqrt{2\pi\sigma^2}}\right)^n \exp\left(-\frac{1}{2\sigma^2}(\boldsymbol{y} - \boldsymbol{X}\boldsymbol{\beta})^\top(\boldsymbol{y} - \boldsymbol{X}\boldsymbol{\beta})\right).$$

Ignoring σ^2 for the moment, maximizing the likelihood is equivalent to minimizing

$$(\boldsymbol{y} - \boldsymbol{X}\boldsymbol{\beta})^\top(\boldsymbol{y} - \boldsymbol{X}\boldsymbol{\beta}) = \boldsymbol{\epsilon}^\top\boldsymbol{\epsilon} = \sum_{i=1}^n \epsilon_i^2,$$

i.e. the least squares criterion that is often also used as the foundation for determining regression coefficients in the linear model without relying on the normal distribution for the error terms. The solution to minimizing the least squares criterion is the ordinary least squares estimator

$$\hat{\boldsymbol{\beta}} = \left(\boldsymbol{X}^\top\boldsymbol{X}\right)^{-1}\boldsymbol{X}^\top\boldsymbol{y}, \tag{1.4}$$

while the maximum likelihood estimator for the variance of the error terms is given by

$$\hat{\sigma}_{\mathrm{ML}}^2 = \frac{1}{n}\left(\boldsymbol{y} - \boldsymbol{X}\hat{\boldsymbol{\beta}}\right)^\top\left(\boldsymbol{y} - \boldsymbol{X}\hat{\boldsymbol{\beta}}\right).$$

Since $\hat{\sigma}_{\mathrm{ML}}^2$ is biased, a commonly used alternative is the unbiased estimator, which can also be derived as a restricted maximum likelihood (REML) estimator:

$$\hat{\sigma}_{\mathrm{REML}}^2 = \frac{1}{n-p-1}\left(\boldsymbol{y} - \boldsymbol{X}\hat{\boldsymbol{\beta}}\right)^\top\left(\boldsymbol{y} - \boldsymbol{X}\hat{\boldsymbol{\beta}}\right).$$

We illustrate a simple linear regression model in action with data from the Fourth Dutch Growth Study Fredriks et al. (2000a,b), a cross-sectional study measuring growth and development in the Dutch population between the ages 0 and 23 years. Figure 1.2(a) shows $n = 3,512$ observations of the body mass index (BMI) and age of boys between 10 and 20 years of age. The response variable is BMI, and there is only

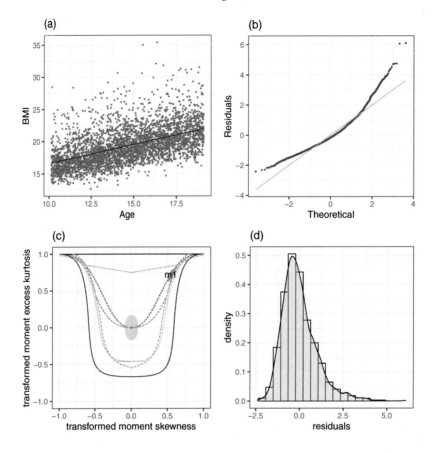

Figure 1.2 Dutch boys' BMI for boys aged between 10 and 20 years: (a) the data and the fitted least squares line; (b) QQ-plot of the residuals from the linear model `m1`; (c) bucket plot for checking the moment transformed skewness and kurtosis of the residuals of model `m1`; (d) histogram and density estimate of the residuals.

one explanatory variable, `age`. The fitted least squares line shown in Figure 1.2(a) is $\hat{y} = 10.87 + 0.577$ `age` and it captures the trend in growth well. We refer to this linear model as `m1`.

Unfortunately, other features in the data are not captured well by the linear model. In Figure 1.2, panels (b) and (c) show residual diagnostics from the linear model. Panel (b) shows a QQ-plot of the normalized quantile residuals (defined in Section 4.7.1) of the linear model, which checks the normality assumption. Most points in the QQ-plot are far from the diagonal line, indicating strong deviations from normality. Panel (c) shows a *bucket* plot, which is a diagnostic graphical tool checking the skewness and kurtosis assumption. (See Section 4.7.3 and also De Bastiani et al. (2022).) The point marked as `m1` in Figure 1.2(c) is the transformed moment skewness (on the x axis) against the transformed moment kurtosis (on the y axis) of the residuals of model `m1`. The cloud of red points around `m1` are 99 bootstrap points of transformed moment skewness and kurtosis points, obtained by bootstrapping the original residuals. This shows the variability of the measures of skewness and kurtosis of m1. The important feature is the shaded area in the middle of the figure around the point $(0, 0)$, which represents the normal distribution. This shaded area is a 95% confidence region based on the Jarque–Bera test for simultaneously testing whether skewness and kurtosis exist in the data. The point m1 is far from the 95% confidence region of the the the Jarque–Bera test, indicating that the residuals of the model m1 show considerable skewness and kurtosis. The model m1 has not taken the skewness and kurtosis observed in the data, into account.

Even without diagnostic tools, one can spot more variation above the fitted line in Figure 1.2(a) than below it. To highlight this, Figure 1.2(d) shows a histogram and density estimate of the residuals of the of m1 model, which highlights considerable skewness. This is unlikely to be modeled adequately by the assumption of normality inherited by the linear model. The fact is, that while we fitted a reasonable model for the location parameter of the data (the mean in this case), the basic assumptions of the linear model are inevitably broken and any inference on the parameters or other features of the model would be affected by this.

Figure 1.2(a) is based on $n = 3,512$ observations for children and teenagers from 10 to 20 years old. The situation becomes more complicated if we consider the original dataset with $n = 7,294$ observations with age range of 0.30 to 22.7 years, shown in Figure 1.3(a). Obviously, the least squares fit of the linear model will fail miserably on the complete data. The curve shown in Figure 1.3(a) was fitted using P-splines (Eilers and Marx, 1996, 2021), one of the techniques we will discuss extensively in Chapter 3. In addition, there are other features in the data which indicate that the assumptions of normally distributed error terms with constant variance, are not appropriate here. There is evidence in Figure 1.3(a) of *heteroscedasticity* (the variance varies with age); *skewness*; and possibly *kurtosis* (since there exist a number of observations further away from the central line, in both directions, suggesting heavier tails than the normal distribution). These features suggest that skewness and kurtosis may also vary with age. The GAMLSS model introduced in Section 1.5 accommodates these features.

Generally, GAMLSS enables us to

- consider a much wider range of response distributions than the normal distribution,

- deal with heterogeneity, possibly covariate-dependent, in various distributional features of the response distribution (not only the mean), and

- relax the assumption of a linear predictor such that various kinds of complex regression relations can be accommodated.

In fact, equation (1.3) provides the simplest example of a GAMLSS, in which normally distributed responses are assumed with purely linear effects on only the mean while the variance is the same for all observations. In Sections 1.4.2 to 1.4.4 we discuss various extensions that preceded the development of GAMLSS. It is convenient at this point to introduce a different notation for Equation (1.3) that emphasizes that the elements of the response vector are independent but not identically distributed. More precisely, we rewrite model (1.3) as

$$\boldsymbol{y} \stackrel{\text{ind}}{\sim} \mathcal{N}(\boldsymbol{\mu}, \sigma^2), \qquad \boldsymbol{\mu} = \boldsymbol{X}\boldsymbol{\beta}, \tag{1.5}$$

where the notation indicates that each element y_i of $\boldsymbol{y}$ is independently distributed as $y_i \sim \mathcal{N}(\mu_i, \sigma^2)$ for $i = 1, \ldots, n$. The mean parameter $\boldsymbol{\mu}$ is a linear function of the explanatory variables constituting the columns of $\boldsymbol{X}$.

1.4.2 Generalized Linear Models

A big step in the development of distributional regression models was the introduction of Generalized Linear Models (GLMs) by Nelder and Wedderburn (1972). The GLM was popularized by McCullagh and Nelder (1989) and Dobson and Barnett (2018) and also by the introduction of the first interactive statistical package **GLIM**; see, for example, Francis et al. (1993) and Aitkin (2018).

In a GLM, the normal response distribution in (1.5) is replaced by the exponential family of distributions such that

$$\boldsymbol{y} \stackrel{\text{ind}}{\sim} \mathcal{E}(\boldsymbol{\mu}, \phi),$$

where $\mathcal{E}$ denotes the exponential family, $\boldsymbol{\mu} = \mathbb{E}(\boldsymbol{y})$ is the expectation of the response and $\phi > 0$ is a scale parameter. The exponential family includes many important distributions such as the normal, Bernoulli, Poisson, gamma, inverse Gaussian and Tweedie distributions, therefore providing a unifying framework for regression analyses in variety of settings. Most importantly, the framework includes the linear model as a special case, but also allows the analysis of binary responses (based on the Bernoulli distribution), count responses (based on the Poisson distribution), nonnegative continuous responses (based on the gamma and inverse Gaussian distributions), and nonnegative continuous responses supplemented with a positive probability of observing zero (based on the Tweedie distribution).

Regression effects are now assumed for the expectation $\boldsymbol{\mu}$, with a further generalization of the linear model, allowing a monotonic *link function* $g(\cdot)$ that relates $\boldsymbol{\mu}$ to the linear predictor $\boldsymbol{\eta} = \boldsymbol{X}\boldsymbol{\beta}$:

$$g(\boldsymbol{\mu}) = \boldsymbol{\eta} = \boldsymbol{X}\boldsymbol{\beta}.$$

This opens up the relationship between $\boldsymbol{\eta}$ and $\boldsymbol{\mu}$ to a variety of shapes not possible under the linear model. For example, if $g(\cdot)$ is the logarithmic function, this implies a multiplicative relationship between the covariates and the mean response since

$$\log(\mu_i) = \beta_0 + \beta_1 x_{i1} + \cdots + \beta_p x_{ip}$$

and therefore

$$\mu_i = \exp(\beta_0) \cdot \exp(\beta_1 x_{i1}) \cdots \exp(\beta_p x_{ip}). \tag{1.6}$$

The inverse of the link function $h(\cdot) = g^{-1}(\cdot)$ is called the response function and maps the linear predictor to the expectation of the response, namely

$$\boldsymbol{\mu} = h(\boldsymbol{\eta}).$$

Since the domain of $\boldsymbol{\mu}$ is often restricted (e.g. to the unit interval in case of the Bernoulli distribution or to the positive half axis in case of Poisson, gamma, and inverse Gaussian distributions), the link function also serves as a convenient way of constraining the distribution parameter $\boldsymbol{\mu}$ to the appropriate range when modeled as a function of the explanatory variables.

Note that unlike in the linear model, GLMs in general entail a non-separable structure where the regression predictor (the structural part of the model) cannot easily be disentangled from the random component (the response distribution). Rather, one specific aspect of this distribution (namely the mean) is related to the structural model component.

Unifying various regression models under the umbrella of the exponential family allows us to derive general principles and implementations for statistical inference. In particular, iteratively weighted least squares (IWLS) estimation provides a convenient way of implementing Fisher scoring iterations for determining the maximum likelihood estimate for the regression coefficients $\boldsymbol{\beta}$ (see McCullagh and Nelder (1989) for details). Furthermore, theoretical properties of the exponential family result in asymptotic normality and the validity of likelihood ratio tests. An important property of GLM models (which is shared with the generalized estimating equation approach (Hardin and Hilbe, 2002)) is that it is always consistent in estimating the population mean.[1] The problem is that if the distribution is not correct it could be a very inefficient way of doing so. Another important theoretical implication of assuming a response distribution in the exponential family is that the variance of the responses is intrinsically linked to the expectation based on a variance function that is specific to the chosen member of the exponential family. More precisely, we find

$$\mathbb{V}(y_i) = V(\mu_i)\phi,$$

that is, the variance is determined by the product of a variance function $V(\cdot)$ and the scale parameter ϕ. For example, in case of the normal distribution, the variance function and scale parameter are given by $V(\mu) = 1$ and $\phi = \sigma^2$, providing one example where indeed the variance does not depend on the expectation μ but only on the scale parameter, which then coincides with the error variance. For other members of the exponential family, variance function and scale parameter are given by, for example, $V(\mu) = \mu$, $\phi = 1$ (Poisson distribution), $V(\mu) = \mu(1 - \mu)$, $\phi = 1$ (Bernoulli distribution), and $V(\mu) = \mu^2$, $\phi > 0$ (gamma distribution).

1.4.3 Generalized Additive Mixed Models

While generalized linear models enable considerable flexibility with respect to the response distribution, they keep the restrictive assumption of a purely linear re-

---

[1] The concept of the population of interest is introduced in Chapter 4.

gression predictor $X\beta$ determining the conditional expectation of the response, via $\mu = h(X\beta)$. Various extensions have been introduced to overcome this limitation, following the advent of generalized additive models (Hastie and Tibshirani, 1986, 1990) that expanded the predictor to include nonlinear effects of continuous covariates, yielding

$$\eta_i = x_i^\top \beta + s_1(x_{i1}) + \cdots + s_J(x_{iJ}), \tag{1.7}$$

where the $s_j(\cdot)$ are nonlinear smooth functions for the explanatory variables x_{ij}. In this book, we will rely on penalized splines for modeling these nonlinear effects, as discussed in more detail in Section 3.1. In the wake of generalized additive models, it became apparent that a multitude of other effects could be integrated into the regression predictor in similar ways. For example, extended model classes include

- spatial effects $s_{\text{spat}}(z_i)$ where z_i denotes information on the spatial allocation of individual i, in terms of either coordinates or administrative regions,

- varying coefficient terms $x_{i1}s(x_{i2})$, where the effect of x_{i1} (the interaction variable) smoothly varies with respect to the value of covariate x_{i2} (the effect modifier), and

- interaction surfaces $s(x_{i1}, x_{i2})$ of two continuous covariates.

Various approaches for defining such terms will be discussed in Chapter 3.

Figure 1.3(a) shows the benefit of using smoothing techniques (otherwise known as smoothers) for modeling the relationship between an explanatory term and the response. The fitted curve for all $n = 7294$ observations of the BMI dataset fits the trend in the data very well and it is hard to imagine we could have achieved the same effect using parametric curve fitting. [2] The GLM/GAM framework provides three distributions appropriate for modeling a continuous response variable such as BMI: the normal, the gamma, and the inverse Gaussian distributions. We have fitted all three distributions; the fitted values for the conditional mean of those distributions were very similar and indistinguishable from the line shown in Figure 1.3(a), which plots the fitted values for the inverse Gaussian model. The inverse Gaussian model had the lowest AIC[3] value of the three GAMs. Figure 1.3(b) shows the QQ-plot of the normalized quantile residuals from the three GLM/GAM fitted models. None of them fits the data well. Figure 1.3(c) shows the bucket plot of the three fitted distribution models. The points N, G, and I represent the transformed skewness and the transformed kurtosis of the normalized quantile residuals of the three GAM fitted distributions. All points are far from the 95% confidence region of the Jarque–Bera test. This provides additional evidence that none of the three GLM/GAM distributions adequately fits skewness and kurtosis in the BMI data. The inability of the exponential family to model skewness and kurtosis of the BMI data is also partially shown in Figure 1.3(d), in which the centile curves[4] at centiles 3, 10, 25, 50, 75, 90, and 97, of the fitted inverse Gaussian distribution model, are plotted.

---

[2] Note that, to improve the fit, we used the transformed variable $x = \texttt{age}^{1/3}$ rather than $\texttt{age}$.

[3] Use of the AIC as a way of choosing between models is discussed in Section 4.4.1.

[4] A centile is a quantile multiplied by 100.

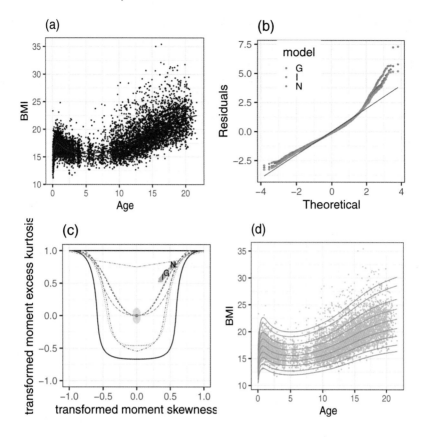

Figure 1.3 Dutch boys' BMI for boys aged between 0 and 23 years: (a) the data and the fitted smooth curve from a GAM model using an inverse Gaussian response distribution; (b) QQ-plots of the residuals from GAMs with normal distribution (N) in blue, gamma distribution (G) in red, and inverse Gaussian distribution (I) in blue; (c) bucket plot of the normalized quantile residuals from the three fitted GAM models: N G, and I; (d) fitted centile curves at centiles values 3, 10, 25, 50, 75, 90, and 97, using the inverse Gaussian GAM model.

In general, we would expect $\alpha\%$ of the data to be below the α centile curve and $(100 - \alpha)\%$ above. For example, in Figure 1.3(d) we would expect 3% of the data to be below the 3 centile curve (the curve at the bottom of the plot). The observed percentage is 1.78%. Above the 97 centile curve (at the top of the plot), we would expect 3% of the data while actually there are 4.2%. The difference does not sound large, but when the centile curves are used for risk stratification, this could be crucial.

Another extension concerns the ability to adjust for potential correlation induced by unobserved heterogeneity associated with grouping structures in the data. In combination with linear predictors, random effect models were introduced by Laird and Ware (1982), and popularized by Pinheiro and Bates (2000). These are models which accommodate correlation between observations through the use of random intercepts and (optionally) random slopes. (See Section 3.6 for details.) A *mixed effects* model in which fixed effect and random effect coefficients coexist is written as

$$\eta_i = \boldsymbol{x}_i^\top \boldsymbol{\beta} + \boldsymbol{z}_i^\top \boldsymbol{\alpha},$$

where $\boldsymbol{x}_i$ contains explanatory variables associated with the linear fixed effect coefficients $\boldsymbol{\beta}$, while $\boldsymbol{z}_i$ contains explanatory variables associated with the random effects (coefficients) $\boldsymbol{\alpha}$. The $\boldsymbol{\alpha}$ are assumed to be normally distributed with zero mean and covariance matrix $\boldsymbol{Q}$, namely $\boldsymbol{\alpha} \sim \mathcal{N}(\boldsymbol{0}, \boldsymbol{Q})$.

The most prominent case for the application of random effects are longitudinal data with repeated observations on the same set of statistical units (subjects, individuals). In this case, within-subject correlation is accounted for by the random effects component. An alternative perspective is that (individual-specific) random effects account for individual-specific, unobserved heterogeneity between the statistical units. However, the application of random effects models is actually much broader since most smoothers $s_j(\cdot)$ in the GAM equation (1.7) (and also most of the extensions mentioned earlier) can be represented as random effects models. This connection of the smoothers to random effect models led to the further understanding and development of smoothers and also to different ways of estimating their smoothing parameters. The family of smoothers which fall into this category were called *structured additive terms* by Fahrmeir et al. (2004).

The combination of random effects with additive model structures leads to generalized additive mixed models, see Ruppert et al. (2003), Wood (2017), and Fahrmeir et al. (2021) for overviews on the state of the art for this model class.

1.4.4 Mean and Dispersion Models

As an important step towards relaxing the common focus on exclusively modeling the mean of the response variable in terms of (possibly complex) regression effects, Aitkin (1987) introduced a model with normally distributed response, in which both the mean and the variance of the model are functions of explanatory variables:

$$
\begin{aligned}
y_i &\sim \mathcal{N}(\mu_i, \sigma_i^2) \\
g_1(\mu_i) &= \boldsymbol{x}_{i1}^\top \boldsymbol{\beta}_1 \\
g_2(\sigma_i) &= \boldsymbol{x}_{i2}^\top \boldsymbol{\beta}_2,
\end{aligned}
\tag{1.8}
$$

where $\boldsymbol{x}_{i1}$ and $\boldsymbol{x}_{i2}$ are design vectors containing explanatory variables associated with the mean and the standard deviation, while the link functions $g_1(\cdot)$ and $g_2(\cdot)$ are taken to be the identity and log functions, respectively. Smyth (1989) extended model (1.8) with the gamma response distribution. Both authors used maximum likelihood for the estimation of model parameters. Rigby and Stasinopoulos (1996) introduced smoothers into model (1.8). Nelder and Pregibon (1987) considered the more general case of the exponential family, namely, $y_i \sim \mathcal{E}(\mu_i, \phi_i)$, using an extended quasi-likelihood function for parameter estimation.

Mean and dispersion models provide one simple special case of generalized additive models for location, scale, and shape (GAMLSS), in which the mean and dispersion are modeled in terms of (linear) predictors. GAMLSS considerably extends this by allowing potentially all parameters characterizing the response distribution to depend on covariate information.

1.5 Generalized Additive Models for Location, Scale and Shape

1.5.1 GAMLSS as a Distributional Regression Model

In a distributional regression model, the relationship between the response y and the covariates $\boldsymbol{x}$ is of a *stochastic* nature. The response y depends on $\boldsymbol{x}$ through the *conditional* distribution $f(y|\boldsymbol{x})$, which is the main subject of interest since it provides rich information about how the covariates $\boldsymbol{x}$ affect various aspects of the (conditional) distribution of y.

GAMLSS provides a parametric framework for *statistical inference* in distributional regression, in which we approximate the conditional distribution $f(y|\boldsymbol{x})$ by a parametric distribution $f(y|\boldsymbol{\theta}(\boldsymbol{x}))$, where $\boldsymbol{\theta}(\boldsymbol{x}) = (\theta_1(\boldsymbol{x}), \theta_2(\boldsymbol{x}), \ldots, \theta_K(\boldsymbol{x}))^\top$ is a K-dimensional vector of (unknown) model parameters which themselves depend on explanatory terms. The basic idea of statistical inference is to use $f(y|\boldsymbol{\theta}(\boldsymbol{x}))$ to say something sensible about the population distribution $f(y|\boldsymbol{x})$, see Chapter 4.

The notation $\boldsymbol{\theta}(\boldsymbol{x})$ emphasizes that any of the model parameters in $\boldsymbol{\theta}$ can be functions of any of the explanatory variables $\boldsymbol{x}$, not only the mean as in equation (1.5). This is one of the main features of the distributional regression model on which we focus in this book. The implication of such an approach is that the shape of the model distribution for y can change according to the values of explanatory variables $\boldsymbol{x}$. By modeling all the parameters of $f(y|\boldsymbol{\theta}(\boldsymbol{x}))$ as functions of the explanatory terms, we explicitly *simultaneously* model all the characteristics of the distribution including location, scale (variability), quantiles, moments, skewness, and kurtosis. Modeling only the mean allows shifts exclusively in the location of the distribution with all other distribution parameters remaining constant. Consequently, modeling all parameters allows for various types of changes in the shape of the distribution of a response variable, based on one or more explanatory variables (e.g., the age of a child).

The distinction between the basic assumptions of the standard regression model and those of a GAMLSS is shown in Figure 1.4, in which we depict simulated samples of a response variable y with a single explanatory variable x. Figure 1.4(a) demonstrates the distributional assumptions of the linear model. The mean of the normal distribution of the response y varies linearly with x; the shape of the distribution remains the same over the range of x, since the variance of y is constant. Figure 1.4(b) illustrates how the assumptions of a GAMLSS model may operate. A GAMLSS model allows a nonlinear (smooth) relationship between x and the location parameter of the distribution, but also allows all the parameters of the distribution to vary with ex-

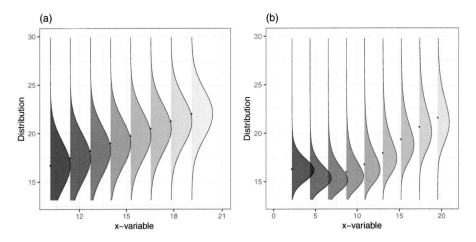

Figure 1.4 Different distributional regression models assumptions: (a) the linear regression model; (b) the GAMLSS model.

planatory terms. It allows the shape of the response distribution to change according to different values of x.

Figure 1.5(a) displays the BMI data and fitted values for the location parameter μ of a GAMLSS model fitted using the Box–Cox t (`BCTo`) distribution, which is a four-parameter distribution introduced by Rigby and Stasinopoulos (2006). The parameters are μ (location parameter, approximately the median), σ (scale parameter, approximately the coefficient of variation), and ν and τ as skewness and kurtosis parameters, respectively (see Section 2.2.2 and Rigby et al. (2019)). The model was fitted using smoothers for all of the parameters as functions of `age`.[5] The QQ-plot of the GAMLSS model, in Figure 1.5(b), shows that the `BCTo` distribution fits the data very well. It is only in the lower tail that a few points deviate from the diagonal line. Given that there are more than 7,000 observations, this behaviour is not unusual. The bucket plot in Figure 1.5(c) shows that the fitted `BCTo` distribution corrects properly for skewness and kurtosis in the data, as the value of the transformed skewness and transformed kurtosis of the residuals from the `BCTo` model fall very close to the origin $(0,0)$ (representing the normal distribution) and within the 95% confidence region of the Jarque–Bera test. The fitted centiles from the `BCTo` distribution shown in Figure 1.5(d) provide further graphical evidence that the distribution fits well.

In the following, we introduce GAMLSS more formally and discuss the different ingredients of a GAMLSS specification.

1.5.2 Response Distributions

In a GAMLSS model, the responses are assumed to be generated from a K-parametric family of distributions with (covariate-dependent) parameters $\boldsymbol{\theta}(\boldsymbol{x}) = (\theta_1(\boldsymbol{x}), \theta_2(\boldsymbol{x}),$

---

[5] The transformed variable $x = \texttt{age}^{1/3}$ was fitted instead of `age`.

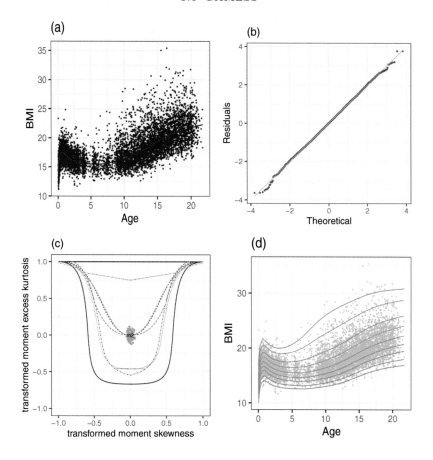

Figure 1.5 Dutch boys BMI for boys aged between 0 and 23 years: (a) the data and the fitted smooth curve from a GAMLSS model with BCT response distribution; (b) QQ-plots of the normalized quantile residuals from the GAMLSS model; (c) bucket plot of the residuals from the GAMLSS models; (d) fitted centile curves fitted at centile values 3, 10, 25, 50, 75, 90, and 97, from the GAMLSS model.

$\ldots, \theta_K(\boldsymbol{x}))^\top$. Subsequently we allow those K parameters to possibly differ for each observation i, for $i = 1, \ldots, n$, so we introduce the notation $\boldsymbol{\theta}_{[i]}(\boldsymbol{x}_i) = (\theta_{i1}(\boldsymbol{x}_i), \theta_{i2}(\boldsymbol{x}_i), \ldots, \theta_{iK}(\boldsymbol{x}_i))^\top$ for $i = 1, \ldots, n$. By suppressing the explicit dependence of θ_{ik} on $\boldsymbol{x}_i$ to keep the notation short, we have $\boldsymbol{\theta}_{[i]} = (\theta_{i1}, \theta_{i2}, \ldots, \theta_{iK})^\top$. We assume that there is one common type of distribution applying to all observations (such as normal, Poisson, etc.) but that the K parameters of this distribution are allowed to vary over the individual observations, that is,

$$y_i \overset{\text{ind}}{\sim} \mathcal{D}(\theta_{i1}, \ldots, \theta_{iK}), \qquad \text{for } i = 1, \ldots, n.$$

We denote the density and the cumulative distribution function of this distribution as $f(y_i|\boldsymbol{\theta}_{[i]})$ and $F(y_i|\boldsymbol{\theta}_{[i]})$, respectively. We note that each θ_{ik} for $k = 1, \ldots, K$ may depend on different subsets of $\boldsymbol{x}_i$.

$\mathcal{D}$ denotes the response distribution: The GAMLSS framework allows a multitude of response types, including (but not limited to) (i) models for continuous responses, enabling us to not only deal with but to systematically study phenomena such as heteroscedasticity and skewness, (ii) models for continuous nonnegative responses, potentially featuring a discrete point mass at zero, (iii) count responses potentially featuring zero inflation and/or overdispersion, (iv) continuous fractional or bounded responses (e.g. proportions), again including the option for discrete point masses at one or both endpoints, and (v) multivariate response distributions. More details on potential choices for response distributions are provided in Chapter 2.

For many univariate continuous distributions defined on the real line $\mathbb{R}$, the first two parameters θ_1 and θ_2 are related to location and scale (or dispersion), but this is not always the case. For $K > 2$, the remaining parameter(s) are generally shape parameters, although they may also capture specialized features such as zero inflation. In special cases, θ_3 and θ_4 are true skewness and true kurtosis parameters (see Rigby et al. (2019) for the definitions of these concepts). Rigby and Stasinopoulos (2005) and subsequent publications by those authors use $K = 4$ parameters with the notation μ, σ, ν, and τ, respectively. Note, however, that neither the original definition of GAMLSS (1.10) nor its original fitting algorithms have restrictions on the number of parameters K.

1.5.3 Link Functions

For each of the distribution parameters θ_k, a monotonic link function $g_k(\cdot)$ and corresponding response function $h_k(\cdot) = g_k^{-1}(\cdot)$ relate the regression predictor η_k with the corresponding parameter, namely,

$$\eta_k = g_k(\theta_k) \quad \text{and} \quad \theta_k = h_k(\eta_k) \ .$$

The response function $h(\cdot)$ is often chosen to map the regression predictors from the real line (where the predictor η_k can take its values) to the correct support for θ_k, ensuring that parameters are appropriately constrained. For example, standard deviation and variance have to be positive while parameters representing probabilities are restricted to the unit interval $[0, 1]$. This is an important and useful feature of the link function, but note that link functions reflect the relationship between parameter and covariates. For example, in a model with an identity link for θ_k, the contribution of each explanatory variable to the distribution parameter θ_k is additive, while for a model with a log link the effect is multiplicative, as shown in equation (1.6).

While default choices for the link functions exist (e.g. the logarithmic link for positive parameters such as variances, or the logit or probit links for parameters restricted to the unit interval), it is important to emphasize that the choice for a link function also implies a modeling decision. This decision determines the exact relation between the covariates and the conditional response distribution and has consequences for both the fit of the model and the interpretation of the estimated regression effects. It therefore makes sense to consider competing specifications for the link function and to include the decision on a specific link function in the model building and model checking process.

An interesting avenue to circumvent the difficulties arising from the need to identify the most appropriate link function, is to consider flexible link functions estimated from the data along with the regression effects of interest. In GLMs, this has, for example, been addressed under the notion of single index models; see Ichimura (1993), where kernel density estimates are used to determine the response function, or Yu and Ruppert (2002), Muggeo and Ferrara (2008), and Yu et al. (2017) who employ penalized splines for the specification of the response function. Estimated link functions have also been combined with additive model specifications, for example in Tutz and Petry (2016) and Spiegel et al. (2019). Another way of relaxing the assumption of one given link function are composite links, where multiple transformations of linear predictors are additively combined to one composite model specification; see Thompson and Baker (1981). In this book, we will not pursue these ideas further but will rather focus on fixed, pre-specified link functions.

1.5.4 Structured Additive Predictors

The simplest case of a GAMLSS is a *fully parametric model*, where a linear predictor is specified for each of the distribution parameters, i.e.

$$\eta_{ik} = \beta_0^{\theta_k} + \beta_1^{\theta_k} x_{i1}^{\theta_k} + \cdots + \beta_{J_k}^{\theta_k} x_{iJ_k}^{\theta_k}$$

leading for n observations to the n-dimensional vector

$$\boldsymbol{\eta}_k = \boldsymbol{X}_k \boldsymbol{\beta}_k$$

of predictor evaluations. While looking rather restrictive at first glance, considerable flexibility can already be achieved in the parametric setting, by considering various types of transformations such as polynomials or interactions. (See also Section 3.8.1.) Still more flexibility is achieved when assuming that each of the regression predictors η_k is additively composed of an intercept $\beta_0^{\theta_k}$ and a sum of J_k functions $s_j^{\theta_k}(\boldsymbol{x}_i)$ (or $s_{jk}(\boldsymbol{x}_i)$ for simplicity), $j = 1, \ldots, J_k$, leading to the structured additive predictor

$$\eta_{ik} = \beta_0^{\theta_k} + s_1^{\theta_k}(\boldsymbol{x}_i) + \cdots + s_j^{\theta_k}(\boldsymbol{x}_i) + \cdots + s_{J_k}^{\theta_k}(\boldsymbol{x}_i) \qquad (1.9)$$

or the more compact variant

$$\eta_{ik} = \beta_{0k} + s_{1k}(\boldsymbol{x}_i) + \cdots + s_{jk}(\boldsymbol{x}_i) + \cdots + s_{J_k k}(\boldsymbol{x}_i) \ .$$

The functions $s_j^{\theta_k}(\boldsymbol{x}_i)$ are used as a generic notation that may represent a variety of different effects, as discussed in more detail in the following, and in Chapter 3. In particular, the functions can simply represent a linear effect or more complex effects such as nonlinear effects of continuous covariates, spatial effects, or random effects. Notationally, we allow each function to depend on the complete covariate vector, although in practice each effect will usually only depend on a small subset of $\boldsymbol{x}_i$. However, to avoid notational complexity, we do not make this explicit. Furthermore, to make the model identifiable, appropriate centering constraints have to be applied to the different functions.

Model equation (1.9) can also include terms that do not fit into the framework of structured additive terms. For example, the local regression smoothers (**loess**) of Cleveland et al. (2017) were part of the original implementation of the GAM models in **Splus** since the early 1990s. Decision trees, neural networks, and the fitting of non-linear terms have been implemented in the **gamlss** package since 2010. The original GAMLSS algorithms of Rigby and Stasinopoulos (2005) allow the inclusion of any statistical regression-type technique which allows prior weights in its implementation. However, while for the structured additive terms there is a strong theoretical justification (see Chapter 3), the justification for the techniques mentioned above comes from the fact that empirically they work well. Dimensionality reduction techniques such as lasso regression (Tibshirani, 1996) and principal component regression have also been implemented within **gamlss** (Stasinopoulos et al., 2022), see Section 5.4.

There is great potential to be gained by merging some of *machine learning* techniques with distributional regression. Machine learning originated in the computer science world, and as a result its language is somewhat different from that of statistical modeling. It generally encompasses algorithms and computational techniques designed to produce a prediction of an *output* (response variable) on the basis of given *inputs* (explanatory variables). In this respect its aim is similar to statistical modeling. The difference arises from the fact that while statistical modeling aims to interpret and understand the underlying structure of relationships, machine learning takes a *black box* approach. Both approaches can be helpful in different circumstances but caution and knowledge of their limitations are crucial.

1.5.5 Basis Function Representation

We use basis function expansions to represent nonlinear effects in the structured additive predictors, i.e. each function is approximated in terms of a linear combination of basis functions such that (after dropping the parameter index θ_k and the function index j for notational convenience) we obtain

$$s(\boldsymbol{x}_i) = \sum_{l=1}^{L} \gamma_l B_l(\boldsymbol{x}_i),$$

where γ_l are the basis amplitudes while $B_l(\boldsymbol{x}_i)$ represent different types of basis functions (discussed in detail in Chapter 3). In matrix notation, each of the predictors can be written for all observations as

$$\boldsymbol{\eta} = \beta_0 \mathbf{1}_n + \boldsymbol{B}_1 \boldsymbol{\gamma}_1 + \cdots + \boldsymbol{B}_J \boldsymbol{\gamma}_J.$$

To enforce specific properties of the function estimates such as smoothness or shrinkage, each parameter vector $\boldsymbol{\gamma}_j$, $j = 1, \ldots, J$, is supplemented by a quadratic penalty term

$$\text{pen}(\boldsymbol{\gamma}_j) = \lambda_j \boldsymbol{\gamma}_j^\top \boldsymbol{K}_j \boldsymbol{\gamma}_j$$

that is, augmented to the likelihood, $\lambda_j \geq 0$ is the smoothing parameter determining the impact of the penalty and $\boldsymbol{K}_j$ is a positive semi-definite penalty matrix. In a Bayesian framework, the penalty is replaced by the equivalent prior distribution

$$f(\boldsymbol{\gamma}_j | \tau_j^2) \propto \left(\tau_j^2\right)^{-\frac{\text{rank}(\boldsymbol{K}_j)}{2}} \exp\left(-\frac{1}{2\tau_j^2} \boldsymbol{\gamma}_j^\top \boldsymbol{K}_j \boldsymbol{\gamma}_j\right) \mathbb{1}(\boldsymbol{A}_j \boldsymbol{\gamma}_j = \boldsymbol{0}),$$

where the prior variance τ_j^2 is related inversely to the smoothing parameter, the penalty matrix $\boldsymbol{K}_j$ plays the role of a prior precision matrix, and $\boldsymbol{A}_j$ is an appropriate constraint matrix that ensures identifiability of the model. In more general cases, the penalty or prior distribution may involve multiple smoothing parameters and/or it may be notationally convenient to absorb the smoothing parameter into the penalty matrix. We then write $\boldsymbol{K}_j(\boldsymbol{\lambda}_j)$ to emphasize that, indeed, the penalty term $\text{pen}(\boldsymbol{\gamma}_j) = \boldsymbol{\gamma}_j^\top \boldsymbol{K}_j(\boldsymbol{\lambda}_j)\boldsymbol{\gamma}_j$ depends on a (possibly vector-valued) hyperparameter λ.

1.5.6 Compact Summary

The GAMLSS model is expressed in matrix notation as

$$\boldsymbol{y} \stackrel{\text{ind}}{\sim} \mathcal{D}(\boldsymbol{\theta}_1, \cdots, \boldsymbol{\theta}_K) \tag{1.10}$$

$$g_k(\boldsymbol{\theta}_k) = \boldsymbol{\eta}_k \tag{1.11}$$

$$\boldsymbol{\eta}_k = \beta_{0k}\boldsymbol{1}_n + \boldsymbol{B}_{1k}\boldsymbol{\gamma}_{1k} + \cdots + \boldsymbol{B}_{J_k k}\boldsymbol{\gamma}_{J_k k} \tag{1.12}$$

$$\boldsymbol{\gamma}_{jk} \sim \mathcal{N}(\boldsymbol{0}, \tau_{jk}^2 \boldsymbol{K}_{jk}^-), \tag{1.13}$$

where $\mathcal{D}(\theta_1, \ldots, \theta_K)$ is a K-parametric distribution and the vectors $\boldsymbol{\theta}_k$ and $\boldsymbol{\eta}_k$ are of length n, i.e. $\boldsymbol{\theta}_k = (\theta_{1k}, \theta_{2k}, \ldots, \theta_{nk})^\top$ for $k = 1, \ldots, K$. Notice the difference between the K-dimensional vector $\boldsymbol{\theta} = (\theta_1, \theta_2, \ldots, \theta_K)^\top$, which represents the distribution parameters in general for any $\boldsymbol{x}$, the K-dimensional vector $\boldsymbol{\theta}_{[i]} = (\theta_{i1}, \theta_{i2}, \ldots, \theta_{iK})^\top$, which represents the distribution parameters for the ith observation, and the n-dimensional vector $\boldsymbol{\theta}_k = (\theta_{1k}, \theta_{2k}, \ldots, \theta_{nk})^\top$, which represents the kth distribution parameter for n observations.[6]

The assumptions of the GAMLSS models defined by equations (1.10)–(1.13) are scrutinized and discussed throughout this book. Equation (1.10) concerns the distributional assumptions of a GAMLSS model: Chapter 2 covers some aspects related to the type of distribution appropriate for the distributional assumption, and gives practical advice for choosing an appropriate response distribution. Chapter 3 covers the different terms appropriate for equation (1.12). Link functions appropriate for equation (1.11) are not particularly targeted in this book and are usually chosen by default to map the predictors $\boldsymbol{\eta}_k$ onto the appropriate support of the parameter $\boldsymbol{\theta}_k$.

---

[6] We use the following terminology: *distribution parameters* for the $\boldsymbol{\theta}$'s; *coefficients* for the β's and γ's; and *hyperparameters* or *smoothing parameters* for the τ^2's or equivalently the λ's.

Chapter 4 introduces general ideas underlying statistical modeling and inference, and some general tools for working with GAMLSS, in particular with respect to model choice and interpretation. Different methods of estimating the parameters of equation (1.12) and the hyperparameters in equation (1.13) are discussed in Chapters 5, 6 and 7.

1.6 Other Distributional Regression Approaches

We outline here some alternative approaches to distributional regression, that is, other approaches of overcoming the focus on mean-based regression analyses. This treatment is by no means exhaustive and focuses on quantile regression and conditional transformation model as specific model classes. More extensive reviews are provided in Kneib (2013) and Kneib et al. (2023)

1.6.1 Quantile Regression

In GAMLSS, the whole distribution of the response is estimated simultaneously, making all its characteristics available to the researcher based on one convenient and coherent model assumption. The downside of this approach is that we are strongly relying on the assumption that we are able to specify a single distribution that fits all the data well. Quantile regression, in contrast, does not aim at inferring all aspects of the conditional distribution of a response variable given covariates, but rather focuses on local features of this conditional distribution, namely conditional quantiles for given quantile levels. As an advantage, it does not require the assumption of a specific response distribution,[7] alleviating the risk of distribution model misspecification.

Under suitable assumptions on the data generating process, quantile regression provides us with consistent and asymptotically unbiased estimates of the underlying population quantiles (given the model is approximately correct). Since a set of quantiles also provides an (indirect) characterization of the conditional response distribution, including the possibility to study features such as variability and skewness, quantile regression is also a distributional regression approach. There is an important general point to be made here. While the distribution-free approach sounds appealing, it makes it more difficult to check the adequacy of the fitted model. It seems that the more assumptions we make, the easier it is to check the adequacy of those assumptions. For example, if we do assume a distribution we can easily define the residuals of the model and through those residuals check the adequacy of the distribution. If we do not specify a distribution, the residuals are more difficult to obtain and therefore it is more difficult to check the model adequacy. Put another way, within statistical modeling there is no free lunch.

In the following, we briefly sketch the basic approach to quantile regression for models with linear predictors, while providing some references for more general model variants at the end of this section.

---

[7] It does assume that the conditional cdf $F(y|x)$. exists, but does not specify the exact form for it.

The classical linear model specifies the conditional mean of the response variable y given covariate $\boldsymbol{x}$ as

$$\mathbb{E}(y|\boldsymbol{x}) = \boldsymbol{x}^\top \boldsymbol{\beta}$$

and estimates for the regression are typically obtained by minimizing the least squares criterion

$$S_2(\boldsymbol{\beta}) = \sum_{i=1}^{n} \left(y_i - \boldsymbol{x}_i^\top \boldsymbol{\beta}\right)^2$$

with respect to the regression coefficients $\boldsymbol{\beta}$. For i.i.d. samples, it is well known that minimizing the sum of absolute deviations from a central tendency measure yields the median, such that it seems natural to define regression medians as the minimizers of the absolute error criterion

$$S_1(\boldsymbol{\beta}) = \sum_{i=1}^{n} \left|y_i - \boldsymbol{x}_i^\top \boldsymbol{\beta}\right|.$$

More generally, considering the asymmetrically weighted absolute error criterion

$$S_q(\boldsymbol{\beta}) = (1-q) \sum_{i:y_i < \boldsymbol{x}_i^\top \boldsymbol{\beta}_q} \left|y_i - \boldsymbol{x}_i^\top \boldsymbol{\beta}_q\right| + q \sum_{i:y_i \geq \boldsymbol{x}_i^\top \boldsymbol{\beta}_q} \left|y_i - \boldsymbol{x}_i^\top \boldsymbol{\beta}_q\right| \qquad (1.14)$$

for $0 < q < 1$ yields regression quantiles, with the special case $q = 0.5$ reducing to the regression median.

An alternative perspective that emphasizes the model structure underlying quantile regression starts from the regression specification

$$y = \boldsymbol{x}^\top \boldsymbol{\beta}_q + \varepsilon_q,$$

where, instead of assuming $\mathbb{E}(\varepsilon_q) = 0$ as in mean-based regression, we assume that $Q_q(\varepsilon_q) = 0$, that is, the q-quantile of the error term ε_q is assumed to be zero. This implies that

$$Q_q(y) = \boldsymbol{x}^\top \boldsymbol{\beta}_q,$$

that is, the regression predictor determines the q-quantile of the response distribution.

Comparing quantile regression to GAMLSS, the two main advantages of quantile regression are the absence of a global distributional assumption; and the robustness with respect to outliers which is inherent to the definition of quantiles. A typical example in which quantile regression can work better than GAMLSS is when the conditional distribution is bimodal while the assumed GAMLSS distribution is unimodal.

While individual quantiles, estimated using quantile regression, are consistent and asymptotically unbiased, a set of estimated quantiles based on the same data may not be. The locality of the model assumed for quantile regression implies that in fact no globally consistent model can be defined except for the trivial case of $\boldsymbol{\beta}_q \equiv \boldsymbol{\beta}$ independent of the quantile level q. Indeed, if the separate quantile regressions are not exactly parallel to each other, the fitted quantiles will inevitably cross at some point. In many cases, this will only happen well outside the range of the observed covariates, but for a dense set of quantile levels and small samples, quantile crossing may also be an issue inside the range of observed covariates.

This is a rather unpleasant feature of quantile regression. There are several attempts in the literature to rectify the "crossing" problem. The joint estimation of multiple quantile regressions called *quantile sheets* was introduced by Schnabel and Eilers (2013) as a valuable alternative, but unfortunately the methodology only applies to models with a single explanatory variable; see also Sottile and Frumento (2021) for a recent attempt. The non-crossing is usually achieved by adding more constraints to the estimating function of quantile regression (see equation (1.14)). It is hard though to see how those added constraints will not affect the consistency and unbiasedness of the resulting estimates.

We advocate a "dual" approach, particularly when the focus of the analysis is on the quantiles of the conditional distribution, as for example in centile estimation for growth curves. This consists of fitting a GAMLSS distribution model to the data and then using different quantile regression curves to check it, or vice versa. Note that quantile regression is necessarily restricted to continuous response distributions, while GAMLSS accommodates continuous, discrete, and mixed discrete–continuous distributions.

The estimation of quantile regression models usually relies on linear programming such that flexible extensions, for example, models including random effects or penalized splines, are more difficult to derive. However, one can use the fact that the quantile estimating function of equation (1.14) is identical to an asymmetric Laplace distribution probability function, and use GAMLSS to fit it.[8] An alternative estimation scheme for quantile regression is statistical boosting (see Chapter 7).

Because of its local nature, quantile regression does not have residuals in the conventional sense, neither is there a general measure for goodness of fit. The only residuals are binary residuals, which indicate whether an observation is above or below the fitted quantile curve. While a GAMLSS model is more difficult to find, when it is found, it provides far more information about the data generating mechanism and its properties, and checking of its assumptions is easier.

In summary, the effect of a covariate on any part of the conditional response distribution can be easily assessed with quantile regression without needing the specify a parametric distribution. We illustrate the method on the BMI measurements for Dutch boys. Figure 1.6 shows quantile curves for the quantile levels $q = 0.03$, 0.1, 0.25, 0.5, 0.75, 0.9, and 0.97. Panel (a) shows data on boys aged between 10 and 20 years. A linear term for age has been fitted. Panel (b) shows the complete data, with a nonparametric smoothing term for age. (As previously, the transformed variable $x = \texttt{age}^{1/3}$ was used instead of $\texttt{age}$ in the fitting process.) The corresponding

---

[8] The asymmetric Laplace distribution is a special case of the GAMLSS distribution SEP3, with fixed parameters $\sigma = \tau = 1$ and $\nu = [(1 - q)/q]^{0.5}$ where q is the quantile value.

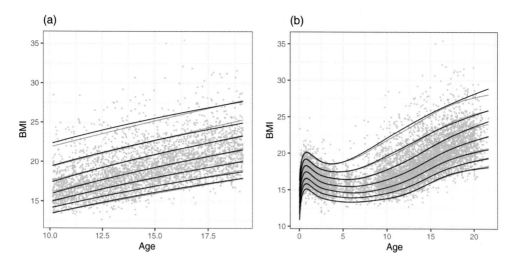

Figure 1.6 Quantile regression performed on the Dutch boys' BMI: (a) boys aged between 10 and 20 years, (b) boys aged between 0 and 23 years. The fitted quantile curves are evaluated at quantiles 0.03, 0.10, 0.25, 0.50, 0.75, 0.90, and 0.97 and are compared in panel (b) to the corresponding centiles of the GAMLSS model fitted using the BCT distribution as in Figure 1.5(d).

centiles of the GAMLSS model fitted using the BCT distribution as in Figure 1.5(d) are also shown, for comparison. As the curves are approximately parallel across the age range, we can conclude that the effect of age on BMI is similar in all regions of the distribution of BMI. The quantile regression was implemented using `gamlss()` with the asymmetric Laplace response distribution, that is, `SEP3` with $\sigma = \tau = 1$ and $\nu = [(1 - q)/q]^{0.5}$. This produced similar results to the `qgam()` function in the **qgam** package (Fasiolo et al., 2021).

An extensive treatment of quantile regression methodology is provided in the classical textbook of Koenker (2005). A more recent overview of current developments is available in the *Handbook of Quantile Regression* (Koenker et al., 2020), which also discusses advances on additive quantile regression approaches (see also Fenske et al., 2011; Waldmann et al., 2013; Fasiolo et al., 2021) and ways of circumventing crossing quantiles (see also Chernozhukov et al., 2009; Bondell et al., 2010; Rodrigues and Fan, 2017).

Although quantile regression was originally developed as a distribution-free approach which does not lend itself well to a Bayesian treatment, Bayesian quantile regression has been suggested utilizing the asymmetric Laplace distribution as working model (Yu and Moyeed, 2001). Since the asymmetric Laplace distribution enjoys a representation as a location–scale mixture of normals (Kozumi and Kobayashi, 2011; Yue and Rue, 2011), efficient Bayesian inference can be implemented; this allows complex predictor structures as in GAMLSS (see for example Waldmann et al., 2013).

An alternative to quantile regression is expectile regression, where instead of considering an asymmetrically weighted absolute error criterion, an asymmetrically weighted squared error criterion is employed. Expectile regression then includes the ordinary least squared (OLS) based mean regression as a special case, but still allows for studying the complete response distribution by varying the asymmetry of the estimation criterion. Expectiles were originally suggested in Newey and Powell (1987) and have regained more interest in recent years due to their ability to accommodate flexible predictor structures (see, for example, Schnabel and Eilers, 2009; Sobotka and Kneib, 2012).

1.6.2 Conditional Transformation Models

Conditional transformation models (CTMs) are another approach to the issue of allowing the conditional distribution to be fully responsive to covariate values. Instead of directly specifying the response distribution of interest, CTMs aim at identifying the required transformation to map the conditional distribution of the responses to a simple reference distribution. This is similar in spirit to earlier attempts such as the Box–Cox transformation, which aims to make the response distribution more normal-like.

In a very general approach, CTMs can be specified as

$$h(Y|\boldsymbol{x}) \overset{\mathcal{D}}{=} Z \sim \mathcal{N}(0, 1),$$

where $h(\cdot|\boldsymbol{x})$ is a covariate-dependent transformation function that is strictly increasing in y and which is chosen such that the conditional distribution of the response is matched to a standard normal. Indeed, for continuous distributions, one can show that a unique transformation of this type always exists, if we are flexible enough with respect to $h(\cdot|\boldsymbol{x})$. Note that here we are explicitly denoting random variables as capital letters to ease the understanding of the model specification.

Due to the monotonicity assumed for the transformation function, the model can be inverted to

$$Y \overset{\mathcal{D}}{=} h^{-1}(Z), \quad Z \sim \mathcal{N}(0, 1).$$

Another perspective on the model is obtained when looking at the conditional cumulative distribution function (cdf) of the response variable, which is given by

$$F_{Y|\boldsymbol{x}}(y) = \mathbb{P}(Y \leq y|\boldsymbol{x}) = \Phi(h(y|\boldsymbol{x})).$$

Thus the CTM allows us to relate the cdf of Y to the cdf of a standard normal evaluated at a transformed argument. From the cdf, we can directly determine the density

$$f_{Y|\boldsymbol{x}}(y) = \phi(h(y|\boldsymbol{x})) \left| \frac{\partial}{\partial y} h(y|\boldsymbol{x}) \right|$$

which then also gives rise to likelihood-based inference.

The main difficulties with turning CTMs into practice are the choice of a suitable parametrization of the transformation function and the interpretation of the resulting models. For the former, ensuring monotonicity of the transformation function is the main obstacle, where solutions based on Bernstein polynomials are particularly attractive since monotonicity constraints can then be enforced via linear constraints; see Hothorn et al. (2018) for details. While the original formulation of CTMs is targeted towards univariate, continuous responses, discrete and multivariate versions have also been suggested, see Siegfried and Hothorn (2020) and Klein et al. (2022). In addition to a likelihood-based treatment of CTMs, Bayesian variants (Carlan et al., 2023) and boosting approaches (Hothorn et al., 2014; Hothorn, 2020) are also conceivable.

2

Distributions

Distributions are essential to the specification of any GAMLSS model: Picking the right response distribution is, in combination with the choice of the parametrization, the response functions, and the specification of regression predictors, one of the central modeling tasks when coming up with a GAMLSS specification for a given dataset. Solid knowledge of the types of distributions that are available for GAMLSS and the possibilities for adapting existing and generating new types of distributions, is therefore very relevant for anyone interested in applying GAMLSS. In this chapter, we

- review important types of distributions available for distributional regression modeling,

- introduce multivariate extensions that have only recently made their way into the GAMLSS paradigm, and

- indicate methods for selecting an appropriate response distribution.

A variety of resources providing a plethora of distributions is available in the **R** environment. Most importantly, the **gamlss.dist** package provides access to over 100 *explicit* response distributions tailored towards applications in the GAMLSS modeling frame and, in addition, facilitates the generation of user-defined *"implicit"* response variable distributions. The available univariate response distributions (both implicit and explicit) are comprehensively documented in Rigby et al. (2019), to which the reader is referred for derivations and detailed properties. Methods for generating many of the distributions are described briefly in Section 2.5, and in chapter 13 of Rigby et al. (2019). Other **R** packages provide additional distributions. For example, the **bamlss** package (Umlauf et al., 2018) includes several distributions which have been used for Bayesian fitting of a GAMLSS, see Chapter 6. The **bamlss** package also allows all explicit distributions of **gamlss** to be imported. The package **GJRM** (Marra and Radice, 2020) fits a range of distributional regression models with a wide choice of response variable distribution, and in particular multivariate distributions based on copulas. The package **VGAM** (Yee, 2019) for vector generalized additive models (VGAMs), is another good source of distributions in **R**, with over 100 available response distributions.

The choice of an appropriate distribution for a given response variable depends on properties of this response, such as continuous vs. discrete vs. mixed discrete–continuous; restrictions on the support such as non-negativity; univariate vs. multivariate distributions, etc.; as well as the question under investigation. Practical advice for choosing the response distribution for a given dataset is given in Section 2.6.

In the next sections, we provide a broad overview of categories of distributions.

2.1 Some Terminology and Notation

We assume the random variable y has density or probability mass function $f(y)$ with support on range $\mathcal{S}$:

- y is a *continuous* random variable if $\mathcal{S}$ is an interval of $\mathbb{R} = (-\infty, \infty)$, or the union of two or more intervals of $\mathbb{R}$ and the probability distribution is characterized by the density function $f(y) \geq 0$ such that $\int_{\mathcal{S}} f(y) = 1$;

- y is a *discrete* random variable if $\mathcal{S}$ is a discrete set, i.e. a set with a finite or countably infinite number of elements and the probability distribution is characterized by the probability mass function $f(y)$, such that $\sum_{\mathcal{S}} f(y) = 1$;

- y is a *mixed* random variable if $\mathcal{S} = \mathcal{S}_1 \cup \mathcal{S}_2$ is the union of interval(s) $\mathcal{S}_1 \subset \mathbb{R}$ and a discrete set $\mathcal{S}_2$ with nonzero probability.

For simplicity we will refer to probability functions as a unifying term for *probability mass functions* for discrete random variables and *probability density functions* for continuous random variables and use the notation $f(y)$ throughout. Typically distributions depend on parameter(s) and where necessary this dependence is made explicit as, for example, $f(y|\boldsymbol{\theta})$ or $f_{\boldsymbol{\theta}}(y)$.

One-parameter distributions are of limited use in flexible regression modeling, as only one dedicated aspect of the response distribution such as the location of the distribution may be modeled while all other distribution parameters are implicitly determined by the single parameter. Two-parameter distributions are capable of modeling more flexible types of response distributions, for example by considering location and scale while any other shape features are implied. Three-parameter and four-parameter distributions provide even more flexibility, at the expense of increased complexity in the distributional form.

In what follows, the **gamlss** names for distributions are given as, for example, NO for the normal distribution.

2.2 Continuous Distributions

The continuous distributions can be subdivided into three categories, according to their support on:

- the real line $\mathbb{R} = (-\infty, \infty)$,
- the positive real line $\mathbb{R}_+ = (0, \infty)$, and

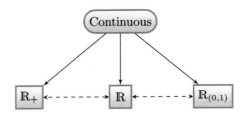

Figure 2.1 The different types of continuous distributions. Note that $\mathbb{R}$ and $\mathbb{R}_+$, and $\mathbb{R}$ and $\mathbb{R}_{(0,1)}$, are connected by dashed arrows to emphasize the fact that within the GAMLSS software, distributions on $\mathbb{R}$ can become distributions on $\mathbb{R}_+$ or $\mathbb{R}_{(0,1)}$, and vice versa, by transformation, censoring, or truncation.

- the unit interval $\mathbb{R}_{(0,1)} = (0,1)$.

The subdivisions are shown diagrammatically in Figure 2.1.

Notice that in general any distribution on $\mathbb{R}$ can be transformed to a distribution defined on $\mathbb{R}_+$ or $\mathbb{R}_{(0,1)}$, and vice versa. For example, the exponential transformation of a distribution on $\mathbb{R}$ results in a distribution on $\mathbb{R}_+$; the logit transformation of a distribution on $\mathbb{R}$ yields a distribution on $\mathbb{R}_{(0,1)}$; and the inverse transformations work in the reverse direction. In addition, any distribution on $\mathbb{R}$ can be truncated or censored to one on $\mathbb{R}_+$ or $\mathbb{R}_{(0,1)}$. In the **gamlss** package there are functions which enable the user to move from distributions on $\mathbb{R}$ to those on $\mathbb{R}_+$ or $\mathbb{R}_{(0,1)}$; for more detailed information see Rigby et al. (2019, sections 5.6 and 6.3).

Distributions for continuous data available in **gamlss** are shown in Table A.1.

2.2.1 Continuous Distributions on $\mathbb{R}$

One-Parameter Distribution on $\mathbb{R}$

Arguably the most well-known continuous distribution after the normal is the t- (or Student's t-) distribution. It was developed famously in the sampling context as the distribution underlying the t-test (Gosset, 1908). In its standard form, it is symmetric about zero and has a single parameter (degrees of freedom, usually denoted as ν) which controls its spread (and kurtosis). As the t-distribution does not have a location parameter, it is not useful as a response distribution. However, by introducing a location and scale parameters to the simple t, the resulting three-parameter t-family distribution is useful in the regression context and is discussed below.

Two-Parameter Distributions on $\mathbb{R}$

The *normal* and *logistic* are well-known symmetric distributions, with the logistic being leptokurtic (more peaked in the middle and with heavier tail) compared with

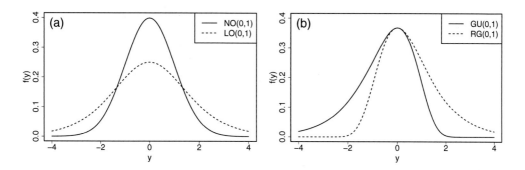

Figure 2.2 Two-parameter distributions on $\mathbb{R}$. (a) Normal NO(0,1) and logistic LO(0,1); (b) Gumbel GU(0,1) and reverse Gumbel RG(0,1) distributions.

the normal. The *Gumbel* and *reverse Gumbel* are left-skewed and right-skewed, respectively. Figure 2.2 compares the probability functions of the standardized versions ($\mu = 0$, $\sigma = 1$) of these distributions.

Three-Parameter Distributions on $\mathbb{R}$

- The *power exponential* distribution is symmetric about its mean μ, with variance σ^2 and parameter ν controlling the kurtosis. Special cases are the Laplace ($\nu = 1$) and normal ($\nu = 2$) distributions; and the uniform distribution is the limiting case as $\nu \to \infty$.

- The *t-family* distribution is an extension of the original Student's t-distribution with symmetry about the location parameter μ, and with scale parameter σ and the original degrees of freedom parameter ν.

- The *skew normal* distribution types 1 and 2 have location parameter μ (in type 2, μ is the mode), scale parameter σ, and skewness parameter ν, where $\nu \in \mathbb{R}$. Left skewness is therefore possible. The limiting case as $\nu \to \infty$ is the half-normal, and as $\nu \to -\infty$ the reflected half-normal.

- The *normal family* distribution is a normal distribution, but with its mean-variance relationship controlled by ν. It is symmetric about its mean μ and mesokurtotic, with variance proportional to a power of the mean: $\mathbb{V}(y) = \sigma^2 \mu^\nu$. Note that ν is not designed to be modeled with covariates, but is used to regulate the mean-variance relationship. Note also that we must have $\mu > 0$.

- The *exponential Gaussian* (or exponentially modified Gaussian) distribution is derived as the distribution of the sum of independent exponential and normal random variates, namely $y = y_1 + y_2$, where $y_1 \sim \text{Exp}(\nu)$ and $y_2 \sim \mathcal{N}(\mu, \sigma^2)$, independently. It is positively skewed, which is inherited from its exponential component, and has mean $\mu + \nu$.

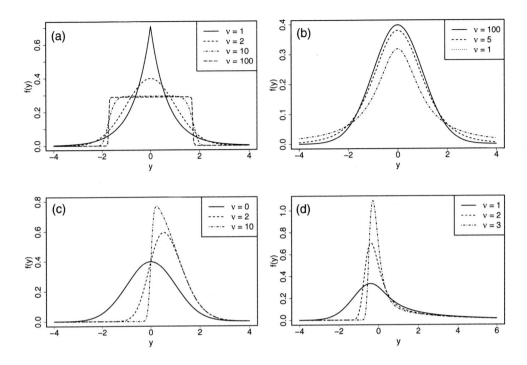

Figure 2.3 Three-parameter distributions on $\mathbb{R}$: (a) power exponential $(\mathtt{PE}(0, 1, \nu))$, (b) t-family $(\mathtt{TF}(0, 1, \nu))$, (c) skew normal type 1 $(\mathtt{SN1}(0, 1, \nu))$, and (d) exponential Gaussian $(\mathtt{exGAUS}(0, 1, \nu))$.

Probability functions of the power exponential, t-family, skew normal, and exponential Gaussian distributions are shown in Figure 2.3.

Four-Parameter Distributions on $\mathbb{R}$

The groups of distributions are as follows.

- Distributions that model skewness and kurtosis:
  - *skew exponential power* distributions (SEP1–SEP4),
  - *sinh-arcsinh* distributions (SHASH, SHASHo, SHASHo2) (Figure 2.4).
- Distributions that model skewness and leptokurtosis:
  - *skew t-*distributions (ST1–ST5, SST),
  - *Johnson's SU* distribution (JSU, JSUo),
  - *exponential generalized beta type 2* distribution (EGB2).
- The *generalized t-*distribution (GT) is symmetric. Kurtosis may be modeled, but not skewness.

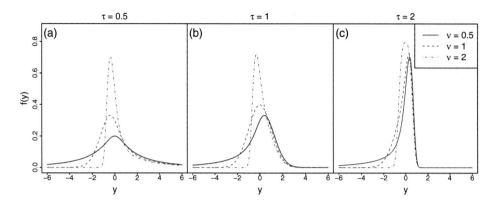

Figure 2.4 Four-parameter distribution on $\mathbb{R}$: sinh-arcsinh `SHASH`$(0, 1, \nu, \tau)$. Panel (a) $\tau = 0.5$, (b) $\tau = 1$, and (c) $\tau = 2$.

- The *normal-exponential-t* distribution (`NET`) has four parameters, but is treated as a two-parameter distribution in **gamlss**, with ν and τ fixed. It is symmetric about μ (which is the mean if it exists) and was introduced for robust estimation of location and scale.

The reader can find more details and properties for continuous distributions defined on $\mathbb{R}$ in Rigby et al. (2019, chapters 4 and 18).

2.2.2 Continuous Distributions on $\mathbb{R}_+$

One-Parameter Distribution on $\mathbb{R}_+$

The *exponential* distribution, a special case of the gamma and the Weibull, is a well-known one-parameter distribution on $\mathbb{R}_+$.

Two-Parameter Distributions on $\mathbb{R}_+$

The *gamma* and *inverse Gaussian* (Figure 2.5) are members of the exponential family of distributions. The inverse Gaussian has a higher positive skewness and a heavier right tail than the gamma. Other two-parameter distributions available for modeling are

- the *log normal*,

- the *Pareto*, the well-known heavy-tailed distribution for modeling extreme events,

- the *Weibull* distribution, used in parametric survival analysis, and

- the *inverse gamma* distribution, much-used as a prior in Bayesian analysis.

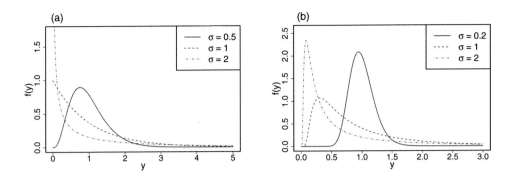

Figure 2.5 Two-parameter distributions on $\mathbb{R}_+$: (a) gamma (`GA`) distribution, with $\mu = 1$ and $\sigma = 0.5, 1$ and 2; (b) inverse Gaussian (`IG`) distribution, with $\mu = 1$ and $\sigma = 0.2, 1$ and 2.

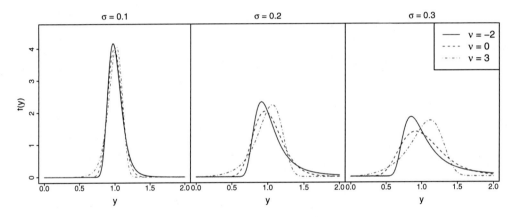

Figure 2.6 Three-parameter distribution on $\mathbb{R}_+$: the Box–Cox Cole and Green (`BCCG`) distribution, with $\mu = 1$, $\sigma = 0.1, 0.2$ and 0.3; and $\nu = -2, 0$ and 3.

Three-Parameter Distributions on $\mathbb{R}_+$

- The *generalized inverse Gaussian* is a flexible three-parameter distribution, of which the inverse Gaussian is a special case.

- The *gamma family* distribution is a three-parameter version of the gamma distribution, in which the third parameter controls the mean-variance relationship.

Other three-parameter distributions on $\mathbb{R}_+$ in the **gamlss** system are derived by transformation:

- the *Box–Cox Cole and Green* (`BCCG`) distribution is defined by assuming that the random variable z has a truncated standardized normal distribution, where z is a Box–Cox transformation of y (Figure 2.6);

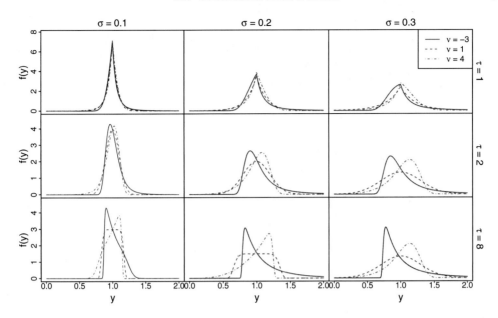

Figure 2.7 Four-parameter distribution on $\mathbb{R}_+$: the Box–Cox power exponential (BCPE), with $\mu = 1$, $\sigma = 0.1, 0.2, 0.3$, $\nu = -3, 1, 4$, and $\tau = 1, 2, 8$.

- the *generalized gamma* distribution is derived by transformation by assuming that $z = (y/\mu)^\nu$ has a gamma distribution with mean 1; and

- the *log normal family* is obtained in a similar way to the BCCG distribution.

Four-Parameter Distributions on $\mathbb{R}_+$

- The *Box–Cox power exponential* (BCPE) distribution is derived by assuming z (as above) has a truncated standard power exponential distribution (Figure 2.7).

- The *Box–Cox t* (BCT) is derived by assuming z (as above) has a truncated t-family distribution.

- The *generalized beta type 2* (GB2) is a flexible distribution which contains various known distributions as special cases. For example, setting $\sigma = 1$ we have a specific form of the Pearson type VI distribution. The Burr XII (or Singh–Maddala) distribution is given when $\nu = 1$. The Burr III (or Dagum) distribution is given when $\tau = 1$. For $\sigma = \tau = 1$ we have the Pareto distribution.

Both BCT and BCPE distributions are flexible, reliable and fit a variety of data well. The GB2 is flexible and can model skewness and both platykurtosis and leptokurtosis.

The reader can find more details and properties for continuous distributions defined on $\mathbb{R}_+$ in Rigby et al. (2019, chapters 5 and 19).

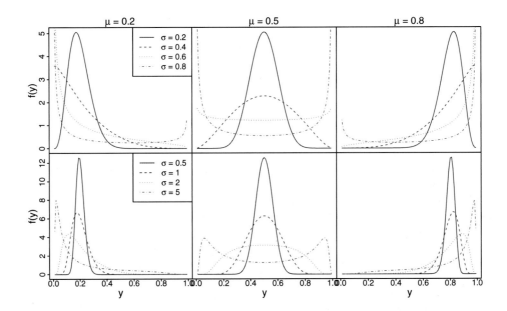

Figure 2.8 Top row: beta (BE) distribution. Bottom row: SIMPLEX distribution.

2.2.3 *Continuous Distributions on* $\mathbb{R}_{(0,1)}$

Typically, a bounded continuous response is a fraction or proportion on $\mathcal{S} = (0,1)$; as bounded responses on other ranges can be scaled to $(0,1)$, we assume $\mathcal{S} = (0,1)$ without loss of generality. In the package **gamlss.dist** there are only four explicitly defined distributions on $\mathbb{R}_{(0,1)}$; however, any distribution on $\mathbb{R}$ can be inverse "logit" transformed to a $\mathbb{R}_{(0,1)}$ distribution. The most well-known of the explicitly defined is the beta distribution (Figure 2.8, top row), being until fairly recently the only candidate for this range. Other, lesser-known, two-parameter bounded continuous distributions are the *logit normal* distribution, derived by the inverse logit transformation of a normal random variable, and the *simplex* distribution (Figure 2.8, bottom row), a special case of the four-parameter generalized simplex distribution. The *generalized beta type I* is a four-parameter bounded continuous distribution, of which the beta is a special case. Location and dispersion may be modeled with the two-parameter distributions; the generalized beta type I has more flexibility. All of these distributions are capable of U- and J-shapes; the simplex can be bimodal, where the modes are inside $(0,1)$ (Figure 2.8, bottom middle panel). More information and examples about $\mathbb{R}_{(0,1)}$ defined distributions is given in Rigby et al. (2019, chapters 6 and 21). A feature of all of the bounded continuous distributions is that the support does not include the endpoints of the range. Section 2.2.4 deals with zero- and one-inflated distributions, in which the endpoints zero and one are allowed.

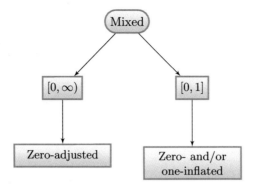

Figure 2.9 The types of mixed distributions.

The Dirichlet distribution is a multivariate generalization of the beta distribution, and is discussed in Section 2.4.2.

2.2.4 Mixed Distributions

There are two types of mixed distributions: zero-adjusted distributions defined on the range $[0, \infty)$ and zero- and/or one-inflated distributions defined on $[0, 1]$ (Figure 2.9). More on mixed distributions can be found in Rigby et al. (2019, chapter 9). The package **gamlss.dist** contains only a few explicitly defined zero-adjusted and zero-inflated distributions; however **gamlss.inf** allows the transformation of all distributions defined on $\mathbb{R}_+$ to zero-adjusted, and all defined on $\mathbb{R}_{(0,1)}$ to zero- and/or one-inflated.

Zero-Adjusted Distributions

Zero-adjusted distributions are continuous distributions on $\mathbb{R}_+$ with a spike or probability mass at zero. These are encountered in situations such as daily rainfall, where the response can be either exactly zero, or a continuous positive quantity. (Min and Agresti (2002) refer to these distributions as *semicontinuous*.) The support is $\mathcal{S} = [0, \infty)$ and the probability function can be expressed as the *mixed probability function*:

$$f(y|\boldsymbol{\theta}, \pi) = \begin{cases} \pi & \text{if } y = 0 \\ (1 - \pi) \, f_c(y|\boldsymbol{\theta}) & \text{if } y > 0, \end{cases}$$

where $f_c(y|\boldsymbol{\theta})$ is the probability density function (pdf) of $y \in \mathbb{R}_+$. The zero-adjusted version of any distribution on $\mathbb{R}_+$ can be constructed in this way. The current implementation of **gamlss** includes the zero-adjusted gamma (`ZAGA`) and zero-adjusted inverse Gaussian (`ZAIG`) distributions (Figure 2.10). Zero-adjusted versions of all other `gamlss.family` distributions on $\mathbb{R}_+$ can be generated using the `gen.Zadj()` function. More on zero-adjusted distributions can be found in Rigby et al. (2019, chapters 9 and 21).

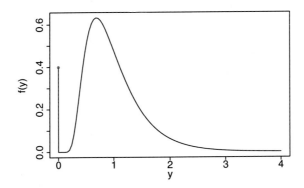

Figure 2.10 Zero-adjusted inverse Gaussian (`ZAIG`) distribution, with $\mu = 1$, $\sigma = 0.5$, and $\nu = \mathbb{P}(y = 0) = 0.4$.

2.2.5 Inflated Distributions on $\mathbb{R}_{(0,1)}$

When we are dealing with a bounded continuous response on $\mathbb{R}_{(0,1)}$, a problem is encountered if the values zero and/or one are valid outcomes. We distinguish between two scenarios:

(1) zero and/or one are included in the range, but are not associated with inflated probabilities;

(2) zero and/or one are included in the range, and have probability spikes.

In the first scenario, a simple scaling proposed by Smithson and Verkuilen (2006) may be used. Assuming n raw observations $y^* \in [0, 1]$, use y defined as

$$y = \frac{(n-1)y^* + 0.5}{n},$$

giving $y \in (0, 1)$ as required.

The second scenario necessitates the use of probability spikes at zero and/or one. Similar to zero-adjusted distributions on $\mathbb{R}_+$, we have the mixed probability function of the zero-inflated distribution on $[0, 1)$:

$$f(y|\boldsymbol{\theta}, \pi) = \begin{cases} \pi & \text{if } y = 0 \\ (1 - \pi)\, f_c(y|\boldsymbol{\theta}) & \text{if } 0 < y < 1, \end{cases} \tag{2.1}$$

and the one-inflated distribution on $(0, 1]$:

$$f(y|\boldsymbol{\theta}, \pi) = \begin{cases} (1 - \pi)\, f_c(y|\boldsymbol{\theta}) & \text{if } 0 < y < 1 \\ \pi & \text{if } y = 1, \end{cases} \tag{2.2}$$

where $\pi \in (0, 1)$ and $f_c(\cdot)$ is the probability function of $y \in (0, 1)$. The mixed probability function of the zero- and one-inflated distribution on $[0, 1]$ can be written similarly as

$$f(y|\boldsymbol{\theta}, \pi_0, \pi_1) = \begin{cases} \pi_0 & \text{if } y = 0 \\ (1 - \pi_0 - \pi_1)\, f_c(y|\boldsymbol{\theta}) & \text{if } 0 < y < 1 \\ \pi_1 & \text{if } y = 1, \end{cases} \tag{2.3}$$

where $\pi_0 \in (0, 1)$, $\pi_1 \in (0, 1)$ and $\pi_0 + \pi_1 \in (0, 1)$. As this last constraint presents implementation difficulties in the regression framework, the following formulation is preferred:

$$f(y|\boldsymbol{\theta}, \xi_0, \xi_1) = \begin{cases} \frac{\xi_0}{1+\xi_0+\xi_1} & \text{if } y = 0 \\ \frac{1}{1+\xi_0+\xi_1} f_c(y|\boldsymbol{\theta}) & \text{if } 0 < y < 1 \\ \frac{\xi_1}{1+\xi_0+\xi_1} & \text{if } y = 1, \end{cases} \tag{2.4}$$

where the parameter constraints are $\xi_0 > 0$ and $\xi_1 > 0$, which are easily handled. A disadvantage of formulation (2.4) is that parameters ξ_0 and ξ_1 are not as readily interpretable as the probabilities π_0 and π_1 in model (2.3).

In **gamlss**, models (2.1), (2.2), and (2.4) are explicitly available for the beta distribution. In addition, zero-inflated, one-inflated, and zero- and one-inflated versions of any distribution on $(0, 1)$ may be created using the function `gen.Inf0to1()`. More details on zero- and/or one-inflated distributions are given in (Rigby et al., 2019, chapters 9 and 21).

2.3 Discrete Distributions

Discrete distributions have support which is either finite or a countably infinite set. In practice, these are generally subsets of the integers, the two most important categories being:

- $\mathcal{S} = \{0, 1, 2, \ldots\}$: (unbounded) count distributions, sometimes called "Poisson type" since the Poisson is the most common amongst them, and

- $\mathcal{S} = \{0, 1, \ldots, n\}$: bounded count distributions, also called "binomial type", since they are often modeled by binomial or related distributions.

These are discussed in Sections 2.3.1 and 2.3.2. Discrete distributions available in **gamlss** are shown in Table B.1.

2.3.1 Unbounded Count Distributions: $\mathcal{S} = \{0, 1, 2, \ldots\}$

Poisson Distribution

Historically, the Poisson (`PO`) distribution, derived from the Poisson point process, has been the starting point for (unbounded) count response variable distributions. It is a member of the exponential family of distributions and hence Poisson regression is a special case of the generalized linear model. The Poisson has a single parameter μ, and equality of its mean and variance: $\mathbb{E}(y) = \mathbb{V}(y) = \mu$. In practice, one frequently encounters count data which do not conform with the restrictive behavior of the Poisson distribution, leading to the problems of *overdispersion* (or less frequently *underdispersion*), *zero-inflation*, and *extreme tail* behaviour.

Figure 2.11 provides a diagram of the different types of (unbounded) count data distributions available within **gamlss**, according to which of the "Poisson" problems they tackle. If none of the available count distributions fits the data adequately, another option is the use of discretized continuous distributions or finite mixture distributions.

We now consider the "Poisson" problems in more detail.

Overdispersion

Overdispersion, in the Poisson context, is the phenomenon of observed variance far exceeding the observed mean. It can, for example, be modeled by assuming that the Poisson parameter is a random variable:

$$y|\lambda \sim \text{PO}(\lambda)$$
$$\lambda \sim \mathcal{D}(\mu, \boldsymbol{\theta}), \qquad (2.5)$$

where $\mathcal{D}$ is a continuous distribution on $\mathbb{R}_+$ having $\mathbb{E}(\lambda) = \mu$. The unconditional distribution of y is then

$$f(y|\mu, \boldsymbol{\theta}) = \int_0^\infty f_{y|\lambda}(y|\lambda) \, f_\lambda(\lambda|\mu, \boldsymbol{\theta}) \, d\lambda \qquad \text{for } y \in \mathbb{N}, \qquad (2.6)$$

where $f_{y|\lambda}(\cdot)$ and $f_\lambda(\cdot)$ are the probability functions of the $\text{PO}(\lambda)$ and $\mathcal{D}(\mu, \boldsymbol{\theta})$ distributions, respectively. The marginal distribution given by $f(y|\mu, \boldsymbol{\theta})$ is called a *mixed Poisson* (or *compound Poisson*) distribution, and $\mathcal{D}$ is called the mixing distribution. From the law of iterated variance, we get

$$\mathbb{V}(y) = \mathbb{E}[\mathbb{V}(y|\lambda)] + \mathbb{V}[\mathbb{E}(y|\lambda)]$$
$$= \mathbb{E}(\lambda) + \mathbb{V}(\lambda)$$
$$= \mu + \mathbb{V}(\lambda) > \mu,$$

such that the variance of a mixed Poisson distribution always exceeds its mean, as desired. Table 2.1 gives the mixed Poisson distributions implemented in **gamlss**, and plots of the Poisson and Poisson–inverse Gaussian distributions over a range of (μ, σ) combinations are shown in Figure 2.12. The variances of the alternative versions of the distributions in Table 2.1 (e.g. **NBI**, **NBII**, **NBF**) vary according to the parametrization of the mixing distribution.

Note that the negative binomial derived as a mixed Poisson distribution with gamma mixing distribution has a different parametrization to its genesis as the distribution of the number of failures till the kth success in independent Bernoulli trials, where the probability of success in each trial is π. The beta negative binomial (**BNB**) is

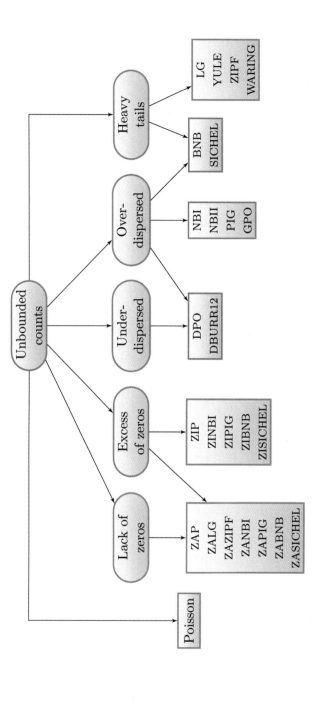

Figure 2.11 Unbounded count distributions available for GAMLSS modeling. Note that continuous distributions can be discretized using censoring and multimodal discrete distributions can be modeled using a finite mixture.

Table 2.1 *Mixed Poisson distributions implemented in* **gamlss**.

Distribution	Mixing distribution $\mathcal{D}$	**gamlss** name(s)
Delaporte	Shifted gamma	DEL
Negative binomial	Gamma	NBI, NBII, NBF
Poisson–inverse Gaussian	Inverse Gaussian	PIG, PIG2
Poisson–shifted generalized inverse Gaussian	Shifted generalized inverse Gaussian	PSGIG
Sichel	Generalized inverse Gaussian	SICHEL, SI

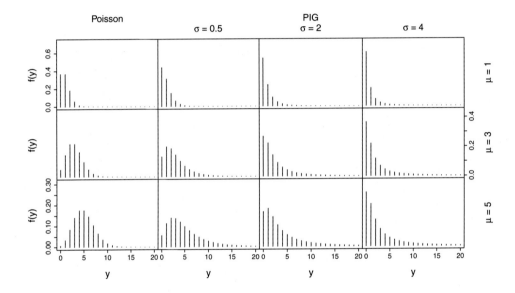

Figure 2.12 The Poisson and Poisson–inverse Gaussian (PIG) distributions, for $\mu = 1, 3, 5$ and $\sigma = 0.5, 2, 4$.

an overdispersed version of this original form, derived by using the beta mixing distribution for the parameter π. The geometric (GEOM) distribution is the special case of the negative binomial with $k = 1$, and the WARING distribution is the corresponding special case of the BNB. The one-parameter YULE distribution is a special case of the WARING. The generalized Poisson (GPO) is another overdispersed Poisson distribution, derived as a special case of a generalized negative binomial distribution.

Underdispersion

Underdispersion, or variance less than expected under the Poisson model, is less commonly observed than overdispersion. In **gamlss** the discrete Burr XII (DBURR12) distribution is capable of modeling underdispersion, and the double Poisson (DPO) of both underdispersion and overdispersion.

Zero-Inflated and Zero-Adjusted Count Distributions

The value zero may display a greater frequency than that expected under a count distribution model $\mathcal{P}(\boldsymbol{\theta})$. We model this phenomenon by assuming a proportion π of the population generates zero with certainty, and the remaining $(1 - \pi)$ generates counts according to $\mathcal{P}(\boldsymbol{\theta})$. An example is the number of claims on insurance policies over, say, a year, when a proportion of policyholders are averse to making claims. We have the zero-inflated probability function:

$$f(y|\boldsymbol{\theta}, \pi) = \begin{cases} \pi + (1 - \pi)f_{\mathcal{P}}(0|\boldsymbol{\theta}) & \text{if } y = 0 \\ (1 - \pi)f_{\mathcal{P}}(y|\boldsymbol{\theta}) & \text{if } y = 1, 2, \ldots, \end{cases} \tag{2.7}$$

where $f_{\mathcal{P}}(\cdot|\boldsymbol{\theta})$ is the probability function of $\mathcal{P}(\boldsymbol{\theta})$.

Zero-inflated count distributions currently implemented in **gamlss** are the zero-inflated Poisson (ZIP, ZIP2), zero-inflated negative binomial (ZINBI), zero-inflated Poisson–inverse Gaussian (ZIPIG), zero-inflated beta negative binomial (ZIBNB), and zero-inflated Sichel (ZISICHEL).

An alternative model for the situation when zero counts are different to that expected under $\mathcal{P}(\boldsymbol{\theta})$, is the *zero-adjusted* model, in which the zero probability can be inflated or deflated relative to that of the parent distribution:

$$f(y|\boldsymbol{\theta}, \pi) = \begin{cases} \pi & \text{if } y = 0 \\ \dfrac{1 - \pi}{1 - f_{\mathcal{P}}(0|\boldsymbol{\theta})} f_{\mathcal{P}}(y|\boldsymbol{\theta}) & \text{if } y = 1, 2, \ldots. \end{cases} \tag{2.8}$$

This is a *hurdle model* (Mullahy, 1986), with the hurdle at zero. Zero-adjusted models implemented in **gamlss** are the zero-adjusted Poisson (ZAP), zero-adjusted negative binomial (ZANBI), zero-adjusted logarithmic (ZALG), zero-adjusted Poisson–inverse Gaussian (ZAPIG), zero-adjusted Zipf (ZAZIPF), zero-adjusted beta negative binomial (ZABNB), and zero-adjusted Sichel (ZASICHEL).

2.3.2 Bounded Count Distributions: $\mathcal{S} = \{0, 1, \ldots, n\}$

The most well-known bounded count distribution is the binomial (BI), which has its genesis as the distribution of the number of events occurring in n independent Bernoulli trials, where the probability of event occurrence at each trial is π. The "binomial denominator" n is considered known and the binomial treated as a one-parameter distribution. The mean and variance of the binomial distribution are $\mathbb{E}(y) = n\pi$ and $\mathbb{V}(y) = n\pi(1 - \pi)$; where the sample variance of binomial-type data exceeds this, the "extra-binomial variation" may be modeled by assuming that

$$y|\pi \sim \text{BI}(n, \pi)$$
$$\pi \sim \mathcal{D}(\mu, \boldsymbol{\theta}),$$

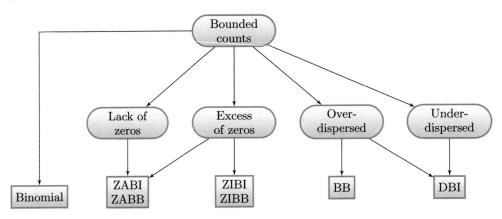

Figure 2.13 Bounded count distributions available for GAMLSS modeling.

where the mixing distribution $\mathcal{D}$ is continuous on $(0,1)$, with mean $\mathbb{E}(\pi) = \mu$. The unconditional distribution of y is then

$$f(y|n,\mu,\boldsymbol{\theta}) = \int_0^1 f_{y|\pi}(y|n,\pi)\, f_\pi(\pi|\mu,\boldsymbol{\theta})\, d\pi \qquad \text{for } y \in \{0,1,\ldots,n\},$$

where $f_{y|\pi}(\cdot)$ and $f_\pi(\cdot)$ are the probability functions of the $\mathtt{BI}(n,\pi)$ and $\mathcal{D}(\mu,\boldsymbol{\theta})$ distributions, respectively. The choice of the beta distribution for $\mathcal{D}$ results in the beta-binomial ($\mathtt{BB}$) distribution, which has $\mathbb{E}(y) = n\mu$ and

$$\mathbb{V}(y) = n\mu(1-\mu)\left[1 + \frac{(n-1)\sigma}{(1+\sigma)}\right],$$

exceeding the variance of the binomial distribution.

For situations where the probability at zero is either inflated or deflated relative to the parent distribution, a zero-inflated or zero-adjusted version of the distribution may be defined analogous to (2.7) and (2.8). Thus we have the zero-inflated and zero-adjusted versions of both the binomial ($\mathtt{ZIBI}$, $\mathtt{ZABI}$) and beta-binomial ($\mathtt{ZIBB}$, $\mathtt{ZABB}$) distributions.

2.4 Multivariate Continuous Distributions

2.4.1 Bivariate Normal Distribution

The normal distribution is unusual amongst statistical distributions, in that it has an obvious and unique multivariate version, whose correlation structure is unrestricted (subject to positive definiteness of its variance–covariance matrix). We restrict our attention to the bivariate case, since this is what we will be using in the chapters that follow. The bivariate normal distribution, denoted as $\mathcal{N}_2(\boldsymbol{\mu}, \boldsymbol{\Sigma})$, has mean parameter $\boldsymbol{\mu} = (\mu_1, \mu_2)^\top$ and variance–covariance matrix

$$\boldsymbol{\Sigma} = \begin{pmatrix} \sigma_1^2 & \sigma_{12} \\ \sigma_{12} & \sigma_2^2 \end{pmatrix}.$$

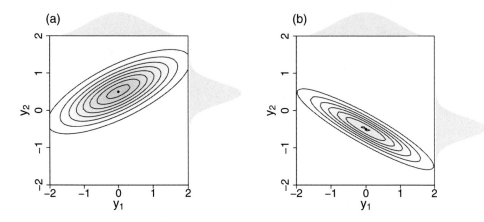

Figure 2.14 Contours of the $\mathcal{N}_2(\boldsymbol{\mu}, \boldsymbol{\Sigma})$ pdf and marginals: (a) $\boldsymbol{\mu} = (0, 0.5)^\top$ and $\boldsymbol{\Sigma} = \left(\begin{smallmatrix} 1 & 0.4 \\ 0.4 & 0.3 \end{smallmatrix}\right)$; (b) $\boldsymbol{\mu} = (0, -0.5)^\top$ and $\boldsymbol{\Sigma} = \left(\begin{smallmatrix} 1 & -0.5 \\ -0.5 & 0.3 \end{smallmatrix}\right)$.

If $\boldsymbol{y} \sim \mathcal{N}_2(\boldsymbol{\mu}, \boldsymbol{\Sigma})$:

- the marginal distributions are $y_j \sim \mathcal{N}(\mu_j, \sigma_j^2)$, for $j = 1, 2$;

- the correlation between y_1 and y_2 is $\rho = \frac{\sigma_{12}}{\sigma_1 \sigma_2}$;

- the conditional distribution of y_1 given y_2 is

$$y_1 | y_2 \sim \mathcal{N}\left(\mu_1 + \rho \sigma_1 \left(\frac{y_2 - \mu_2}{\sigma_2} \right), (1 - \rho^2) \sigma_1^2 \right),$$

which has the features of linear and homoscedastic regression, since its mean can be written as

$$\mathbb{E}(y_1 | y_2) = \beta_0 + \beta_1 y_2$$

and its variance does not involve y_2.

Figure 2.14 shows contours of the $\mathcal{N}_2(\boldsymbol{\mu}, \boldsymbol{\Sigma})$ pdf and marginals. Panel (a) shows $\boldsymbol{\mu} = (0, 0.5)^\top$ and $\boldsymbol{\Sigma} = \left(\begin{smallmatrix} 1 & 0.4 \\ 0.4 & 0.3 \end{smallmatrix}\right)$, that is, with correlation $\rho = 0.73$; panel (b) shows $\boldsymbol{\mu} = (0, -0.5)^\top$ and $\boldsymbol{\Sigma} = \left(\begin{smallmatrix} 1 & -0.5 \\ -0.5 & 0.3 \end{smallmatrix}\right)$, that is, with correlation $\rho = -0.91$.

2.4.2 Dirichlet Distribution

The Dirichlet distribution (Gupta and Richards, 2001) is a multivariate generalization of the beta distribution in which the components sum to 1. It is of interest because of its use as a prior distribution in Bayesian estimation, and also as the response distribution for compositional data, which is illustrated in Chapter 12. The Dirichlet is based on the original parametrization of the beta distribution (denoted

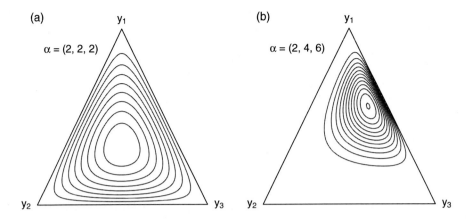

Figure 2.15 Contours of the three-dimensional Dirichlet distribution, on the 2-simplex, for (a) $\boldsymbol{\alpha} = (2, 2, 2)$ and (b) $\boldsymbol{\alpha} = (2, 4, 6)$.

in **gamlss** as $\text{BEo}(\alpha, \beta)$):

$$f(y|\alpha, \beta) = \frac{1}{B(\alpha, \beta)} y^{\alpha-1} (1-y)^{\beta-1} \qquad \text{for } y \in (0,1) \,,$$

where $\alpha > 0$ and $\beta > 0$. The k-dimensional Dirichlet distribution with parameter vector $\boldsymbol{\alpha} = (\alpha_1, \ldots, \alpha_k)^\top, \alpha_i > 0$, has support $y_i \in (0,1), i = 1, \ldots, k$ with $\sum_{i=1}^k y_i = 1$. Its pdf is given by

$$f(\boldsymbol{y}|\boldsymbol{\alpha}) = \frac{\Gamma(\alpha_0)}{\prod_{i=1}^k \Gamma(\alpha_i)} \prod_{i=1}^k y_i^{\alpha_i - 1},$$

where $\alpha_0 = \sum_{j=1}^k \alpha_j$. Its marginal distributions are $y_i \sim \text{BEo}(\alpha_i, \alpha_0 - \alpha_i)$, with means $\mathbb{E}(y_i) = \alpha_i/\alpha_0$ and variances

$$\mathbb{V}(y_i) = \frac{\alpha_i (\alpha_0 - \alpha_i)}{\alpha_0^2 (\alpha_0 + 1)},$$

for $i = 1, \ldots, k$. The correlation is necessarily negative because of the constraint that the y_i's sum to one:

$$\text{Cor}(y_i, y_j) = -\sqrt{\frac{\alpha_i \alpha_j}{(\alpha_0 - \alpha_i)(\alpha_0 - \alpha_j)}} \qquad \text{for } i \neq j.$$

Figure 2.15 shows contours of the three-dimensional Dirichlet distribution, on the 2-simplex.

2.4.3 Copula Models

Copulas (Joe, 1997; Nelsen, 2007) provide a convenient method for constructing multivariate continuous distributions as functions of their marginal distributions and a copula function. As in Section 2.4.1, we restrict our attention to the bivariate case. Consider a bivariate distribution with joint cdf $F(y_1, y_2)$ and marginal distributions $F_1(y_1)$ and $F_2(y_2)$. The copula function that links the joint cdf with its marginals is $C(u_1, u_2; \boldsymbol{\xi})$ where $u_i \in [0, 1]$ and

$$F(y_1, y_2) = C(F_1(y_1), F_2(y_2); \boldsymbol{\xi}). \tag{2.9}$$

$C(u_1, u_2; \boldsymbol{\xi})$ is a bivariate cdf with $\mathcal{U}(0, 1)$ marginals. Its *dependence parameter(s)* $\boldsymbol{\xi}$ control the dependence structure of the joint distribution. For example, the bivariate Clayton copula has the form

$$C(u_1, u_2; \xi) = \left(u_1^{-\xi} + u_2^{-\xi} - 1 \right)^{-1/\xi} \qquad \text{for } u_i \in [0, 1], i = 1, 2; \quad \xi > 0.$$

Dependence can, for example, conveniently be quantified by the rank correlation coefficient Kendall's τ (Puka, 2011) since the rank correlation is independent of the marginal distributions. For the Clayton copula, we find that Kendall's τ is related to the copula parameter ξ via $\tau = \xi/(\xi + 2)$, which is nonnegative, implying a monotonically positive relation between the two random variables. When not both of the marginals are continuous, $F(y_1, y_2)$ can still be expressed via equation (2.9) but in this case the copula is not unique.

Thus one can "mix and match" continuous marginal distributions with copula functions to obtain desired multivariate distributions. The top and middle rows of Figure 2.16 show contours of four bivariate densities, all with the same marginal distributions, one normal and one gamma, different copula functions, and dependence given by Kendall's $\tau = 0.6$. The Gumbel, Frank, Clayton, and Joe copulas give bivariate densities with quite different shapes. Not all copula functions admit negative dependence; the bottom row of Figure 2.16 shows the bivariate densities with the same marginals as in the top and middle rows, but with negative dependence (Kendall's $\tau = -0.6$), given by the Frank and normal copulas.

In the regression framework, copula models provide a convenient mechanism for the modeling of correlated outcomes. In addition to the specification of models for some or all of the parameters of the marginal distributions, covariates may also act on the dependence parameter(s) $\boldsymbol{\xi}$.

2.5 Generated Distributions

Distributions may be generated by several methods. We briefly describe some which are implemented in **gamlss**. These are discussed in more detail in Rigby et al. (2019, chapter 13).

- *Transformation of a single random variable* may be derived in various ways. For example, the log-normal distribution is derived as the distribution of $y = \exp(z)$, where $z \sim \mathcal{N}(\mu, \sigma^2)$. Let $y = g(z)$, and the inverse function of $g(\cdot)$ be $z = g^{-1}(y) = h(y)$. The log-normal and logit-normal distributions are explicit distributions in the `gamlss.family`, derived as above with $z \sim \mathcal{N}(\mu, \sigma^2)$ and $h(y) = \log(y)$ for the log-normal (`LOGNO`), and $h(y) = \text{logit}(y)$ for the logit-normal

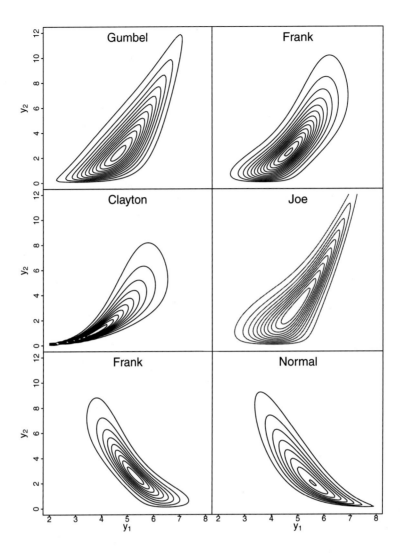

Figure 2.16 Contours of bivariate density functions with marginals
$y_1 \sim \mathcal{N}(5, 1)$ and $y_2 \sim \mathtt{GA}(1, 0.7)$. Top and middle rows: Gumbel, Frank,
Clayton, and Joe copula functions, all with Kendall's $\tau = 0.6$. Bottom row:
Frank and normal copula functions, both with Kendall's $\tau = -0.6$.

(`LOGITNO`). In addition, **gamlss** has the functionality for transforming any continuous distribution on $\mathbb{R}$ to one on either $\mathbb{R}_+$ (with $h(y) = \log(y)$), or on $(0, 1)$ (with $h(y) = \mathrm{logit}(y)$), using the function `gen.Family()`. For example, the log-logistic distribution, which is an explicit distribution in **gamboostLSS**, may be constructed in **gamlss** by using `gen.Family()` and specifying the logistic distribution for z and $h(y) = \log(y)$.

- *Transformation of two or more random variables.* For example, the Student *t-*

family $y \sim \text{TF}(\mu, \sigma, \nu)$ is defined as

$$y = \mu + \sigma t; \quad t = z(w/\nu)^{-1/2}$$
$$z \sim \mathcal{N}(0, 1) \text{ independently of } w \sim \chi_\nu^2 .$$

- *Convolution.* For example, the definition of the exponential Gaussian $y \sim \text{exGAUS}(\mu, \sigma, \nu)$ is

$$y = z_1 + z_2$$
$$z_1 \sim \mathcal{N}(\mu, \sigma^2) \quad \text{independently of } z_2 \sim \text{EXP}(\nu).$$

- *Azzalini-type methods.* These allow the generation of families of skew distributions (e.g. skew normal, skew t) but have the computational disadvantage that their cdf is, in general, not available in closed form. For details see Rigby et al. (2019, chapter 13).

- *Splicing.* Skewness is introduced into symmetric distribution families by defining $f(y)$ as

$$f(y) = \pi_1 f_{y_1}(y) \mathbb{1}\,(y < \mu) + \pi_2 f_{y_2}(y) \mathbb{1}\,(y \geq \mu),$$

where $f_{y_1}(y)$ and $f_{y_2}(y)$ are pdfs symmetric about μ, and π_1 and π_2 are constrained to ensure that $f(y)$ is a valid pdf. Examples include the skew normal type 2 (SN2) and the skew t type 3 (ST3).

- *Mixtures.* Mixed Poisson distributions were discussed in Section 2.3.1. Mixture distributions can, in principle, be derived for any parent distribution $f_{y|\lambda}(y|\lambda)$, where λ is a random variable with pdf $f_\lambda(\lambda)$ and suitable support. Then, similar to equation (2.6), the marginal distribution of y is

$$f(y) = \int f_{y|\lambda}(y|\lambda) f_\lambda(\lambda)\, d\lambda$$

for continuous λ, and a similar expression involving summation for discrete λ. For example, another definition of the t family is:

$$y|\lambda \sim \mathcal{N}(\mu, \lambda^2)$$
$$\lambda \sim \text{GG}(\sigma, [2\nu]^{-0.5}, -2),$$

where GG is the generalized gamma distribution.

- *Truncation.* Any distribution may be truncated on the left, right, or on both sides.

2.6 Choice of Response Distribution

The issue of finding an appropriate response distribution consists of three main tasks and decisions to be made by the analyst. The first task concerns the choice from the available distributions with their corresponding distribution parameters. The second task is the choice of explanatory terms in each of the distribution parameters once a response distribution has been chosen. Those explanatory terms could affect different aspects of the distributions, for example location, scale, shape. This manifests by affecting the corresponding distribution parameters. As the third task, we also have the added complexity of how the explanatory terms affect the parameters, for example linearly or nonlinearly; do we need the main effects only or also the interactions?

It would be convenient to have an automatic method to choose the distributions and model terms. While this may be possible for specific cases it is, unfortunately, an impossible task for the general GAMLSS model. In Section 4.5.1, we discuss selection of the response distribution in more detail, in the larger context of model selection for distributional regression. Furthermore, statistical boosting as introduced in Chapter 7 provides us with a convenient and flexible approach for combining fitting GAMLSS models with variable and effect selection.

In summary, the flexibility of GAMLSS, by allowing a large variety of different distributions, comes with a price. We have to think harder about which aspects of the distribution we need to pay attention to, and whether these aspects are adequately modeled. We illustrate model building in several real-world applications in Part III.

3

Additive Model Terms

GAMLSS rely on regression predictors specified for each of the distribution parameters θ_k characterizing the response distribution. In this book, we consider structured additive predictors that add further flexibility to the GAMLSS regression framework, beyond the flexibility induced by the possibility to consider complex types of response distributions. Structured additive predictors provide a rich and unifying framework for various types of regression effects, as briefly discussed in Section 1.5.4. Here, we

- introduce various types of effects that can be utilized as building blocks in GAMLSS models,

- motivate their use by relating them to applications considered in Part III, and

- embed the different terms in a generic setup that facilitates the description of inferential approaches in Part II.

Generically, each of the predictors in a GAMLSS model is assumed to be of the additive form

$$\eta_{ik} = \beta_0^{\theta_k} + s_1^{\theta_k}(\boldsymbol{x}_{i1}) + \cdots + s_j^{\theta_k}(\boldsymbol{x}_{ij}) + \cdots + s_{J_k}^{\theta_k}(\boldsymbol{x}_{iJ_k}),$$

that is, composed of an intercept term $\beta_0^{\theta_k}$ and J_k potentially nonlinear functions $s_j^{\theta_k}(\boldsymbol{x}_{ij})$, where $\boldsymbol{x}_{ij}$ denotes the covariate vector for the jth effect in distribution parameter θ_k (while notationally suppressing its potential dependence on θ_k). In the following, we will summarize several special cases that can be used within the GAMLSS framework for building complex regression specifications. For this purpose and to simplify notation, we will typically rely on a simple setup such as

$$y_i = s(\boldsymbol{x}_i) + \varepsilon_i, \quad \varepsilon_i \sim \mathcal{N}(0, \sigma^2),$$

since this considerably facilitates the discussion and visualization. However, all of the additive terms discussed in this chapter can be included in larger additive predictors η_{ik} for the different distribution parameters θ_k.

3.1 Penalized Splines

Penalized splines provide a flexible and easy way of accounting for the potentially nonlinear effect of a continuous covariate x. While simple, fully parametric specifications such as the linear model $\beta_0 + \beta_1 x$ or polynomial expansions $\beta_0 + \beta_1 x + \cdots + \beta_l x^l$

are straightforwardly implemented in standard software and convenient to interpret, they are often not flexible enough. Moreover, in complex model specifications such as GAMLSS, it is typically hard to decide a priori whether the shape of a covariate effect can safely be reduced to a polynomial form or whether more flexibility is indeed warranted. Penalized splines offer a flexible, data-driven way of determining the amount of nonlinearity in the effect of x.

The starting point for penalized spline smoothing is to derive a convenient approximation of nonlinear effects $s(x)$ by combining a local polynomial approximation with global smoothness assumptions. More precisely, let $a = \kappa_1 < \cdots < \kappa_m = b$ be a sequence of m knots decomposing the interval $[a, b]$, corresponding to the domain of covariate x, into disjoint intervals. Then $s(x)$ is assumed to be a polynomial spline of degree D, namely

- $s(x)$ is a polynomial of degree D on each of the intervals $[\kappa_l, \kappa_{l+1})$ and

- $s(x)$ is $D - 1$ times continuously differentiable, where $D = 1$ implies continuity without being differentiable and $D = 0$ does not make any smoothness assumptions.

This specification combines flexibility achieved via the local polynomial approximation with overall smoothness as determined by the differentiability condition. The overall smoothness is therefore dictated by the degree D of the spline. A convenient standard is cubic polynomial splines, since these lead to twice continuously differentiable, visually smooth functions. Flexibility of the function approximation, on the other hand, is driven by the number of knots since decomposing the domain of the covariate into more intervals allows for the incorporation of more and more localized features.

It can then be shown that the space of polynomial splines for a given degree D and a given sequence of m knots is a vector space of dimension $L = m + D - 1$. As a consequence, every polynomial spline $s(x)$ can be represented as a linear combination of L basis functions, leading to

$$s(x) = \sum_{l=1}^{L} \gamma_l B_l^{(D)}(x).$$

This implies that a polynomial spline model

$$y_i = s(x_i) + \varepsilon_i = \sum_{l=1}^{L} \gamma_l B_l^{(D)}(x_i) + \varepsilon_i$$

is simply a large linear model with transformed covariates corresponding to the basis function evaluations. Therefore, polynomial splines can still be conveniently estimated by least squares (or maximum likelihood for generalized model variants). More precisely, the model can be written in matrix notation as

$$y = B\gamma + \varepsilon$$

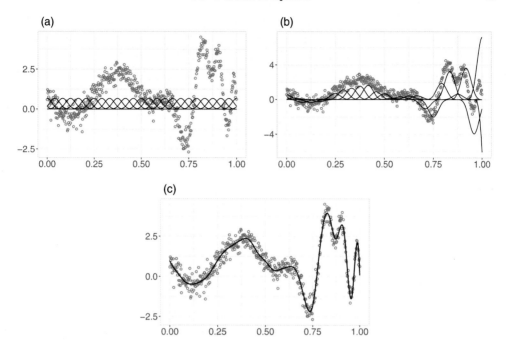

Figure 3.1 Illustration of the fitting process for polynomial splines: (a) simulated data and B-spline basis functions, (b) B-spline basis functions scaled with the estimated basis coefficients, and (c) sum of the scaled basis functions (bottom) leading to the final model fit.

with design matrix

$$\boldsymbol{B} = \begin{pmatrix} B_1^{(D)}(x_1) & \dots & B_L^{(D)}(x_1) \\ \vdots & & \vdots \\ B_1^{(D)}(x_n) & \dots & B_L^{(D)}(x_n) \end{pmatrix}$$

and the basis coefficients $\boldsymbol{\gamma}$ can conveniently be estimated as

$$\hat{\boldsymbol{\gamma}} = (\boldsymbol{B}^\top \boldsymbol{B})^{-1} \boldsymbol{B}^\top \boldsymbol{y}.$$

Note that we are basically just using $\boldsymbol{B}$ to replace the common design matrix $\boldsymbol{X}$ representing the different covariates in classical linear regression, where $\hat{\boldsymbol{\beta}} = (\boldsymbol{X}^\top \boldsymbol{X})^{-1} \boldsymbol{X}^\top \boldsymbol{y}$. In the case of a polynomial spline, the function estimate is then given by

$$\hat{f}(x) = \sum_{l=1}^{L} \hat{\gamma}_l B_l^{(D)}(x),$$

which can be evaluated at arbitrary points x within the support $[a, b]$. The vector of function evaluations for the observed covariate values is simply given by $\hat{\boldsymbol{f}} = \boldsymbol{B}\hat{\boldsymbol{\gamma}}$. The fitting process is visualized in Figure 3.1 for simulated data.

Different bases for representing the space of polynomial splines exist, but we will rely exclusively on B-splines. The corresponding basis functions are defined recursively as

$$B_l^{(D)}(x) = \frac{x - \kappa_{l-D}}{\kappa_l - \kappa_{l-D}} B_{l-1}^{(D-1)}(x) + \frac{\kappa_{l+1} - x}{\kappa_{l+1} - \kappa_{l+1-D}} B_l^{(D-1)}(x)$$

with

$$B_l^{(0)}(x) = \mathbb{1}(\kappa_l \leq x < \kappa_{l+1}) = \begin{cases} 1 & \kappa_l \leq x < \kappa_{l+1} \\ 0 & \text{otherwise} \end{cases} \quad l = 1, \ldots, L-1.$$

The main advantages of B-splines are their numerical stability combined with their local support.

While polynomial splines are very flexible and therefore have the inherent ability to approximate a variety of nonlinear shapes, the main issue in their practical application is the dependence on the number of knots. (See Figure 3.2 for a graphical illustration.) Choosing too few may lead to function estimates that are not flexible enough and therefore miss out important aspects of the true effect structure. On the other hand, choosing the number of basis functions too large may lead to overfitting such that the function estimate is overly wiggly and therefore also not reliable. Penalized splines overcome this issue by resorting to a regularized estimation approach. The basic idea is to use a moderately large number of basis functions (typically 20 to 40 basis functions will be sufficient) based on equidistant knots, enabling enough potential flexibility to represent a large variety of effects. To control the overall flexibility, a penalty term representing the complexity of the function estimate is then augmented to the fitting criterion.

A very general way of measuring the complexity of a nonlinear function is the integrated squared second derivative $\int (s''(x))^2 dx$, which can also be interpreted as the L_2 norm in the space of twice continuously differentiable functions. Instead of performing least squares estimation, we then minimize the penalized least squares criterion

$$\sum_{i=1}^{n} (y_i - s(x_i))^2 + \lambda \int (s''(x))^2 dx,$$

where $\lambda \geq 0$ is a smoothing parameter determining the relative preference we assign to the fit to the data on the one hand (represented by the least squares criterion) and the desired smoothness of the function estimate on the other hand (represented by the penalty term). Setting $\lambda = 0$ yields the standard polynomial spline estimate without any penalty, while $\lambda \to \infty$ places heavy weight on the penalty such that the minimizer is a linear function in x (that leads to $s''(x) = 0$). Instead of choosing the number (and potentially location) of the knots, we optimize the smoothing parameter to achieve a data-driven amount of smoothness.

This technique hence leads to the single parameter λ which controls whether the resulting spline is rather wiggly (small λ – trying to fit the data well), or rather smooth (larger λ – trying to generalize the fit). The choice of λ hence represents the

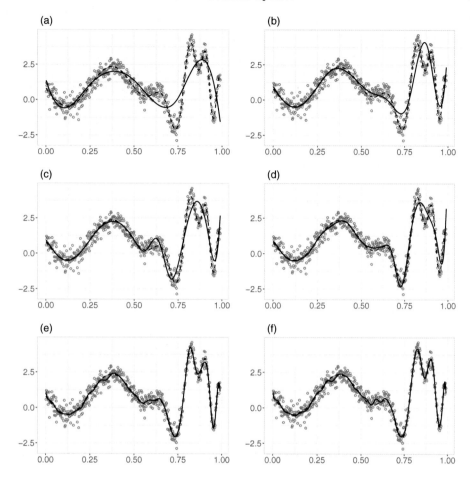

Figure 3.2 Impact of the number of knots on the estimated polynomial spline: (a) 5 knots, (b) 10 knots, (c) 15 knots, (d) 20 knots, (e) 30 knots, and (f) 40 knots. The true curve is represented as the dashed line and the resulting fit for a given number of knots is represented as the solid line.

common trade-off between bias and variance in statistical modeling: Larger values of λ lead to a model with small variance but larger bias, while for small values of λ the variance increases and the bias is small. We will return to this point shortly.

To implement penalized spline estimation for a given value of the smoothing parameter, we derive a more explicit representation of the penalty term as

$$\int (s''(x))^2 dx = \sum_{l_1=1}^{L} \sum_{l_2=1}^{L} \gamma_{l_1} \gamma_{l_2} \int B''_{l_1}(x) B''_{l_2}(x) dx = \boldsymbol{\gamma}^\top \boldsymbol{K} \boldsymbol{\gamma},$$

where $\boldsymbol{K}$ is a positive semidefinite matrix with elements

$$\boldsymbol{K}[l_1, l_2] = \int B''_{l_1}(x) B''_{l_2}(x) dx.$$

In matrix notation, the penalized least squares criterion is then given by

$$(\boldsymbol{y} - \boldsymbol{B}\boldsymbol{\gamma})^\top (\boldsymbol{y} - \boldsymbol{B}\boldsymbol{\gamma}) + \lambda \boldsymbol{\gamma}^\top \boldsymbol{K}\boldsymbol{\gamma}$$

and the penalized least squares estimate is determined as

$$\hat{\boldsymbol{\gamma}} = (\boldsymbol{B}^\top \boldsymbol{B} + \lambda \boldsymbol{K})^{-1} \boldsymbol{B}^\top \boldsymbol{y}.$$

A simpler, approximate alternative to the integrated squared second derivative penalty for B-splines can be derived by using finite differences of adjacent spline coefficients to construct the penalty

$$\text{pen}(\boldsymbol{\gamma}) = \lambda \sum_{l=r+1}^{L} (\Delta^r \gamma_l)^2 = \lambda \boldsymbol{\gamma}^\top \boldsymbol{K}\boldsymbol{\gamma},$$

where the rth-order difference operator Δ^r is defined recursively as

$$\Delta^1 \gamma_l = \gamma_l - \gamma_{l-1}$$
$$\Delta^2 \gamma_l = \Delta^1 \Delta^1 \gamma_l = \Delta^1 \gamma_l - \Delta^1 \gamma_{l-1} = \gamma_l - 2\gamma_{l-1} + \gamma_{l-2}$$
$$\vdots$$
$$\Delta^r \gamma_l = \Delta^{r-1} \gamma_l - \Delta^{r-1} \gamma_{l-1}.$$

The general intuition is that if neighboring basis coefficients do not deviate too much, the resulting fit will be smooth. In the case of equidistant knots, there is an exact equivalence between the integrated squared second derivative penalty and the second-order difference penalty. More generally, rth-order differences approximate integrated squared rth-order derivatives such that for $\lambda \to \infty$, a polynomial of degree $r - 1$ remains unpenalized.

In matrix notation, the penalty matrix is now given by

$$\boldsymbol{K} = \left(\boldsymbol{D}_r^{(L)}\right)^\top \boldsymbol{D}_r^{(L)}$$

with first-order difference matrix

$$\boldsymbol{D}_1^{(L)} = \begin{pmatrix} -1 & 1 & & & \\ & -1 & 1 & & \\ & & \ddots & \ddots & \\ & & & -1 & 1 \end{pmatrix}$$

of dimension $(L - 1) \times L$ and the recursion

$$\boldsymbol{D}_r^{(L)} = \boldsymbol{D}_1^{(L-r+1)} \boldsymbol{D}_{r-1}^{(L)}.$$

The resulting penalized least squares estimate is then of exactly the same form as before.

The difference-based penalty also enables establishment of a Bayesian perspective on penalized splines, in which a random walk assumption leads to an equivalent prior

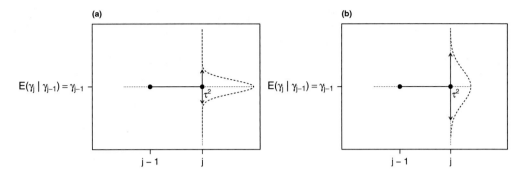

Figure 3.3 Illustration of the random walk prior considered in Bayesian penalized splines for (a) a small and (b) a large value of the variance for the random walk innovations.

structure. More specifically, rth-order differences correspond to rth-order random walk priors

$$\Delta_r \gamma_l | \tau^2 \sim \mathcal{N}(0, \tau^2)$$

such that, for example, for a first-order random walk we have

$$\gamma_l - \gamma_{l-1} | \tau^2 \sim \mathcal{N}(0, \tau^2).$$

The variance parameter τ^2 of the random walk innovations then plays the role of an inverse smoothing parameter, since large values of the variance enable large deviations between neighboring basis coefficients, while small variances force rth-order differences to be close to zero, as illustrated in Figure 3.3.

To complete the prior, we assume flat priors for the initial values $\gamma_1, \ldots, \gamma_r$. The joint prior is then given by a partially improper multivariate normal distribution with density

$$f(\boldsymbol{\gamma}|\tau^2) \propto \left(\frac{1}{\tau^2}\right)^{(L-r)/2} \exp\left(-\frac{1}{2\tau^2}\boldsymbol{\gamma}^\top \boldsymbol{K} \boldsymbol{\gamma}\right),$$

which exactly resembles the form of the quadratic penalty from penalized least squares. Note that $\boldsymbol{K}$ is inherently rank deficient, which represents the fact that a polynomial of degree $r - 1$ remains unpenalized. As a consequence, the joint prior cannot be integrated to one and is only defined upon proportionality. The rank of $\boldsymbol{K}$ is given by $\text{rank}(\boldsymbol{K}) = L - r$.

So far we have introduced penalized splines as a convenient tool for determining nonlinear effects. To make use of the full potential of this approach, it is important to include a data-driven determination of the smoothing parameter λ, or smoothing variance τ^2, so that the appropriate amount of smoothness is achieved, dependent on the requirements of the data. Possible solutions include the optimization of an appropriate fit criterion such as cross-validated least squares (or cross-validated log-likelihood scores) or the Akaike information criterion. In the Bayesian framework, assigning a suitable hyperprior to τ^2 allows us to make determination of the smoothing variance τ^2 an immediate part of the inferential procedure. We refrain from a full discussion at this point since different inferential approaches will be discussed in more detail in Chapters 5, 6, and 7. Penalized splines are illustrated, in the Bayesian modeling context, in Chapters 11 and 12.

There are also a number of variations of the basic penalized splines setup that can be useful in specific applications. For example, cyclic splines enable smoothing on circular domains. This is often desirable when the covariate x represents temporal information such as the hour of a day or the day of a year. In this case it seems plausible to assume that the effect at the end of the day or year should connect smoothly with the effect at the beginning of the day or year. A simple modification of the basis and the penalty allows the integration of this constraint (see, for example, Wood, 2017, chapter 5.3). This is illustrated with monthly, weekday, and hourly effects, in Chapter 10.

Shape constraints on the effect $s(x)$ are another topic of applied interest to ensure, for example, monotonicity, convexity, or unimodality of effects. Such constraints can be particularly useful in sparse data situations where the constraints can lead to more stable estimates that concur with theoretical assumptions on the effect of interest. There is a whole variety of approaches that implements different types of shape constraints, for example by modifying the parametrization of the spline basis or by augmenting additional penalties; see Pya and Wood (2015) or Hofner et al. (2016b) for details.

Finally, the assumption of one global smoothing parameter may be too restrictive if the nonlinearity of the function of interest changes considerably over the covariate domain. In such cases, adaptive smoothing may be more appropriate and there have been a number of suggestions to account for this. Several of these approaches make the smoothing parameter λ (or equivalently the smoothing variance τ^2) a function of the covariate itself such that $\lambda = \lambda(x)$. The concrete specification can take different forms, for example by setting $\lambda(x) = \exp(g(x))$ and assuming $g(x)$ to follow a penalized spline as well (Krivobokova et al., 2008) or by assuming that $\lambda(x)$ is a step function with breakpoints and step sizes estimated from the data (Scheipl and Kneib, 2009).

3.2 Generic Representation

Penalized spline smoothing provides us with a blueprint for a rather generic class of smoothing approaches that cover a variety of different regression terms, including interaction surfaces, varying coefficients, spatial smoothing and random effects. To facilitate their discussion in the next sections and to highlight their similarity, we now introduce a generic representation that provides us with the framework to discuss general inferential principles in Chapters 5, 6, and 7.

We start by assuming that the generic effect $s(\boldsymbol{x}_i)$ can be represented in terms of L

basis functions such that

$$s(\boldsymbol{x}_i) = \sum_{l=1}^{L} \gamma_l B_l(\boldsymbol{x}_i), \tag{3.1}$$

and the vector of function evaluations $\boldsymbol{s} = (s(\boldsymbol{x}_1), \ldots, s(\boldsymbol{x}_n))^\top$ at the observed covariate values is given by

$$\boldsymbol{s} = \boldsymbol{B}\boldsymbol{\gamma},$$

where $\boldsymbol{B}$ is the design matrix obtained from the basis function evaluations and $\boldsymbol{\gamma}$ is the corresponding vector of basis coefficients. Since typically the number of basis functions employed will be large to achieve sufficient flexibility, we regularize estimation by adding to the fit criterion a quadratic penalty term of the following form:

$$\mathrm{pen}(\boldsymbol{\gamma}) = \lambda \boldsymbol{\gamma}^\top \boldsymbol{K} \boldsymbol{\gamma}$$

with a positive semidefinite penalty matrix $\boldsymbol{K}$ and smoothing parameter $\lambda \geq 0$. Assuming likelihood-based inference, we obtain a penalized likelihood criterion of the form

$$\ell_{\mathrm{pen}}(\boldsymbol{\gamma}) = \ell(\boldsymbol{\gamma}) - \mathrm{pen}(\boldsymbol{\gamma})$$

such that (depending on the smoothing parameter), we balance between fidelity to the data (as reflected by the log-likelihood $\ell(\boldsymbol{\gamma})$) and smoothness of the estimate (quantified via the penalty $\mathrm{pen}(\boldsymbol{\gamma})$).

Different types of smoothness (or shrinkage) assumptions are obtained by specifying $\boldsymbol{K}$ appropriately and we will discuss several options in the sections to come. Note that the quadratic form of the penalty is mostly chosen for convenience to make the penalty twice continuously differentiable with respect to the basis coefficients $\boldsymbol{\gamma}$. This allows the construction of, for example, penalized versions of Fisher scoring iterations. To achieve sparsity of the basis coefficients, other types of penalties based on, for example, absolute values of the regression coefficients have received considerable attention. The least absolute selection and shrinkage operator (lasso) is the most well-known of these; see, for example, Efron et al. (2004). These typically require more complex optimization algorithms but can also be recast into our framework via quadratic approximations of the penalty term (Oelker and Tutz, 2017).

As for the penalized spline setting, $\lambda \to 0$ approaches an unpenalized fit while for $\lambda \to \infty$ the penalty starts to overwhelm the information in the data. Optimizing a penalized least squares criterion then leads to a solution that is within the null space of the penalty matrix and optimizes the class of functions $s(\boldsymbol{x}_i)$ that can be represented in the basis expansion (3.1), subject to the restriction $\boldsymbol{\gamma}^\top \boldsymbol{K} \boldsymbol{\gamma} = 0$. Indeed, in many cases, $\boldsymbol{K}$ will be rank deficient such that $L = \dim(\boldsymbol{\gamma}) < \mathrm{rank}(\boldsymbol{K})$, which implies that there is a nontrivial null space of $\boldsymbol{K}$. Therefore, even for $\lambda \to \infty$, some parts of the function will not be penalized such that the resulting fit still depends on the data. For example, in the case of penalized splines with rth-order difference penalty, polynomial effects of degree $r-1$ in the covariate of interest remain unpenalized and $\lambda \to \infty$ implies estimating such an $(r-1)$-degree polynomial from the data.

Note that the penalty matrix $\boldsymbol{K}$ may contain additional unknowns to make the approach more flexible. Making $\boldsymbol{K}$ a function of unknown parameters also leads to a kind of adaptive smoothing where the exact amount and form of smoothness depends on the unknown parameters, which in turn makes computations more difficult. We will focus on the standard setting with fixed penalty matrix $\boldsymbol{K}$.

In a Bayesian framework, instead of assuming a penalty we assign an informative prior distribution

$$f(\boldsymbol{\gamma}|\tau^2) \propto \left(\frac{1}{\tau^2}\right)^{0.5\,\mathrm{rank}(\boldsymbol{K})} \exp\left(-\frac{1}{2\tau^2}\boldsymbol{\gamma}^\top \boldsymbol{K}\boldsymbol{\gamma}\right) \tag{3.2}$$

to express our prior beliefs in smoothness as defined by the penalty matrix $\boldsymbol{K}$. This corresponds to a multivariate normal prior distribution with expectation zero and precision matrix $\boldsymbol{K}/\tau^2$. If $\boldsymbol{K}$ is rank deficient, the prior is partially improper and can therefore only be defined upon proportionality. Furthermore, there is no unique inverse of the precision matrix which would give rise to the covariance matrix. Still, convenient methodology for working with such partially improper normal distributions has been developed, providing numerical recipes that we will employ in Bayesian inference in Chapter 6. (See Rue and Held (2005) for an overview.) The precision matrix itself also provides us with interesting information on a priori conditional independence patterns in the basis coefficients $\boldsymbol{\gamma}$ since two elements of $\boldsymbol{\gamma}$ (γ_{l_1} and γ_{l_2}, say) are conditionally independent (given all other elements in $\boldsymbol{\gamma}$) if and only if the corresponding element $\boldsymbol{K}[l_1, l_2]$ is equal to zero.

The prior for $\boldsymbol{\gamma}$ is specified conditional on the smoothing variance τ^2 which plays a similar role to the smoothing parameter λ in the frequentist approach, albeit on an inverse scale. To obtain a data-driven amount of smoothness, a hyperprior $f(\tau^2)$ will usually be added to the model specification. We will discuss this in more detail in Chapter 6.

Another perspective on the prior distribution (3.2) is to consider the basis coefficients $\boldsymbol{\gamma}$ as random effects in a mixed model where (3.2) specified the corresponding random effects distribution. Unlike the classical case of group-specific random effects typically employed with longitudinal or clustered data (see Section 3.6), the random effects in (3.2) are "global" in the sense that they apply to all observations simultaneously and they are not i.i.d. but correlated to represent smoothness assumption on the effect of interest. Casting smoothing approaches as mixed models has received considerable interest in the statistical literature since it provides efficient ways of determining the smoothing parameters (treated as random effects variances) via (restricted) maximum likelihood (see, for example, Ruppert et al., 2003; Fahrmeir and Kneib, 2011). It also provides the foundation for various other statistical tools including specification tests (see Crainiceanu et al., 2005) and model selection (see Greven and Kneib, 2010).

3.3 Tensor Product Penalized Splines

To include flexible interaction effects of two continuous covariates x_1 and x_2, we now discuss a bivariate extension of penalized spline smoothing to determine estimates for the surface $s(x_1, x_2)$. This requires the definition of corresponding bivariate basis functions as well an appropriate adaptation of the penalization concept. For the former, we rely on tensor product basis functions that are obtained by taking all pairwise interactions of univariate bases. More precisely, let

$$s_1(x_1) = \sum_{l_1=1}^{L_1} \gamma_{l_1} B_{l_1}(x_1) \qquad \text{and} \qquad s_2(x_2) = \sum_{l_2=1}^{L_2} \gamma_{l_2} B_{l_2}(x_2)$$

be polynomial spline approximations of the univariate effects of x_1 and x_2. Then the tensor product spline for the interaction surface is given by

$$s(x_1, x_2) = \sum_{l_1=1}^{L_1} \sum_{l_2=1}^{L_2} \gamma_{l_1 l_2} B_{l_1 l_2}(x_1, x_2)$$

with tensor product basis functions

$$B_{l_1 l_2}(x_1, x_2) = B_{l_1}(x_1) B_{l_2}(x_2),$$

that is, all pairwise interactions of basis functions are considered as tensor product basis functions. (See Figures 3.4 and 3.5 for visualizations.) This principle is of course not limited to the bivariate case or polynomial splines; interactions of arbitrary order and composed of various different types of univariate effects can be constructed. For example, spatio-temporal models can be determined by taking the tensor product of a temporal effect (e.g. a spline) with a spatial effect (which may itself by represented as a tensor product of two univariate splines in the coordinates). See Kneib et al. (2019) for a more general discussion.

As before, the vector of function evaluations can now be written as

$$s = B\gamma,$$

where B contains the evaluations of the tensor product basis functions. However, the construction of the tensor product approach also allows representation of the resulting design matrix as $B = B_1 \odot B_2$, where B_1 and B_2 are the design matrices corresponding to the univariate effects $s_1(x_1)$ and $s_2(x_2)$ respectively, and $\odot$ denotes the "dot" product of two matrices which is obtained by computing all pairwise products of columns in the two respective matrices.

To construct the penalty for a tensor product interaction surface, it is helpful to interpret the regression coefficients $\gamma_{l_1 l_2}$ as being arranged on a lattice where l_1 is interpreted as indexing row orientation while l_2 indexes column orientation. Then an overall smooth tensor product is obtained by assuming smoothness along both the rows and the columns of the coefficient lattice. With the help of Kronecker products, this leads to the penalty matrix

$$K = K_1 \otimes I_{L_2} + I_{L_1} \otimes K_2,$$

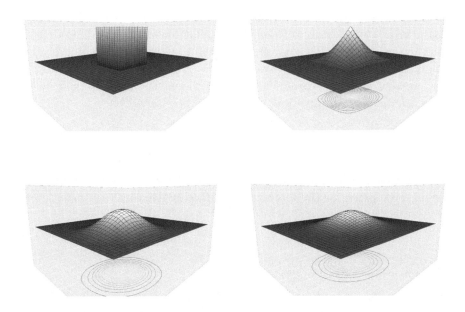

Figure 3.4 Tensor product spline basis functions for different spline degrees. Top left: $D = 0$, top right: $D = 1$, bottom left: $D = 2$, bottom right: $D = 3$.

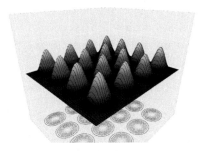

Figure 3.5 A partial set of cubic tensor product splines.

where K_1 and K_2 are the penalty matrices associated with the main effects $s_1(x_1)$ and $s_2(x_2)$ that form the basis of the interaction, and I_L is the identity matrix of dimension L. The Kronecker products then extend the main effects penalties such that they are applied multiple times along the rows and columns of the coefficient lattice.

If both K_1 and K_2 are rank deficient, then K will be rank deficient as well, with the rank deficiency given by the product of the rank deficiencies of the main effects penalties. In the case of tensor product splines, this implies that with first-order difference penalties (rank deficiency of one), the overall rank deficiency of the interaction surface is given by one which corresponds to not penalizing an overall constant. For second-order difference penalties, we end up with a rank deficiency of four for the tensor product that corresponds to a constant, linear main effects and a linear interaction term that are not penalized.

Summing the terms $\boldsymbol{K}_1 \otimes \boldsymbol{I}_{L_2}$ and $\boldsymbol{I}_{L_1} \otimes \boldsymbol{K}_2$ to build the complete penalty matrix $\boldsymbol{K}$ assumes an isotropic penalty structure with one single, joint penalty parameter for the interaction. Conceptually, formulating anisotropic penalties

$$\text{pen}(\boldsymbol{\gamma}) = \lambda_1 \boldsymbol{\gamma}^\top (\boldsymbol{K}_1 \otimes \boldsymbol{I}_{J_2})\boldsymbol{\gamma} + \lambda_2 \boldsymbol{\gamma}^\top (\boldsymbol{I}_{J_1} \otimes \boldsymbol{K}_2)\boldsymbol{\gamma}$$

with two separate smoothing parameters $\lambda_1 > 0$ and $\lambda_2 > 0$ for the two penalties derived from the main effects is straightforward, but the separate determination of the two smoothing parameters can be algorithmically challenging. Similarly, it can be interesting to decompose the overall interaction effect $s(x_1, x_2)$ into the sum of main effects and interaction, that is,

$$s(x_1, x_2) = s_1(x_1) + s_2(x_2) + s_{1|2}(x_1, x_2),$$

where s_1 and s_2 are main effects while $s_{1|2}$ is an interaction with the main effects removed. This is often referred to as the smoothing spline analysis of variance (smoothing spline ANOVA) decomposition of tensor products. Different possibilities exist for its implementation, for example, embedding the model into the framework of reproducing kernel Hilbert spaces (Gu, 2002), using the mixed model representation to achieve an explicit reparametrization (Wood, 2017), or supplementing the parameters of the tensor product with linear constraints to remove certain portions of the interaction (Kneib et al., 2019).

3.4 Radial Basis Functions and Geostatistical Smoothing

Tensor product splines provide a straightforward, convenient extension of univariate splines and can also be considered for the construction of more general types of interactions. However, there are two drawbacks: Firstly, tensor product splines are designed for rectangular domains in which the data are more or less uniformly distributed over the rectangular domain. This relates to the fact that univariate bases are used to construct the interaction. As a consequence, the expanded grid obtained by combining the knots in both univariate dimensions serves as the bivariate knot set. This has the disadvantage that many knots may be placed in areas where no data are available. Secondly, tensor product splines are not rotationally invariant. As a consequence, they may be considered less appropriate, for example in the case of spatial smoothing, where longitude and latitude have quite an arbitrary meaning and we are interested in obtaining spatial effect estimates that remain the same under rotational changes of the coordinate system. On the other hand, tensor products work naturally when constructing interactions of two variables that are measured in different units.

An alternative to tensor product interactions for surface estimation is provided by radial basis functions. Here, the set of knots $\kappa_1, \ldots, \kappa_L$ are chosen arbitrarily from the domain of the bivariate covariate $x = (x_1, x_2)^\top$ (usually as a subset of the observed covariate values). The basis functions are then constructed by centering a radial function around the knots, such that

$$B_l(r) = B(||x - \kappa_l||) \, ,$$

where $r = ||x - \kappa_l|| = \sqrt{(x - \kappa_l)^\top (x - \kappa_l)}$ is the Euclidean distance and $B(r)$ defines the generic form of the radial basis function. One common choice is

$$B(r) = r^2 \log(r),$$

which has a close connection to bivariate smoothing splines. In contrast to tensor product splines, radial basis functions are (by construction) invariant under rotations since they are only a function of the Euclidean distance, which itself is invariant under rotation. On the other hand, radial basis functions do not work well with covariates that are measured in different units, because the Euclidean distance assumes that differences in the covariate dimensions can be compensated for each other.

In the general radial basis function setup, a possible analogue to the integrated squared second derivative penalty discussed in Section 3.1 is given by the biharmonic differential operator

$$\int \int \left[\left(\frac{\partial^2}{\partial^2 x_1} + 2 \frac{\partial^2}{\partial x_1 \partial x_2} + \frac{\partial^2}{\partial^2 x_2} \right) s(x_1, x_2) \right]^2 dx_1 dx_2.$$

While different differential operators are conceivable, they all lead to quadratic penalty terms with entries obtained by applying the penalty operator to pairs of radial basis functions. Of course, derivative-based penalties could also be constructed for the tensor product approach discussed in Section 3.3.

Radial basis functions also enable the inclusion of spatial effects modeled via Gaussian random fields. Assume that the bivariate surface $s(x_1, x_2)$ is indeed a spatial effect defined based on longitude and latitude and that we assume this spatial effect to follow a zero mean Gaussian random field, as in geostatistics (Diggle et al., 1998). Under the assumptions of stationarity (i.e. invariance of the spatial dependence structure under spatial shifts) and isotropy (invariance of the dependence structure under rotations), such a Gaussian random field is completely characterized by its variance τ^2 and its correlation function $\rho(r)$, where r is the Euclidean distance between two points in space. One can then show (Kammann and Wand, 2003) that, for a finite set of spatial locations, the Gaussian random field can be interpreted as a basis function approach where the correlation functions play the role of (radial) basis functions. For example, the power exponential correlation function leads to radial basis functions

$$B(r) = \exp\left(- \left| \frac{r}{\phi} \right|^k \right)$$

with range parameter $\phi > 0$ and shape parameter $0 < k \leq 2$. The power exponential correlation includes the standard exponential ($k = 1$) as well as the Gaussian ($k = 2$)

correlation functions as special cases. Usually k is fixed in applications a priori, based on assumptions on the smoothness of the spatial field.

For the special case of radial basis functions based on correlation functions, there is also a more immediate construction of the penalty matrix $\boldsymbol{K}$ as

$$\boldsymbol{K}[l_1, l_2] = B(||\boldsymbol{\kappa}_{l_1} - \boldsymbol{\kappa}_{l_2}||).$$

Note that in this case the resulting penalty matrix is of full rank due to the positive definiteness of the correlation function employed for the definition of the basis functions.

3.5 Markov Random Fields

While both tensor product splines and radial basis functions can be employed for spatial smoothing when continuous coordinate information is available, one often faces situations where only discrete spatial information is available, for example. via the assignment of spatial observations to different regions. This may, for example, be the case for ensuring data confidentiality or because the quantities of interest indeed only exist on a regional level.

Assume that the assignment of observation i to one of the spatial regions is represented by a spatial indicator $r_i \in \{1, \ldots, L\}$ where, for simplicity, the L regions are numbered consecutively. Then we can assign separate regression coefficients γ_l, $l = 1, \ldots, L$, to each of the L regions which basically corresponds to dummy coding for the regions. Based on indicator basis functions, we can then connect the spatial effect $\gamma_{r_i} = s(r_i)$ of an individual observation i collected in region r_i with the corresponding effect via

$$s(r_i) = \sum_{l=1}^{L} \gamma_l B_l(r_i),$$

where

$$B_l(r_i) = \begin{cases} 1 & \text{if } r_i = l, \\ 0 & \text{otherwise.} \end{cases}$$

In matrix notation this yields the $(n \times L)$ design matrix $\boldsymbol{B}$ with entries

$$\boldsymbol{B}[i, l] = \begin{cases} 1 & \text{if } r_i = l, \\ 0 & \text{otherwise} \end{cases}$$

and the complete vector of spatial effects is given by $\boldsymbol{\gamma} = (\gamma_1, \ldots, \gamma_L)^\top$.

Since the number of regions may be large relative to the sample size and only a small number of observations located in some of the regions, smoothing across regions that are close by is often desirable. Owing to the discrete nature of the spatial information, however, there is no direct distance measure available and we have to rely on indirect distances determined by neighborhood relationships. Let $N(r)$ denote the set of indices of all neighbors of region r (e.g. defined by a common border between the regions). One possible spatial smoothness penalty is then given by

$$\text{pen}(\boldsymbol{\gamma}) = \lambda \sum_{l=1}^{L} \sum_{r \in N(l), r < l} (\gamma_r - \gamma_l)^2,$$

that is, the sum of squared differences in spatial effects between all pairs of neighboring regions. Assigning a large smoothing parameter to this penalty will then make all spatial effects similar, with the limiting case of a constant spatial effect obtained with $\lambda \to \infty$. In matrix notation, the penalty can again be written as $\lambda \boldsymbol{\gamma}^\top \boldsymbol{K} \boldsymbol{\gamma}$, where now the penalty matrix is an adjacency matrix with entries

$$\boldsymbol{K}[l, r] = \begin{cases} -1 & l \neq r, l \sim r, \\ 0 & l \neq r, l \nsim r, \\ |N(l)| & l = r, \end{cases}$$

where $l \sim r$ indicates that l and r are neighbors and $|N(l)|$ is the number of neighbors of region l.

Considerable flexibility can be achieved by modifying the definition of neighborhoods beyond simple geographic adjacency. In addition, weights can be introduced to differentiate between the importance of the neighbors. In this case, the simple adjacency matrix is replaced by

$$\boldsymbol{K}[l, r] = \begin{cases} -\omega_{rl} & l \neq r, l \sim r, \\ 0 & l \neq r, l \nsim r, \\ \omega_{l+} & l = r, \end{cases} \tag{3.3}$$

where $\omega_{rl} = \omega_{lr}$ are symmetric weights assigned to the relation between regions l and r and $\omega_{l+} = \sum_{r \in N(l)} \omega_{rl}$.

From a Bayesian perspective, the penalty is equivalent to a Gaussian Markov random field prior, where the conditional distribution of γ_l given all the neighboring effects is specified as

$$\gamma_l \,|\, \gamma_r, r \neq l \sim \mathcal{N}\left(\frac{1}{|N(l)|} \sum_{r:r \sim l} \gamma_r, \frac{\tau^2}{|N(l)|} \right),$$

that is,

- the prior expectation for the spatial effect in region l is given by the average of all spatial effects of neighboring regions,

- the effect in region l is conditionally independent of all non-neighbors, and

- the variance of the conditional prior distribution in region l is inversely proportional to the number of neighbors.

It can then be shown that these conditional distributions imply a multivariate Gaussian joint distribution for $\boldsymbol{\gamma}$ given by

$$f(\boldsymbol{\gamma}\,|\,\tau^2) \propto \left(\frac{1}{\tau^2}\right)^{(\mathrm{rank}(\boldsymbol{K}))/2} \exp\left(-\frac{1}{2\tau^2}\boldsymbol{\gamma}^\top \boldsymbol{K}\boldsymbol{\gamma}\right),$$

where $\boldsymbol{K}$ is the same adjacency matrix as (3.3). If each region has at least one neighbor and the map is fully connected, the rank of the spatial adjacency matrix is given by $\mathrm{rank}(\boldsymbol{K}) = L - 1$.

When considering the weighted version of a Markov random field, the conditional distribution is altered to

$$\gamma_l \,|\, \gamma_r, r \neq l \sim \mathcal{N}\left(\frac{1}{\omega_{l+}}\sum_{r:r\sim l}\omega_{rl}\gamma_r, \frac{\tau^2}{\omega_{l+}}\right).$$

The use of Gaussian Markov random fields for the modeling of a spatial effect on childhood undernutrition is illustrated in Chapter 11, and for a spatial effect on voting response is illustrated in Chapter 12.

3.6 Random Effects

Random intercepts are typically included in a regression specification to account for unobserved heterogeneity arising from grouping structures in the data. A common example is longitudinal data, in which repeated measurements are available on the same set of individuals. If not all individual-specific characteristics can be explained by covariates, this leads to dependence between the observations on the same individual. Random effects are one way of taking these into account. A stylized regression model could then be

$$y_{it} = \beta_0 + \alpha_i + \beta_1 x_{it} + \cdots + \varepsilon_{it}, \quad i = 1, \ldots, n, \ t = 1, \ldots, T,$$

where, for simplicity, we assume a balanced design with all individuals (indexed by i) being observed at the same time points (indexed by t). While the usual i.i.d. $\mathcal{N}(0, \sigma^2)$ assumption is made for the error terms ε_{it} (both within and between observations for the different individuals and time points), the individual-specific intercepts α_i are also assumed to be random. More precisely, the classical assumption is $\alpha_i \overset{\text{iid}}{\sim} \mathcal{N}(0, \tau^2)$. Since the same random intercept is shared by all observations on the same individual, this induces dependence between such observations, while observations on different individuals remain independent. The amount of dependence is quantified by the intra-class correlation coefficient $\tau^2/(\sigma^2 + \tau^2)$, which can also be interpreted as the ratio between group-specific heterogeneity (quantified by the random effects variance τ^2) and the total heterogeneity (quantified by $\sigma^2 + \tau^2$).

In extended model variants, the regression coefficients associated with covariates can also vary over the individuals, leading to a model such as

$$y_{it} = \beta_0 + \alpha_{0i} + \beta_1 x_{it} + \alpha_{1i} x_{it} + \cdots + \varepsilon_{it}, \quad i = 1, \ldots, n, \ t = 1, \ldots, T,$$

where now, in addition to the individual-specific random intercept α_{0i}, individual-specific random slopes α_{i1} also account for heterogeneity between individuals. The random effects vector $\boldsymbol{\alpha}_i = (\alpha_{0i}, \alpha_{i1})^\top$ for individual i follows the bivariate normal distribution:

$$\boldsymbol{\alpha}_i \sim \mathcal{N}_2(\mathbf{0}, \boldsymbol{Q}).$$

Some structure can be imposed on $\boldsymbol{Q}$, leading for example to independence between random intercept and random slope. Conceptually, it is also easy to extend the model to comprise multiple random slopes, but models with several random effects are notoriously difficult to estimate, in particular without imposing additional structure on $\boldsymbol{Q}$.

To cast random intercepts into our generic framework and to make them applicable in the more complex case of GAMLSS specifications, we consider the more abstract situation where data are grouped into L disjoint sets of observations. For longitudinal data, L represents the number of individuals; more generally, the grouping may also represent regions in a spatial analysis, family membership, or any other kind of grouping structure. Similar to the case of discrete regional data, we then define group-specific regression coefficients γ_l, $l = 1, \ldots, L$. If group membership is represented by the indicator $g_i \in \{1, \ldots, L\}$, the group-specific effects can be represented as

$$s(g_i) = \sum_{l=1}^{L} \gamma_l B_l(g_i)$$

with indicator basis functions

$$B_l(g_i) = \begin{cases} 1 & \text{if } g_i = l, \\ 0 & \text{otherwise.} \end{cases}$$

This more abstract way of representing the grouping structure has the advantage that we avoid the explicit double indexing, which also facilitates working with multiple levels of grouping.

In matrix notation, the indicator basis functions imply a dummy-coded design matrix $\boldsymbol{B}$ with elements

$$B[i, l] = \begin{cases} 1 & \text{if } s_i = l, \\ 0 & \text{otherwise.} \end{cases}$$

The assumption of i.i.d. random intercepts translates to $\boldsymbol{\gamma} \sim \mathcal{N}(\mathbf{0}, \tau^2 \boldsymbol{I}_L)$ for the complete vector of random effects, and therefore the corresponding penalty matrix is given by $\boldsymbol{K} = \boldsymbol{I}_L$. Random slopes can be interpreted as a special case of varying coefficient terms, as discussed in the next section, where the covariate of interest is the interaction variable and the random effect is the effect modifier.

In small area statistics, region-specific random effects are often employed for regularized estimation of region-specific effects. In contrast to Markov random fields, however, random intercepts do not take spatial smoothness into account but rather assume independent effects across the regions. This can be interpreted as assuming that unobserved, regionally varying heterogeneity is caused by covariates that themselves do not have a spatial structure. Since it is hard to decide a priori whether this is a reasonable assumption, one can also include both an i.i.d. random intercept and a Markov random field in the same model. However, both effects are only distinguishable by their different prior assumptions (or penalties) such that the separation is not overly reliable. Still, combining both effect types may be helpful in recovering the complete regional heterogeneity, namely the sum of both effects (Fahrmeir and Lang, 2001).

3.7 Varying Coefficient Terms

Varying coefficients provide a general framework for building interactions of the form $s(\boldsymbol{x}) = x_1 \tilde{s}(x_2)$ in which the effect of the interaction variable x_1 varies over the range of values for the effect modifier x_2. While x_1 is usually either binary or continuous, any of the previously discussed modeling possibilities can be used for $\tilde{s}(x_2)$. The most common situation is to rely on a single continuous x_2 such that $\tilde{s}(x_2)$ can be represented by a penalized spline. More general versions are perfectly conceivable, for example to represent spatially varying coefficients. The special case of having a random intercept as effect modifier can also be interpreted as including a group-specific random slope for x_1 in the model.

If the vector of effect modifier function evaluations is given by $\tilde{\boldsymbol{s}} = \tilde{\boldsymbol{B}}\boldsymbol{\gamma}$, the complete design matrix is obtained by premultiplying the design matrix of the effect modifier with a diagonal matrix of interaction variable observations such that

$$\boldsymbol{s} = \boldsymbol{B}\boldsymbol{\gamma} = \mathrm{diag}(x_{11}, \ldots, x_{n1})\tilde{\boldsymbol{B}}\boldsymbol{\gamma}.$$

The penalty matrix is simply the one of the effect modifier.

3.8 Other Effect Types

The list of effects discussed in the previous paragraphs is by no means exhaustive. Here we briefly summarize a number of additional possibilities.

3.8.1 Parametric Linear Effects

So far we have focused on flexible effects for covariates in our regression specifications, but of course it is still perfectly legitimate to include simple, parametric linear effects. This will, for example, be the case for the intercept of the model and for categorical covariates represented in dummy or effect coding. To cast linear effects $\boldsymbol{x}_i^\top \boldsymbol{\gamma}$ into our generic framework, the design matrix $\boldsymbol{B}$ is simply given by the collection of covariates in $\boldsymbol{x}$ and $\boldsymbol{\gamma}$ comprises the corresponding regression coefficients. Unless the coefficient vector is of considerable dimension (see the next subsection), one usually would not include any kind of regularization, which translates into a flat prior in the Bayesian perspective. Technically, this can be achieved by setting $\boldsymbol{K} = \boldsymbol{0}$ or fixing $\tau^2 = 0$.

3.8.2 Regularized Linear Effects

If the vector of parametric linear effects, L, grows large and approaches (or even exceeds) the number of observations n, it may be helpful to enforce shrinkage of the L regression coefficients, for example by a ridge penalty where

$$\text{pen}(\boldsymbol{\gamma}) = \lambda \boldsymbol{\gamma}^{\top} \boldsymbol{\gamma} = \lambda \sum_{l=1}^{L} \gamma_l^2,$$

that is, the penalty collects the sum of squared coefficients. This directly fits into our generic framework by setting $\boldsymbol{K} = \boldsymbol{I}_L$. The effect of this action (when $r < n$) is that the linear coefficients of the linear model shrink towards zero, but they do not disappear altogether. This approach is known as *ridge* regression.

If sparsity of the coefficient vector is desired, for example, by pushing some of the coefficients to zero, other types of penalty terms such as the least absolute selection and shrinkage operator (lasso; see Tibshirani, 1996)

$$\text{pen}(\boldsymbol{\gamma}) = \lambda \sum_{l=1}^{L} |\gamma_l|$$

may be employed. A practical disadvantage is that such nonquadratic penalties require specialized optimization approaches. They are therefore more difficult to combine with the other effect types discussed in this chapter even though mixing, algorithmically, quadratic penalties with other penalties within the penalized likelihood approach can be achieved via the algorithms described in Section 5.2.1. One promising route to overcome this limitation is quadratic approximation of the penalty term (Oelker and Tutz, 2017). The `gamlss()` additive term function `ri()` uses this approach.

Regularization is discussed in more detail in Section 5.4, and use of the lasso is illustrated in Chapter 8.

3.8.3 Functional Effects

It is increasingly common to observe not only single values of covariates but rather complete curves $x_i(t)$, $t \in \mathcal{T}$ that may serve as explanatory variables where $\mathcal{T}$ denotes the index variable of the function, for example time in a continuous-time setting. Some examples of functional covariates are remote sensing data, growth curves, and spectral information. Combining these with scalar response values is often achieved by a model specification such as

$$y_i = \int_{\mathcal{T}} s(t) x_i(t) \, dt + \varepsilon_i,$$

that is, a varying relevance $s(t)$ is assigned to the covariate $x_i(t)$ and the overall effect is accumulated by integrating over the function domain $\mathcal{T}$ (that may represent time but could also represent other domains such as frequencies). This is termed a *functional regression model* and the topic *functional data analysis* (Ramsay and

Silverman, 2005). While not immediately fitting into our framework of basis function smoothing, functional effects can still be expressed in a similar way if the relevance function $s(t)$ is approximated in terms of basis functions as

$$s(t) = \sum_{l=1}^{L} \gamma_l B_l(t).$$

This then leads to a design matrix $\boldsymbol{B}$ with elements

$$\boldsymbol{B}[i, l] = \int_{\mathcal{T}} B_l(t) x_i(t) \, dt,$$

which are typically determined by numerical integration. If a penalized spline is employed for representing $s(t)$ then the penalization approaches discussed in Section 3.1 can be utilized to achieve smoothness of the functional effect.

Part II

Statistical Inference in GAMLSS

4

Inferential Methods for Distributional Regression

In this chapter we discuss the part of the data analysis circle (see Section 1.1) in which we analyze the data and interpret the results within the framework of GAMLSS and

- introduce the general ideas underlying statistical modeling and estimation, that is, the concepts of population, sample, and model;

- discuss the Kullback–Leibler divergence as a risk function suitable for distributional regression approaches; and

- describe some general tools for working with GAMLSS, in particular with respect to model choice, model fit, and model interpretation.

4.1 Introduction

Analyzing data in the context of regression models in general involves the following steps:

- Make appropriate assumptions for the model. In GAMLSS, this mostly relates to the choice of a response distribution, the relevant set of covariates for each predictor in the model, and specific term structures for these covariates.

- Estimate the parameters of the model (that is, using one of the approaches outlined in Chapters 5, 6, and 7).

- Check the adequacy of the model and its assumptions (using, e.g. graphical tools) and make necessary changes.

- Compare the current model with other competing models. These could either be alternative GAMLSS specifications or completely different modeling approaches.

- Interpret the final model (or models) with respect to the research question of interest.

Like all statistical models, GAMLSS is based on assumptions about the data generating mechanism. In Section 4.2, we discuss the notions of the *population*, the *sample* and the *model*, which are essential for understanding the relevance of assumptions on the data generating mechanism for statistical modeling. Section 4.3 contains a discussion of risk functions that may serve as the basis for estimating the model parameters. The GAMLSS definition provided in equations (1.10) to (1.13) assumes an explicit theoretical distribution for the response as well as specific choices for all regression predictors. These are rather strong assumptions but there are several diagnostic tools, most of them graphical, to validate and check them. Borrowing the classical statistical terminology of "within" and "between" groups variation, we will use the terms *within* and *between* model diagnostic tools to differentiate between different tasks. The former relates to checking how well a specific assumption of a fitted model is performing, while the latter refers to a comparison between models. A discussion on comparing different GAMLSS models is given in Section 4.4 while Section 4.5 gives advice for GAMLSS model selection strategies. In Section 4.6, we discuss the impact of parameter (non-)orthogonality on estimating GAMLSS models. Section 4.7 introduces various tools for model diagnostics relevant for GAMLSS. Finally, Section 4.8 presents visualization and interpretation of the effects for an estimated GAMLSS specification.

4.2 The Population, the Sample and the Model

Within distributional regression models such as GAMLSS, the relationship between the response y and the covariates $\boldsymbol{x}$ is assumed to be of a *stochastic* nature. More precisely, we assume that there is a underlying *population*, $\mathcal{P}$, which contains all subjects we would like to study. Whether the population has a finite or infinite number of subjects, and whether it exists or is purely conceptual, is not important in what follows. A *sample* is then a subset (hopefully suitably randomized to avoid bias) from $\mathcal{P}$. The sample is the dataset available to us. If we assume that we can obtain i.i.d. samples from a population, the population distribution can also be thought of as the data generating mechanism that provides us with the sample as the observed data. For more complex sampling schemes, the data result from the combination of the sampling mechanism and the population.

This process is represented diagrammatically in Figure 4.1. Data samples are generated through the (in general unknown) data generating mechanism (left path in Figure 4.1). Afterwards, we use the data to infer something about the population (right path in Figure 4.1). Statistical *inference* is the process of going from the sample (the data) to the population. We achieve this using a statistical model which is based on assumptions on how the population and the data generating mechanism work.

The likelihood of observing the response y and the explanatory variables $\boldsymbol{x}$ in a sample is determined by their joint population distribution $g(y, \boldsymbol{x}) = g(y|\boldsymbol{x})g(\boldsymbol{x})$. The response y depends on $\boldsymbol{x}$ through the *conditional* population distribution $g(y|\boldsymbol{x})$. In distributional regression, $g(y|\boldsymbol{x})$ is the main subject of interest since it provides information on how $\boldsymbol{x}$ affects y; unfortunately it is generally unknown. What we do know is the observed data sample $(y_i, \boldsymbol{x}_i)$ for $i = 1, 2, \ldots, n$. In parametric statistical inference, we approximate the population conditional distribution $g(y|\boldsymbol{x})$ using a *model* $\mathcal{M}$, which assumes a theoretical parametric distribution $f(y|\boldsymbol{\vartheta}(\boldsymbol{x}))$

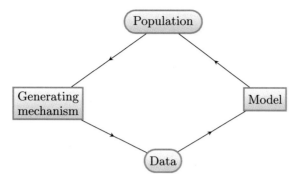

Figure 4.1 The relationship between population and sample or data. The observed data are generated by collecting information on the population via some data generating mechanism representing the sampling scheme (left path). The right path shows the process of statistical *inference*, that is, how to infer something about the population given data that we observed. We achieve this by using a model which is based on assumptions on how the population and the data generating mechanism work.

where $\boldsymbol{\vartheta}(\boldsymbol{x})$ are (covariate-dependent) distribution parameters; see Chapter 2 for various candidate models.

The model density $f(y|\boldsymbol{\vartheta}(\boldsymbol{x}))$ is then employed to provide us with an explanation or mechanism of what the population distribution looks like. The basic idea of statistical inference is to fit the specific model distribution to the data, that is, determining $f(y|\hat{\boldsymbol{\vartheta}}(\boldsymbol{x}))$, where the notation $\hat{\boldsymbol{\vartheta}}$ denotes that the unknown parameters $\boldsymbol{\vartheta}$ have been estimated from the data. Based on this estimated distribution, we attempt to infer something sensible about the population distribution $g(y|\boldsymbol{x})$. If the model choice $f(y|\boldsymbol{\vartheta}(\boldsymbol{x}))$ is "bad", that is, the approximation $f(y|\boldsymbol{\vartheta}(\boldsymbol{x}))$ is far from the true model $g(y|\boldsymbol{x})$, this will in general result in wrong conclusions and this will happen irrespective of how the parameters $\boldsymbol{\vartheta}$ have been estimated. Remember that any model $\mathcal{M}$ is created by making assumptions, and within a distributional regression model the assumption about the distribution $f(y|\boldsymbol{\vartheta}(\boldsymbol{x}))$ is a rather crucial one. The GAMLSS framework provides a wide range of alternative distributions, and diagnostics for checking the adequacy of this assumption, which mitigates the risk of model misspecification.

In GAMLSS, by assuming that the theoretical parametric distribution $f(y|\boldsymbol{\vartheta}(\boldsymbol{x}))$ is a good approximation of the population distribution $g(y|\boldsymbol{x})$, the entire "characteristics" of the population distribution are modeled simultaneously. Usually the question we are trying to answer determines which part of the conditional distribution we need to concentrate on. Linear models, GLMs, and GAMs concentrate on the mean of the conditional distribution, quantile regression on the quantiles (centiles) of the conditional distribution, and GARCH models (Engle, 1982, 2001) on the variance (volatility) of the conditional distribution. In Section 4.3 we show that the *risk* function is the instrument that determines which part of the conditional distribution is focused on. Different risk functions lead us to different characteristics of the conditional distribution.

In GAMLSS, all the characteristics of the conditional distribution $g(y|\boldsymbol{x})$ are modeled simultaneously. Given the estimated conditional distribution $f(y|\hat{\boldsymbol{\vartheta}}(\boldsymbol{x}))$, all information about the conditional distribution is available, namely the center of the distribution, its tails, mean, variance, skewness, kurtosis, quantiles, etc., can be determined from the conditional distribution. In contrast, in quantile regression for example, only quantiles are estimated and each quantile is estimated separately (at least in standard approaches).

The *empirical* conditional distribution function $g_E(y|\boldsymbol{x})$ assigns an equal probability of $1/n$ to each value of the response in the sample, conditional on the predictors. This is an important tool in statistical inference because the empirical cdf $G_E(y|\boldsymbol{x})$ is, under suitable regularity conditions, a consistent estimate of the "true" population conditional cumulative distribution $G(y|\boldsymbol{x})$.

The notions of the *population* distribution, the *empirical* distribution (the observed sample), and the assumed *model* distribution are depicted in Figure 4.2. In this schematic presentation, we concentrate on the concepts by ignoring technical details. For example, to define the population properly, we would need an "infinite" (or very high) dimensional space, for the sample we would need an n-dimensional space, and for the model (to be of any use) we would need far fewer dimensions than n. Our schematic plot presents them all in two dimensions.

The population distribution $g(y|\boldsymbol{x})$ is placed in the center of the figure. Different samples (of equal sample size n) are shown as dots. The gray circle around the population represents samples with a fixed equal probability of being observed. The actually observed sample, represented by its own empirical distribution $g_E(y|\boldsymbol{x})$, is shown on the right of the population distribution.

In order to explain how the conditional population distribution $g(y|\boldsymbol{x})$ is working, we use the model $f(y|\boldsymbol{\vartheta}(\boldsymbol{x}))$, denoted in the figure as a line to emphasize the fact that we are dealing with a parametric model and that for different values of the parameters $\boldsymbol{\vartheta}$ the model can be closer to or further away from the true population distribution $g(y|\boldsymbol{x})$ (or for this matter the empirical distribution $g_E(y|\boldsymbol{x})$).

Some important observations in connection with Figure 4.2 are as follows.

- A model, $f(y|\boldsymbol{\vartheta}(\boldsymbol{x}))$ in Figures 4.2 (a), (b), (c), and (d), will be "good" if the model line passes close to the population distribution $g(y|\boldsymbol{x})$. This will ensure that the model offers enough flexibility to potentially provide a good approximation of the true data generating mechanism. However, it still depends on the observed data and the estimates determined for the parameters $\boldsymbol{\vartheta}(\boldsymbol{x})$ whether, for the given data, conclusions from the model provide accurate inferences about the population of interest. A possible way of measuring "closeness" mathematically more formally is defined in Section 4.3, using the Kullback–Leibler divergence.

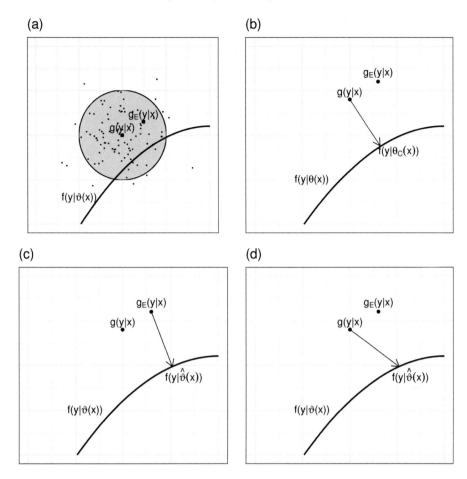

Figure 4.2 Schematic plots of the true population distribution, $g(y|\boldsymbol{x})$, the observed sample represented by its empirical distribution $g_E(y|\boldsymbol{x})$, and the model $f(y|\boldsymbol{\vartheta}(\boldsymbol{x}))$, shown as a line resulting from different parameter values, that is, different values of the parameter $\boldsymbol{\vartheta}$ are represented by different points on the line. Panel (a) shows the population and the samples generated from it, where different samples are represented as dots and the gray circle represents samples of equal probability. Panel (b) shows the model $f(y|\boldsymbol{\vartheta}_c(\boldsymbol{x}))$, that is, closest to the population $g(y|\boldsymbol{x})$ and therefore minimizes the risk (see equation (4.5)). Panel (c) shows the model $f(y|\hat{\boldsymbol{\vartheta}}(\boldsymbol{x}))$ which is closest to the empirical distribution $g_E(y|\boldsymbol{x})$ and therefore minimizes the empirical risk. Panel (d) shows a distance measure useful for selecting between models. It minimizes the distance between the population $g(y|\boldsymbol{x})$ and the fitted model $f(y|\hat{\boldsymbol{\vartheta}}(\boldsymbol{x}))$.

- Among the different values of the parameters $\boldsymbol{\vartheta}$ of the model $f(y|\boldsymbol{\vartheta}(\boldsymbol{x}))$ there is usually one, say $\boldsymbol{\vartheta}_c$, for which the specific model is closest to the population in some sense, see Figure 4.2(b). The value $\boldsymbol{\vartheta}_c$ is then sometimes called the "true" parameter even though there is nothing true about the model or the parameter values $\boldsymbol{\vartheta}_c$ themselves unless we are willing to believe that $g(y|\boldsymbol{x}) = f(y|\boldsymbol{\vartheta}(\boldsymbol{x}))$, that is, the true data generating mechanism indeed coincides with the parametric model chosen to approximate it. This will rarely be the case in practice,[1] but $\boldsymbol{\vartheta}_c$ determines the best approximation that can be achieved based on the model class $f(y|\boldsymbol{\vartheta}(\boldsymbol{x}))$.

- Figures 4.2 (a), (b), (c), and (d) present only one assumed model, $f(y|\boldsymbol{\vartheta}(\boldsymbol{x}))$. A different assumed model, say model $f(y|\tilde{\boldsymbol{\vartheta}}(\boldsymbol{x}))$, would be represented with a different line. Models passing closer to the population will in general be more desirable (that is, more realistic) than models which lie further away from it. In reality, however, we of course do not know which model is closer. Particularly regarding the flexibility of GAMLSS, to reach a sensible decision, we will need the right goodness of fit statistics for comparing competitive models and the right diagnostic tools for checking their assumptions.

- An *overfitted* model will have its model line passing too close to the empirical $g_E(y|\boldsymbol{x})$, but further away from the population $g(y|\boldsymbol{x})$. Too much flexibility in the model is not necessarily good if this means going too close to the sample $g_E(y|\boldsymbol{x})$ but not to the population $g(y|\boldsymbol{x})$.

A different perspective on our discussion around Figure 4.2 can be achieved by disentangling different sources of uncertainty arising when working with samples of empirical data. The *aleatoric* (or stochastic) uncertainty is the population randomness (produced by the data generating mechanism on the left side of Figure 4.1). It is the variability in the data which comes from the fact that the data is a sample from a wider population and unless we are observing the complete population, there will always be aleatoric uncertainty left. The aleatoric uncertainty is therefore something that cannot be reduced by the choice of the model but is an inherent property of sample-based inference. The *epistemic* uncertainty, on the other hand, comes from the lack of knowledge about the true data generating mechanism (the right side of Figure 4.1). A more realistic model or a more efficient estimation approach can reduce epistemic uncertainty by efficiently turning the data available into a model estimate. Within GAMLSS modeling, improved information about, for example, the assumed distribution will reduce the epistemic uncertainty. Aleatoric and epistemic uncertainty are concepts widely used in machine learning; see for example Hüllermeier and Waegeman (2021).

Assuming a model $f(y|\boldsymbol{\vartheta}(\boldsymbol{x}))$ is the first step of statistical modeling. In the next step, the parameters $\boldsymbol{\vartheta}$ of the model have to be estimated from the data in order to be able to interpret the model. At this point, we are using $\boldsymbol{\vartheta}$ as a generic term involving all parameters. In practice, the role of the parameters can be of different nature: in classical parametric statistics, the notion of *parameters of interest* as opposed to *nuisance parameters*[2] exists. Such a division is not helpful for the Bayesian

---

[1] "All models are wrong but some are useful" (Box, 1979)

[2] Nuisance parameters are those parameters needed in the model formulation but without relevance in the investigation of a particular problem.

approach, where all parameters are treated equally, and of very little interest within the GAMLSS framework where the division of parameters into *linear coefficients*, *smooth coefficients* (or random effects), and *hyperparameters* makes more sense; see Chapter 5. Hyperparameters are the *smoothing* or *tuning* parameters when smoothing terms are included in the model. In the next section, we discuss how to go about estimating the parameters $\boldsymbol{\vartheta}$ in general.

4.3 Risk, Empirical Risk, and the Likelihood Function

We now use arguments from statistical decision theory, where *loss* functions are defined as a way to make decisions on the basis of minimizing the associated cost. A *risk* function is an "expected" loss, where the expectation is over all random variables in the model. A typical risk function is the squared error risk (referred to as the L_2 loss), which for regression models takes the form

$$
\begin{aligned}
\mathbb{E}_{Y,\boldsymbol{X}}(g) &= \mathbb{E}_{Y,\boldsymbol{X}}\left[Y - s(\boldsymbol{X})\right]^2 \\
&= \mathbb{E}_{\boldsymbol{X}}\,\mathbb{E}_{Y|\boldsymbol{X}}\left[(Y - s(\boldsymbol{X}))^2\,|\,\boldsymbol{X}\right],
\end{aligned}
\tag{4.1}
$$

where $s(\cdot)$ represents the regression models of interest taking the covariates $\boldsymbol{X}$ as inputs and approximating the response variable Y. The function $s(\cdot)$ should then be chosen such that it minimizes (4.1). Using arguments similar to Hastie et al. (2009, Chapter 2), to minimize (4.1), it is sufficient to minimize the following function pointwise for given values of the covariates:

$$
s(\boldsymbol{x}) = \operatorname*{argmin}_{c}\mathbb{E}_{Y|\boldsymbol{X}}\left((Y - c)^2\,|\,\boldsymbol{X} = \boldsymbol{x}\right),
\tag{4.2}
$$

that is, the random variables representing the covariates $\boldsymbol{X}$ are conditioned on their realized data values $\boldsymbol{x}$. The solution for $s(\boldsymbol{x})$ is the conditional expectation of Y given $\boldsymbol{X}$, which is also known as the regression function, namely,

$$
s(\boldsymbol{x}) = \mathbb{E}_{Y|\boldsymbol{X}}\left(Y|\boldsymbol{X} = \boldsymbol{x}\right).
\tag{4.3}
$$

Therefore, minimizing the risk in equation (4.1), and identifying which function $s(\boldsymbol{X})$ is best for modeling the relationship between Y and $\boldsymbol{X}$, results in the conditional mean $\mathbb{E}_{Y|\boldsymbol{X}}\left(Y|\boldsymbol{X}\right)$ under the L_2 loss. Similarly, if instead of (4.1) we use the absolute risk (referred to as the L_1 loss):

$$
\mathbb{E}_{\boldsymbol{X}}\,\mathbb{E}_{Y|\boldsymbol{X}}\left[|Y - s(\boldsymbol{X})|\,|\,\boldsymbol{X}\right]
\tag{4.4}
$$

results in the conditional median of Y given $\boldsymbol{X}$, that is,

$$
s(\boldsymbol{x}) = \operatorname{median}\left(Y|\boldsymbol{X} = \boldsymbol{x}\right).
$$

Quantile regression can be derived by slightly modifying equation (4.4) using quantile-specific weights; see Section 1.6 for an example. In summary, different risk functions lead to different quantities of interest of the population conditional distribution $g(y|\boldsymbol{x})$.

For GAMLSS models, a sensible measure representing the *risk* (or loss) of using the assumed model $f(y|\boldsymbol{\vartheta}(\boldsymbol{x}))$ rather than the unknown true population $g(y|\boldsymbol{x})$ is given by the Kullback–Leibler (KL) divergence (Kullback and Leibler, 1951). The KL divergence is based on the notion of entropy[3] between the true and theoretical distributions, which is defined as

$$
\begin{aligned}
R(\boldsymbol{\vartheta}) &= \mathbb{E}_{Y|\boldsymbol{X}}\left[\log\left(\frac{g(Y|\boldsymbol{X})}{f(Y|\boldsymbol{\vartheta}(\boldsymbol{X}))}\right)\right] \\
&= \int_y f(y|\boldsymbol{X})\log\left(\frac{g(y|X)}{f(y|\boldsymbol{\vartheta}(\boldsymbol{X}))}\right)dy \\
&\propto -\int_y g(y|\boldsymbol{X})\log f(y|\boldsymbol{\vartheta}(\boldsymbol{X})), dy
\end{aligned}
\tag{4.5}
$$

since $\int_y \log g(y|\boldsymbol{X})f(y|\boldsymbol{X})dy$ is a constant.[4] In order to minimize the risk, we need to identify the value of $\boldsymbol{\vartheta}(\boldsymbol{x})$ which minimizes equation (4.5), denoted as $\boldsymbol{\vartheta}_c(\boldsymbol{x})$. At $\boldsymbol{\vartheta}_c(\boldsymbol{x})$, the model distribution $f(y|\boldsymbol{\vartheta}(\boldsymbol{x}))$ is the closest to the population distribution $g(y|\boldsymbol{x})$ according to the risk defined by equation (4.5). In Figure 4.2(b), the risk $R(\boldsymbol{\vartheta})$ is represented as the length of the directed line from $g(y|\boldsymbol{x})$ to $f(y|\boldsymbol{\vartheta}_c(\boldsymbol{x}))$.

As we do not know the true distribution $g(y|\boldsymbol{x})$, we minimize the *empirical* risk function R_E, in which the population distribution in (4.5) is replaced by the empirical distribution. Note that as the empirical distribution is only defined on a finite number of observed values, the integral in equation (4.5) is replaced by summation, leading to

$$
\begin{aligned}
R_E(\boldsymbol{\vartheta}) &= \sum_{i=1}^n \left(\frac{1}{n}\right)\log\frac{1/n}{f(y_i|\boldsymbol{\vartheta}(\boldsymbol{x}_i))} \\
&\propto -\sum_{i=1}^n \log f(y_i|\boldsymbol{\vartheta}(\boldsymbol{x}_i)),
\end{aligned}
\tag{4.6}
$$

where in the proportionality we ignore both multiplicative and additive constants not depending on the parameter of interest.

The empirical risk is a consistent estimator of the risk since the empirical distribution converges to the true population as the sample size increases to infinity. Figure 4.2(c) represents $R_E(\boldsymbol{\vartheta})$ as an arrow going from the empirical distribution $g_E(y|\boldsymbol{x})$ to the point $\hat{\boldsymbol{\vartheta}}$ in the model line $f(y|\boldsymbol{\vartheta}(\boldsymbol{x}))$ where $\hat{\boldsymbol{\vartheta}}(\boldsymbol{x})$ is the value of $\boldsymbol{\vartheta}(\boldsymbol{x})$ minimizing the empirical risk R_E.

For independent observations, the likelihood is defined as

$$
L(\boldsymbol{\vartheta}) = \prod_{i=1}^n f(y_i|\boldsymbol{\vartheta}(\boldsymbol{x}_i)).
\tag{4.7}
$$

---

[3] *Entropy* is one of the most fundamental concepts of information theory and is widely used in physics, communication theory, economics, computer science, mathematics, probability and statistics; see Cover and Thomas (2007).

[4] We have abused $\propto$ to denote equality up to additive constants here. We also use integration assuming that y is continuous but it can be replaced by summation for discrete responses.

Minimizing the empirical risk $R_E(\boldsymbol{\vartheta})$ in equation (4.6) with respect to the parameters $\boldsymbol{\vartheta}$ is therefore equivalent to maximizing the log-likelihood function $\ell(\boldsymbol{\vartheta}) = \log L(\boldsymbol{\vartheta})$. Rigby et al. (2019, Chapter 11) provides more information on maximum likelihood estimation.

The concept of the likelihood function is of vital importance in parametric statistical inference. It is based on the reasoning that "parameter values which cause the model to suggest that the observed data are probable, are more 'likely' than parameter values that suggest that what was observed was improbable" (Wood, 2017). The three model fitting approaches used in this book, namely penalized likelihood, Bayesian methodology and gradient descent boosting, all rely on the likelihood function as the main source of information coming from the data.

4.4 Model Comparison with GAMLSS

Different GAMLSS models differ with respect to at least one of the following aspects:

- the type of distribution assumed for the response,

- the covariate sets used to explain the parameters of this distribution, and

- the way the covariates enter the predictors for the distribution parameters.

A crucial task in statistical modeling is the comparison between different models, potentially including the investigation of why certain model differences are more important than others. In order to conduct such comparisons, we need a way of evaluating the performance of rival model specifications. How to do this properly depends heavily on the aims of the analysis and particularly on whether the underlying analysis is confirmatory, exploratory or predictive (see Section 1.1). While for confirmatory analyses we are usually interested in the causal interpretation of the model or some of the coefficients, exploratory analyses incorporate models as an interpretable approximation of the data generating mechanism with the aim of identifying interesting associations. Predictive models aim at accurate prediction for new or unobserved observations, therefore sometimes sacrificing interpretability in favor of improved predictive accuracy.

In the following, we distinguish three principal approaches for comparing models: (i) utilizing summary statistics derived from a fitted model, (ii) comparing models via their predictive performance on hold-out (or test) data, and (iii) cross-validation approaches as a kind of hybrid of the two former approaches. As a reference for the models we are considering as candidates, we introduce two benchmark models: the *null model* and the *saturated model*.

The null model is the minimal, that is, most simplistic model that can be fitted under a given distributional assumption. In GAMLSS, where $y_i \sim \mathcal{D}(\boldsymbol{\theta})$, this corresponds to the situation where the distribution parameters $\boldsymbol{\theta}$ are common for all observations, namely $\boldsymbol{\theta}(\boldsymbol{x}_i) \equiv \boldsymbol{\theta}$ for $i = 1, \ldots, n$. In this case, the *marginal* distribution is fitted to the data, and there is no conditioning on explanatory terms and covariates.

At the other end of the spectrum of statistical modeling are saturated models, which have as many parameters as there are observations in the data (and are therefore somehow useless as a model information reduction mechanism). In the schematic presentation of Figure 4.2(a), (c) and (d), the model line of any saturated model (and there is more than one) will pass exactly through the point representing the observed sample. Note that saturated models only represent *full* models in situations where we have at least as many observations available as explanatory terms. In cases in which there are fewer explanatory terms than observations we cannot reach a saturated model. Note that in high-dimensional statistical modeling, it is even possible to use models with more parameters than observations when other restrictions (penalizations) are imposed.

In statistical modeling, a "good" model is typically somewhere between the null and saturated models. It is generally accepted that an *overfitted* model, that is, a model close to a saturated model but far from the population, is very poor for prediction purposes. On the other hand, an *underfitted* model, that is, an overly simplistic model close to the null model, will usually fail to capture important features in the data.

4.4.1 Information Criteria for Evaluating and Comparing Model Fits

Comparing different GAMLSS models can be achieved by using *goodness of fit summary statistics*, which are obtained from fitted models and facilitate comparison of the models while taking their different complexities into account. Since they reduce all aspects of a model fit to a single number, by construction they can only provide a part of the whole picture of how well the fitted model is performing. It is therefore a sensible strategy to complement those statistics either by other (graphical) diagnostics or by additional information about their distributions; see, for example, Aitkin (2010) for ways of judging the performance of different models based on the distribution of the summary measures rather than a single value.

In regression analysis, information criteria are frequently employed as summary statistics. Depending on the inferential approach taken for estimating the model, various variants of information criteria exist and we will discuss those in more detail in the later individual inference chapters. To illustrate the basic idea, we are relying on Akaike's information criterion, which is based on general maximum likelihood principles (Akaike, 1983). The generalized Akaike information criterion (GAIC) is defined as

$$\mathrm{GAIC}(\kappa) = -2\,\hat{\ell} + \kappa \cdot \mathrm{df} \tag{4.8}$$

where κ is a "penalty" constant, the (effective) degrees of freedom df represent a measure of model complexity, and $\hat{\ell}$ is the log-likelihood evaluated at the parameter estimates, representing a measure for the model fit. The GAIC therefore implements a compromise between the ability of the model to explain the data (model fit, measured by the log-likelihood) and the complexity of the model (measured by the degrees of freedom). All information criteria discussed in the remainder of this book follow this idea but differ in the way they quantify model fit and model complexity.

For a GAMLSS model, the model fit part is given by

$$\hat{\ell} = \sum_{i=1}^{n} \log \left[f(y_i | \hat{\boldsymbol{\theta}}(\boldsymbol{x}_i)) \right]. \tag{4.9}$$

The quantity $-2\,\hat{\ell}$ was named the *global deviance* (GD) by Rigby and Stasinopoulos (2005) to distinguish it from the deviance of a GLM/GAM which is defined slightly differently. Increments $\log \left[f(y_i | \hat{\boldsymbol{\theta}}(\boldsymbol{x}_i)) \right]$ in the deviance are important because they quantify how well an individual observation is fitted using the current assumed probability function for the response.

Different values of κ in equation (4.8) define different criteria, for example, $\kappa = 2$ corresponds to the classical Akaike information criterion (AIC) originally suggested in Akaike (1973). The Bayesian information criterion (BIC; Schwarz, 1978), also known as the Schwarz Bayesian criterion (SBC), results for $\kappa = \log(n)$. Other interesting values are $\kappa = 3.841$ corresponding to the χ^2-test with one degree of freedom; and $\kappa = (2 + \log(n))/2$ as a compromise between AIC and BIC, suggested by Ramires et al. (2021). From our experience, values within the range $2.5 < \kappa < 4$ seem to work well in practice.

The AIC can also be related to our discussions on Figure 4.2. How well a model $f(y | \boldsymbol{\vartheta}(\boldsymbol{x}))$ performs in predicting future values depends on how close the fitted model $f(y | \hat{\boldsymbol{\vartheta}}(\boldsymbol{x}))$ is to the population distribution $g(y | \boldsymbol{x})$. If the fitted model approximates the population distribution well, then the model will also predict well. The distance between the population, $g(y | \boldsymbol{x})$, and the fitted model, $f(y | \hat{\boldsymbol{\vartheta}}(\boldsymbol{x}))$, is therefore a measure of how good a model is for prediction. This distance is shown as an arrow in Figure 4.2(d), where the arrow starts from the population distribution and ends at the fitted model. Unfortunately, the evaluation of this distance involves the unknown population distribution. Under regularity conditions assumed for the data generating process, the AIC is an unbiased estimator of this distance.

When the distributional regression model incorporates random effects,[5] use of the model selection criteria is more problematic. More specifically, there are two distinct approaches for the AIC: conditional AIC (cAIC) and marginal AIC (mAIC), depending on whether the AIC is based on the conditional likelihood given the random effects (cAIC), or on the marginal likelihood after integrating out the random effects (mAIC). In the former case, we can express (4.9) explicitly in terms of fixed and random effects as

$$\hat{\ell} = \sum_{i=1}^{n} \log \left[f(y_i | \hat{\boldsymbol{\beta}}(\boldsymbol{x}_i), \hat{\boldsymbol{\gamma}}_i) \right] \tag{4.10}$$

and for given variance (or smoothing) parameters of the random effects use the effective degrees of freedom to quantify model complexity. However, usually there is uncertainty involved in the estimation of the variance (or smoothing) parameters $\boldsymbol{\gamma}$. The most naïve approach is then to ignore this, but this has severe negative

---

[5] In fact, this is also true for mean-based regression models.

side effects, see Säfken et al. (2021). For the linear mixed model, the generalized linear mixed model (GLMM) and the generalized additive mixed model (GAMM), Säfken et al. (2021) present a method for stable computation of the cAIC that takes uncertainty in the hyperparameters into account. A general approach that has also been validated in some special cases of GAMLSS is presented in Wood et al. (2016).

For the mAIC, the random effects are integrated out; this is discussed by Müller et al. (2013) in their review of model selection criteria for the linear mixed model. However, integration of the random effects for the general GAMLSS model is complex and we are not aware of a solution to this. As to the BIC, the fact that the observations are not independent reduces the effective sample size, which is used explicitly in the BIC. Several approaches exist for estimation of the effective sample size in the case of the linear mixed model, for example Lorah and Womack (2019); Cho et al. (2022); however, again we are not aware of a solution to this for the general distributional regression model. In Chapter 9, we perform model selection on a model with zero- and one-inflated simplex response distribution, incorporating random effects, using the GAIC. In the light of the lack of solutions to accounting for variance parameter estimation uncertainty, we use the naïve approach mentioned above, despite its difficulties.

Since the construction and definition of suitable summary statistics usually depends on the mode of estimating the parameters, we will discuss more measures such as the deviance information criterion (DIC) and the widely applicable information criterion (WAIC) in later chapters; see in particular Section 6.8.3. Claeskens and Hjort (2008) is a good reference on information criteria used for statistical modeling in general, including the focused information criterion (FIC) which allows the analyst to focus on specific aspects of the distribution.

Measures other than information criteria have also been used as model fit criteria. For example, there are various generalizations of the R^2 (coefficient of determination) from the linear model; see, for example Nagelkerke (1991), where an R^2 is defined for GLMs. In addition to the conceptual difficulties in deriving generalized versions of R^2, these generalizations all suffer from the same shortcoming as the classical R^2: They measure only the model fit and do not take model complexity into account. As a consequence, they rarely find application in GAMLSS modeling.

Another alternative for comparing models that is often considered in applications are graphical tools such as QQ-plots of normalized quantile residuals or bucket plots; see Section 4.7. Note, however, that information criteria and graphical tools take quite distinct approaches to comparing models, and disagreements in the evaluation of model adequacy are sometimes observed, especially for larger datasets. See De Bastiani et al. (2022) for an in-depth discussion of the underlying reasons.

4.4.2 Predictive Accuracy of Models

Information criteria aim at providing an (unbiased) estimate of the distance between the population model and the estimated model, which is then expected also to result in selecting models with good prediction performance. A different approach is to evaluate directly the prediction accuracy on new data that arose from the same population model. If, indeed, a large dataset is available for the analysis, it may be possible to set parts of the data aside for comparing rival model specifications based on their predictive ability on these hold-out (or test) data. However, in many situations the data available to the analyst will not be large enough to follow this strategy and alternatives such as cross-validation or resampling approaches (e.g. randomly assigning observations to a training or test set), discussed in more detail in the next section, will be needed.

The main question in comparing models with respect to their predictive ability is now related to potential ways of assessing the quality of predictions. In mean-based analyses, a commonly used criterion is the mean squared error of prediction (MSEP), that is,

$$\text{MSEP} = \frac{1}{m} \sum_{k=1}^{m} (\hat{y}_{kM}^* - y_k^*)^2, \qquad (4.11)$$

where y_k^*, $k = 1, \ldots, m$ refer to the observations in the hold-out sample or the left out fold for cross-validation and $\hat{y}_{kM}^*$ are our predictions for these observations based on a model specification M. In GAMLSS, we are rather interested in modeling the complete distribution of the response variable of interest, such that more general measures that allow us to evaluate the quality of a predictive distribution based on the realizations for hold-out data, are needed.

One general option for doing this is the predictive log-likelihood, which can be employed for both discrete and continuous data. If $\boldsymbol{x}_k^*$ denotes the covariate information for the new observation y_k^*, then $\hat{\boldsymbol{\theta}}_k^* = \hat{\boldsymbol{\theta}}(\boldsymbol{x}_k^*)$ are the parameters estimated for the new observation obtained by evaluating the regression predictors at $\boldsymbol{x}_k^*$ and plugging them into the response functions to arrive at the corresponding distribution parameters. We can then use the log-likelihood evaluated at the new data $(\boldsymbol{x}_k^*, y_k^*)$, that is,

$$\text{LS} = \sum_{k=1}^{m} \log \left[f(y_k^* | \hat{\boldsymbol{\theta}}(\boldsymbol{x}_k^*)) \right],$$

as a criterion for evaluating the model fit in a distributional sense (sometimes also referred to as the logarithmic score or log-score).

Other criteria can be constructed from the notion of proper scoring rules. (See Gneiting and Raftery (2007) for a detailed introduction.) The main idea of proper scoring rules is to establish a framework to distinguish between possible evaluations of predictive distributions and favor those that enforce the analyst to report their true beliefs about predictive uncertainty. We will not discuss the underlying theory in detail here, but only present some additional scoring rules for discrete, and continuous data. The case of mixed discrete-continuous data is discussed in, for example, Klein et al. (2015c).

For continuous data, the most common scoring rules (in addition to the logarithmic score) are the Brier or quadratic score (QS), the spherical score (SPS) and the continuous ranked probability score (CRPS), defined as

$$\text{QS} = \sum_{k=1}^{m} 2f(y_k^*|\hat{\boldsymbol{\theta}}(\boldsymbol{x}_k^*)) - \int \left[f(y|\hat{\boldsymbol{\theta}}(\boldsymbol{x}_k^*)) \right]^2 dy$$

$$\text{SPS} = \sum_{k=1}^{m} \frac{f(y_k^*|\hat{\boldsymbol{\theta}}(\boldsymbol{x}_k^*))}{\left(\int \left[f(y|\hat{\boldsymbol{\theta}}(\boldsymbol{x}_k^*)) \right]^2 dy \right)^{1/2}}$$

$$\text{CRPS} = -\sum_{k=1}^{m} \int \left(F(y|\hat{\boldsymbol{\theta}}(\boldsymbol{x}_k^*)) - \mathbb{1}\{y \geq y_k^*\} \right)^2 dy$$

where $F(y|\hat{\boldsymbol{\theta}}(\boldsymbol{x}_k^*))$ denotes the cdf corresponding to the model density $f(y_k^*|\hat{\boldsymbol{\theta}}(\boldsymbol{x}_k^*))$. For the CRPS, an alternative representation is

$$\text{CRPS} = -2\sum_{k=1}^{m} \int_0^1 \left(\mathbb{1}\{y_k^* \leq Q(\alpha|\hat{\boldsymbol{\theta}}(\boldsymbol{x}_k^*))\} - \alpha \right) \left(Q(\alpha|\hat{\boldsymbol{\theta}}(\boldsymbol{x}_k^*)) - y^* \right) d\alpha$$

where integration is now over different quantile levels $\alpha \in [0,1]$ rather than the domain of the response y; and $Q(\alpha|\hat{\boldsymbol{\theta}}(\boldsymbol{x}_k^*))$ denotes the quantile function, than is, the inverse of the cdf $F(y|\hat{\boldsymbol{\theta}}(\boldsymbol{x}_k^*))$. As a major advantage, the second representation of the CRPS allows one to focus on specific quantile levels rather than the overall score to detect a lack of fit in specific parts of the distribution. All three scoring rules can be shown to be strictly proper (as is the logarithmic score), that is, they enforce the analyst to report their true predictive beliefs. Still the logarithmic score is more sensitive to extreme realizations since it evaluates the predictive density at only a single point.

For discrete data, the situation is slightly more complicated since the availability of scores depends on additional assumptions on the scaling of the responses. For count data, the scores for continuous data can still be used, replacing integration by summation. For ordinal responses, there is – to the best of our knowledge – no immediate solution that takes the ordinal scaling into account since the scores inherently rely on an appropriate quantification of the distance between response values. For the case of unordered categorical responses, that is, when the different discrete outcomes are only labels without a specific numerical meaning, the predictive distribution is characterized by the probabilities

$$\hat{\pi}_{kr}^* = f\left(r|\hat{\boldsymbol{\theta}}(\boldsymbol{x}_k^*) \right)$$

where we assumed a total of R potential values such that $y_k^* \in \{1, \dots, R\}$ for simplicity. One scoring rule that is very popular in practice for this type of data is the "hit rate"

$$\text{HR} = \frac{1}{m} \sum_{k=1}^{m} \mathbb{1}\{\hat{\pi}_{ky_k^*} = \max\{\hat{\pi}_{k1}, \dots, \hat{\pi}_{kR}\}\}$$

that represents the fraction of observations for which the actually observed value was associated with the largest probability in the predictive distribution. While the hit rate is a proper scoring rule, it is not strictly proper, that is, the true predictive beliefs are not the unique maximizer of the hit rate. Intuitively, this reflects the fact that the hit rate looks only at the location of the maximum probability within the predictive distribution, but not at the size of the actual probability. Two other (strictly proper) scoring rules for discrete data are analogues to the continuous Brier score:

$$\text{BS} = - \sum_{k=1}^{m} \sum_{r=1}^{R} \left(\mathbb{1}\{y_k^* = r\} - \hat{\pi}_{kr} \right)^2$$

which compares the optimal predictive distribution with point mass 1 in the true category to the predictive distribution; and the continuous spherical score

$$\text{SPS} = \frac{\hat{\pi}_{ky_k^*}}{\sqrt{\sum_{r=1}^{R} (\hat{\pi}_{kr})^2}}$$

which relates the probability of the observed response to the norm of the predictive distribution.

We close this section with the discussion of some general points on evaluating predictions. It may be tempting to evaluate the predictive ability not on hold-out data or in a cross-validation scheme, but on the same data that have been used for estimating the model (training data). However, this will usually lead to overly optimistic evaluations owing to using the same data twice, once for estimation and once for evaluation. It is therefore important to indeed split the data; see also Section 4.4.3 for options based on cross-validation. In some cases, one may also need to split the data into more than two parts. For example, hold-out data are also sometimes used to determine hyperparameters, that is, to tune the performance of a model. In GAMLSS, this may relate to the determination of the smoothing parameters, but it could also relate to other hyperparameters such as the optimal number of iterations in boosting approaches as in Chapter 7. For predictive evaluation, the data would then have to split in three parts, one for estimating the model ("training data"), one for tuning the model ("validation data"), and one for evaluating the predictive performance ("test data").

Finally, all the scores discussed in this section (except for the MSEP) do not have an absolute interpretation but only provide comparative evidence, with *higher* scores indicating better predictive ability.[6] They can be used to compare and order models but there is no absolute meaning to the value of a score. This is a property shared with information criteria such as the AIC. We illustrate the use of some of these scores in Sections 10.3 and 11.2.3.

---

[6] Note, however, that sometimes the scores are also defined with a negative orientation such that smaller scores indicate better predictive ability.

4.4.3 Cross-validation

Since, in many situations, the data available to the analyst will not be large enough to set parts of the data aside for evaluating predictive performance using the approaches discussed in the previous section, one often resorts to cross-validation approaches. The basic idea is randomly to split the data into k parts (or folds) and to use $k-1$ folds of the data to estimate the model and the remaining fold for evaluation. Iterating and averaging over the k folds then provides an estimate for the cross-validated predictive performance.

Actual implementations of cross-validation differ in their exact design, in particular in the construction of the folds.

- If $k = n$, that is, the number of folds coincides with the number of observations, each fold consists of only one observation such that $n-1$ observations are used for estimation while one is set aside for evaluation. This results in leave-one-out cross-validation (LOO-CV). As a major advantage, no assignment of observations to the folds is needed, leaving less room for artefacts resulting from random assignments. On the downside, LOO-CV usually requires n additional model fits and is therefore quite expensive for large datasets.

- In mutually exclusive k-fold cross-validation, the data are split into k mutually exclusive subsets of roughly the same size, that is, each observation is randomly assigned to one of the k folds. This offers effective control over the computational burden by keeping k at a moderate size (often $k = 5$ or $k = 10$), but the results are inherently dependent on the random assignment.

- An alternative that reduces this dependence is to use a larger number of potentially overlapping folds. More precisely, one fixes the size of the folds at a pre-specified fraction of the sample size (that is, 10%) and then determines k folds by randomly picking this fraction of observations. In this way, observations can show up in multiple folds and, as k increases, the result will get less and less dependent on the precise assignments.

4.5 Model Selection Strategies for GAMLSS

Model choice in GAMLSS addresses two main (interlinked) questions:

- How do we find the right distribution for a given dataset?

- How do we specify the predictors for the different parameters of this distribution?

The latter consists of both the selection of the relevant subset of covariates as well as deciding how to best include them in a regression predictor. For example, for continuous covariates we might include them with linear or nonlinear effects or we might also consider different types of transformations. Similarly, for regional data we could include the regional information in the form of an i.i.d. random effect or as a Markov random field (or both). Furthermore, we might also consider different types of interactions. We refer to the task of selecting the covariates as well as the form(s) in which they appear in the predictors, alternatively as *variable selection* or *term selection*. *Effect selection* and *feature selection* are equivalent terminology arising from the machine learning world.

The two tasks of selecting a response distribution and specifying the predictors for its parameters cannot, unfortunately, be solved independently since the fit of a specific type of distribution crucially depends on the chosen predictor structure and, vice versa, the predictors cannot be determined without choosing a response distribution in the first place. We nonetheless discuss the two aspects separately in the following sections 4.5.1 and 4.5.2 with specific emphasis on one problem at a time.

4.5.1 Selection of a Distribution in GAMLSS

In many cases, a first pre-selection of appropriate response distributions can be made based on stylized properties of the response. One important aspect in this regard is whether the response is continuous, discrete, or mixed discrete–continuous, which allows the elimination of distributions that do not adhere to this. On a related note, domain restrictions on the response, such as non-negativity or the restriction to a certain interval can serve as guiding principles for a first selection of distributions. Unfortunately, the distinction is not as clear-cut as it seems at first glance. In some cases, it might still make sense to approximate a discrete distribution by a continuous one, or to construct a discrete distribution from a continuous one, by some categorization. Similarly, a distribution with support on the whole real line may be used for a response with restricted domain if the observed values are not too close to the boundaries of the domain. Nonetheless, the scaling and the domain of the response gives us first hints on suitable candidate distributions (as also reflected in the organization of Chapter 2 according to stylized classes of distributions). Figure 4.3 shows the four most important classes of response variable, and their subdivisions.

- For continuous response variables, the choice is whether the support is $\mathbb{R}$, $\mathbb{R}_+$, or $\mathbb{R}_{(0,1)}$. In practice $\mathbb{R}_+$ or $\mathbb{R}_{(0,1)}$ response variables can be modeled using an $\mathbb{R}$ distribution, if the observed response has values far from the boundaries of zero or zero and one, respectively.

- For count responses the difference is whether we are dealing with "Poisson" or "binomial" type responses: that is, whether the discrete count distribution is unbounded or bounded.

- For mixed responses the most common choice is between zero-adjusted with support $\mathcal{S} = [0, \infty)$, or zero- and/or one-inflated with support $[0, 1)$, $(0, 1]$ or $[0, 1]$.

- When the response is a factor (categorical) we have two choices: *ordered* or *unordered*.[7] Within the GAMLSS framework, unordered categorical responses can be modeled using the *multinomial* distribution. The special case where the classification factor has only two levels, for example "dead" or "alive," "yes" or "no,"

---

[7] In machine learning, models with unordered categorical responses are called *classification* models.

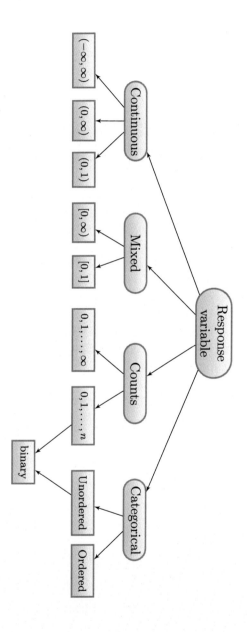

Figure 4.3 Showing how the different ranges of the response variables are associated with different response variable distributions.

can be treated as binary responses and therefore can be fitted within GAMLSS by assuming a binomial response. For binary and ordered responses, GAMLSS does not offer extra advantages over the more classical GLM or GAM models, but these can still be treated within the GAMLSS framework.

Using other stylized features such as skewness or kurtosis of the response distribution is more difficult, since these may change radically once starting to adjust for covariates. While one quite often sees marginal visualizations of the response variable in histograms or kernel density estimates, these are in general not too helpful for making specific choices for the response distribution. This is analogous to the linear model, where the normality assumption should be verified from the error terms and not the marginal distribution of the response.

If not too many covariates are present, one possibility to compare different candidate distributions is to define a close-to-saturated model with a fairly complex, flexible predictor formed by these covariates which is then used for all distribution parameters of any candidate distribution. If the number of covariates is larger, through exploratory analysis and (if possible) prior subject matter knowledge, a "reasonable" small set of explanatory terms for each distribution parameter should be specified. These models can then be compared based on some of the criteria discussed in the previous section (cross-validated predictive performance scores, information criteria, etc.)[8] or using the graphical diagnostics discussed in Section 4.7 (quantile residuals, worm plots, etc.). The saturated approach is used in Chapter 9, in which the number of predictors is small ($p = 4$), and the "reasonable" small set strategy in Chapter 10. If computation times permit it, one can also do a full predictor optimization for each candidate distribution using one of the stepwise procedures discussed in the next section, and then finally compare the best-fitting specification of each.

Whatever strategy is followed, it is important to consider the impact of possibly extensive model searches on the validity of the inferences derived from the final model. Ignoring the model choice decisions will, in general, lead to overly optimistic uncertainty assessments. In an exploratory analysis, this may still be acceptable but for confirmatory, causal analyses, proper quantification of post model selection uncertainty is needed.

4.5.2 Selection of Covariates and Terms in GAMLSS

For the selection of covariates and terms in GAMLSS, given a specific type of response distribution, similar approaches as in other forms of regression models can be used but additional complications arise from the fact that GAMLSS models usually involve multiple structured additive predictors, and in these predictors covariates can enter in various forms. Various approaches for tackling this problem have been developed by relying on a specific mode of statistical inference. For example, the boosting approach discussed in Chapter 7 directly targets variable and term selection in the context of early stopping and indirect penalization. In Bayesian approaches, suitable

---

[8] Facilitated by the **gamlss** function `chooseDist()`.

prior structures can be chosen that enable model choice decisions; see Section 6.7. In frequentist approaches, shrinkage and regularization approaches relying on specific penalties have been suggested; see Section 5.4 for details.

In this section, we focus on some general strategies that can be used with any inferential approach that fits a GAMLSS model. These are basically variations of stepwise procedures that systematically compare a sequence of models by some kind of model choice criterion. Of course, one could also try to simply fit all possible model specifications for a given type of distribution and then choose the "best" according to a specific criterion, but this is only possible when the number of terms available for selection is small. Otherwise, the *curse of dimensionality* soon takes effect and the task becomes almost impossible. For example, let us consider the analysis of ultrasound data that we discuss in detail in Chapter 8, in which we have only seven explanatory terms. For a linear main effect model with no interactions, there are $2^7 = 128$ different models to consider, and that for modeling only the location parameter. When modeling also the scale parameter, the number of models is $2^{2 \times 7}$. More generally, for r terms and K distribution parameters, we would need 2^{Kr} different models to consider. For fitting a normal distribution with $K = 2$ and $r = 10$, we already have 1,048,576 models to consider and for $r = 30$ the number of models exceeds 10^{17}. In addition, if we need to consider first-order interactions, we have to add $Kr(r-1)/2$ interaction terms.

Stepwise procedures create a subset of models by iteratively including and/or excluding effects in forward, backward, or combined stepwise fashion. In contrast to the usual approaches for regression models for a single distribution parameter, stepwise procedures for GAMLSS require an additional loop for the distribution parameters. This additional loop can either be included by looping over all potential effects for all distribution parameters simultaneously and picking or removing one term only once, or by applying the stepwise procedure separately within a loop over the distribution parameters.

Even though a considerable reduction of computing time can be achieved with stepwise selections as compared to a full model search, with a larger number of terms and distribution parameters they may still become too costly, in terms of computation times. One can impose some hierarchy among the parameters in the selection: For example, determine the most important parameter first (usually the location parameter) and proceed sequentially through the other parameters, holding the rest of the parameters fixed.

Stepwise procedures in general require an explicit list of all potential effects that could be included in the final model, including a list of candidate interactions if needed. This list can also comprise mandatory effects that should be considered in any model, competing model terms where different modeling alternatives are available for the same covariate, or decompositions of effects, such as, the decomposition of nonlinear effects into a linear effect and the nonlinear deviation from this.

Note that automated model selection using stepwise variable selection procedures has been roundly criticized by many authors, most notably Harrell (2015). Some of the problems with the method cited by Harrell (2015) are: (i) underestimated standard errors and confidence intervals, (ii) erroneously small p-values, (iii) regression coefficients biased high in absolute value, and (iv) lack of accounting for collinearity. In addition, Smith (2018) states that "big data exacerbates the failings of stepwise regression (...) the larger the number of potential explanatory variables, the more likely stepwise regression is to be misleading." Despite these biases, and the drop in popularity of stepwise selection in the statistical world, we still consider it as a pragmatic solution to a hard covariate selection problem in distributional regression. In any case, it is important to base stepwise procedures on comparisons with appropriate criteria rather than relying on p-values from significance tests that were designed for prespecified hypotheses.

4.5.3 A Recipe for Model Selection

As discussed in Sections 4.5.1 and 5.5.2, there are different paths a data analyst can take to model a dataset within GAMLSS. Despite the difficulties with the naïve application of stepwise procedures discussed previously, our experience indicates that a pragmatic recipe is as follows, but we emphasize that this recipe usually requires adaptation for specific applications.

1) Use the information displayed in Figures 2.1, 2.9, 2.11 and 4.3, and Tables A.1 and B.1 to select the type of distribution, and therefore the set of distributions, most appropriate for the response variable. Within this set of appropriate distributions, usually one can be considered as the standard. For example, for a continuous response variable defined on the real line, the standard distribution is the normal, for continuous nonnegative distributions it is the gamma, or for continuous responses restricted to the unit interval, it is the beta distribution. Of course, given specific considerations depending on the application of interest, one could also contemplate alternative standard distributions.

2) Find a "reasonable" small set of explanatory terms and fit the standard distribution to the data using this set of explanatory terms. At this stage, since the number of terms is small, all parameters of the distribution can be modeled using the small set of explanatory variables.

3) Use the model fitted in (2) above to select the "best" distributions (say the best three) within the class of all available (appropriate) distributions for the response.

4) Having a small set of candidate distributions for the response, we can now select appropriate terms for all the parameters of the distributions using, for example, stepwise selection approaches.

5) Finally, check the final models, for example with residual diagnostics and diagnostics for influential observations. Be guided by ease of interpretation when selecting the final model.

This procedure is of course by no means unique and there is no guarantee that it will always lead to the best model, or even just a good one. For example, an alternative procedure could start by selecting terms for the location model, assuming a standard distribution such as the normal. In a second step, this part of the model is fixed and one proceeds by selecting terms for all the remaining parameters of the distribution. There are many alternative paths a practitioner can take and our advice is to take the path with which one is most comfortable.

Note, however, that there is a *hierarchy* when selecting terms for the parameters. It is better to start from a good location model in order to avoid overfitting the scale parameter model. A bad location model may exaggerate the residuals and therefore will result in overfitting the scale parameter model. The same applies to parameters associated with skewness and kurtosis. A good location and scale parameter model is needed before attempting to fit models for the skewness and kurtosis parameters. For example, a bad model for the scale parameters could affect the model for the kurtosis badly, since large variations not accounted for by the scale model are seen as evidence of kurtosis. The reader should keep this hierarchy of parameters associated with properties of the distribution of the response in mind when selecting a GAMLSS model.

4.6 Parameter Orthogonality

Parameter orthogonality (or lack thereof) is an issue of which users of distributional regression should be aware. It affects both the fitting of a model, and the interpretation of a fitted model. Within GAMLSS there are two types of parameter orthogonality: the first is distribution parameter orthogonality, that is, whether the distribution parameters θ_k for $k = 1, 2, \ldots, K$ are information orthogonal or not; the second concerns the regression coefficients in the predictors.

To investigate distribution parameter orthogonality, consider any two-parameter distribution having log-likelihood $\ell(\theta_1, \theta_2 | \boldsymbol{y})$. It is well known that the asymptotic variance–covariance matrix of the MLEs $(\hat{\theta}_1, \hat{\theta}_2)^\top$ is the inverse of the Fisher information matrix $\mathcal{I}$, which is defined as

$$\mathcal{I} = -\mathbb{E} \begin{pmatrix} \frac{\partial^2 \ell}{\partial \theta_1^2} & \frac{\partial^2 \ell}{\partial \theta_1 \partial \theta_2} \\ \frac{\partial^2 \ell}{\partial \theta_1 \partial \theta_2} & \frac{\partial^2 \ell}{\partial \theta_2^2} \end{pmatrix}.$$

The distribution parameters θ_1 and θ_2 are *orthogonal* if the off-diagonal element of $\mathcal{I}$ is zero, which implies that $\hat{\theta}_1$ and $\hat{\theta}_2$ are asymptotically independent. The concept of distribution parameter orthogonality was first discussed by Huzurbazar (1950) and later by, amongst others, Cox and Reid (1987), generally in the context of a marginal distribution setting, that is, no model equations for the distribution parameters; and with θ_1 being the parameter of interest (location, usually the mean) and θ_2 a "nuisance parameter." In this setting, the advantage of orthogonal parametrization is stability and speed of maximum likelihood estimation; non-orthogonal parametrizations can result in "banana-shaped" log-likelihoods on which ML iterations tend to be slow to converge, or not to converge.

Consider now the simple case of a GAMLSS model with two parameters and a linear model for both of them:

$$g_1(\theta_1) = \boldsymbol{x}_1^\top \boldsymbol{\beta}_1; \quad g_2(\theta_2) = \boldsymbol{x}_2^\top \boldsymbol{\beta}_2,$$

in which $\boldsymbol{x}_1$ and $\boldsymbol{x}_2$ are of length p_1 and p_2, respectively. If the parameters θ_1 and θ_2 are orthogonal, then the $(p_1 + p_2) \times (p_1 + p_2)$-dimensional Fisher information matrix is block-diagonal, that is, the elements of $\boldsymbol{\beta}_1$ and the elements of $\boldsymbol{\beta}_2$ are orthogonal. This type of between-parameter orthogonality simplifies the interpretation of the model. Very rarely we can also achieve within-parameter orthogonality, that is, the variance–covariance matrix of $\boldsymbol{\beta}_j$ is diagonal ($j = 1, 2$). This simplifies the interpretation since any conclusion of any of the $\boldsymbol{\beta}_j$ can be drawn independently of the behaviour of the rest of the parameters. As discussed in Section 1.4.2, GLMs have response distributions in the exponential family $\mathcal{E}(\mu, \phi)$, in which μ and ϕ are the mean and scale parameters, respectively. Since μ and ϕ are orthogonal for all exponential family distributions (Barndorff-Nielsen, 2014), GLMs (including their important special case normal linear models) do not suffer from the problems of parameter non-orthogonality. However, in distributional regression, response distributions outside the exponential family do not typically have orthogonal parameters, even when $K = 2$. The result of this is that

- ML iterations can sometimes be slow or not converge, owing to badly shaped log-likelihoods; and

- misspecification of the model for one parameter can have a biasing effect on estimates of the coefficients of other parameters, even if the models for those other parameters are correctly specified (Heller et al., 2019). When a response distribution has orthogonal parameters, the parameters are essentially modeled independently; this is not the case for non-orthogonal parameters.

As pointed out in Rigby et al. (2019, section 18.1) almost all continuous distributions with support on $\mathbb{R}$ are location–scale families of distributions, which have the advantage that the location μ and the scale σ of parameters are orthogonal if the distribution is symmetric. Unfortunately this does not usually apply to the other parameters of the distribution. It is very common that the scale parameter of a continuous distribution is correlated with its kurtosis parameter.

We demonstrate parameter orthogonality and non-orthogonality for the case of the t-family (TF) distribution (see section 2.5), which is a symmetric, three-parameter continuous distribution with support on $\mathbb{R}$. Distribution parameters are $\mu =$ location (mean, median, mode), $\sigma =$ scale, and $\nu =$ kurtosis. In Figure 4.4 the results of two simulations are shown, both of which are based on $\text{TF}(\mu = 0, \sigma = 1, \nu = 10)$.

- The plots above the diagonal show the log-likelihood contours for a single simulated sample of size $n = 1000$.

- In the plots below the diagonal, we show bivariate density estimates for the MLEs $(\hat{\mu}, \hat{\sigma}, \hat{\nu})$ of 500 samples, each of size $n = 1000$. The correlations between these estimates are also shown.

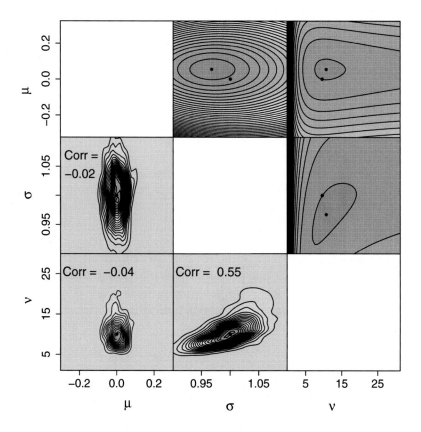

Figure 4.4 Log-likelihood contours (top right) and bivariate density estimates of MLEs (bottom left), of the *t*-family distribution, $\text{TF}(\mu = 0, \sigma = 1, \nu = 10)$. The figures were obtained by sampling $n = 1000$ values from the distributions, and fitting a GAMLSS model, once in the top right plots and 500 times in the bottom left plots. In each plot the black point is the true (simulated) value; in the top right plots the red points are the MLEs.

- From the log-likelihood contours for (μ, σ) and correlation of $(\hat{\mu}, \hat{\sigma})$, it appears that μ and σ are orthogonal parameters (as they should be).

- From the log-likelihood contours for (μ, ν) and correlation of $(\hat{\mu}, \hat{\nu})$, it also appears that μ and ν are orthogonal parameters (or close to orthogonal).

- The log-likelihood contours for (σ, ν) and correlation of $(\hat{\sigma}, \hat{\nu})$ demonstrate strong evidence of non-orthogonality between σ and ν.

An easy way to check distribution parameter orthogonality, for specific parameter values, is to generate a sufficiently large random sample from that distribution, fit a GAMLSS "null" model (no covariates) and compute the correlation of the MLEs as given by the observed Fisher information matrix. These correlations for the sample

Table 4.1 *Asymptotic correlations between MLEs of the TF($\mu = 0, \sigma = 1, \nu = 10$)* *distribution, as estimated by the observed Fisher information matrix based on a simulated* *sample of size $n = 1000$.*

	σ	ν
μ	0.021	0.026
σ		0.758

above are given in Table 4.1, and confirm the orthogonality between μ and σ and between μ and ν, and the strong non-orthogonality between σ and ν.

A non-orthogonal two-parameter distribution may, in theory, be reparametrized to an orthogonal version, although the orthogonalization may not always be mathematically tractable. For example, the more commonly used version of the Poisson-inverse Gaussian distribution (Dean et al. (1989), `PIG` in **gamlss**) is non-orthogonal; its reparametrization to an orthogonal version is straightforward (Heller et al. (2019), `PIG2` in **gamlss**). However, there are many examples of non-orthogonal two-parameter distributions which are not amenable to straightforward orthogonalization; Stadlmann et al. (2023) give a method for a numerical solution of the problem. Beyond two parameters, orthogonalization between all parameters cannot in general be achieved.

The second type of parameter orthogonality concerns the regression coefficients. Orthogonality in these is affected by whether the explanatory variables associated with the particular coefficients are correlated or not. This problem in linear regression is termed *multicollinearity*. It does affect the interpretation of the associated $\boldsymbol{\beta}$ because the fitted coefficients are very sensitive to small changes in the covariates involved. Checking the correlation coefficients of the explanatory terms in advance and eliminating highly correlated covariates is a possible solution to avoid the problem.

Orthogonality in coefficient vectors $\boldsymbol{\gamma}$ of smooth effects is based on whether the linear manifold of the ith smoother, $\mathcal{M}(\boldsymbol{B}_i)$, is close to the linear manifold associated with the jth smoother, $\mathcal{M}(\boldsymbol{B}_j)$, that is, concurvity is present. (See Section 5.2.1.) Concurvity can affect the fitting since highly concurved subspaces make the back-fitting algorithm difficult to converge. The modified backfitting algorithm described in Section 5.2.1, where all the linear terms are fitted separately, provides a solution to the fitting problem. However, the problem of interpreting the smoothers is still present because the shape of one smoother is sensitive to the values of the other one. Fitting a smooth surface for the two associated variables, that is, $s(x_i, x_j)$, is a possible solution provided that the two variables are not too dependent.

4.7 Diagnostic Tools for GAMLSS

In this sectiom, we discuss a number of tools, mainly graphical, that assist checking how well a given model specification, including both the response distribution and the regression predictors, fits to a given dataset. As a general word of caution, we emphasize that – while often agreeing in practice – there are cases where the graphical assessment with any of the approaches discussed below may be in conflict with the evaluation by tools such as information criteria discussed in Section 4.4.1. The underlying reason is often the fact that the fit in the tails has a larger impact on the graphical assessment than on information criteria. An example of this issue can be found in Section 11.3.2, where we analyze childhood malnutrition and find a disagreement in the evaluation by quantile residuals and Bayesian information criteria. A more detailed discussion of this issue can be found in De Bastiani et al. (2022).

4.7.1 Normalized Quantile Residuals

The linear model definition of residuals, as the difference between the observed values of the response, $\boldsymbol{y}$, and its fitted values, $\hat{\boldsymbol{y}} = \hat{\boldsymbol{\mu}}$, does not extend to distributional regression models. Even in the case where $\boldsymbol{\mu}$ is a location parameter, the difference between observed and fitted values tells us only a very incomplete story about the fit since it includes only location aspects but does not reflect the fit with respect to other distributional characteristics. A general and easy method is to define residuals based on the fitted cumulative distribution function (cdf) $\hat{F}(y_i) = F(y_i|\hat{\boldsymbol{\theta}}_{[i]})$. We rely on the well-known fact that, for the continuous random variable y_i with true cdf $F(\cdot)$, we have

$$F(y_i) \sim \mathcal{U}(0, 1),$$

for example, plugging the random variable y_i into its own cdf yields a standard uniform distribution. This is also referred to as the probability integral transform.

In the context of GAMLSS, we can use the probability integral transform to evaluate the estimated cdf at the observed responses and then compare the resulting values to the uniform distribution, for example based on a histogram. Alternatively, the values can be normalized to become *normalized quantile* residuals by additionally applying the inverse standard normal cdf.

Formally, for continuous response distributions, the probability integral transform leads to the residuals

$$\hat{u}_i = F\left(y_i|\hat{\boldsymbol{\theta}}_{[i]}\right)$$

and the normalized quantile residuals are then given by

$$\hat{r}_i = \Phi^{-1}(\hat{u}_i),$$

where $\Phi^{-1}(\cdot)$ is the inverse cdf of the standard normal distribution.

For discrete or censored response variables, we have to take a "randomization" step since in this case the probability integral transform is not directly applicable. In contrast, the cdf is a step function such that the residuals $\hat{u}_i$ will necessarily be discrete. The randomization step jitters the discrete $\hat{u}_i$'s such that they are uniformly distributed over the corresponding step heights of the cdf (Stasinopoulos et al., 2017, section 12.2). Afterwards, one applies the normalization step.

Normalized quantile residuals are useful as observation-based diagnostics (as opposed to variable-based diagnostics), having the ability to identify individual observations in which the response variable has unusual values. They can be plotted individually against the observation index of the data, or against explanatory terms, to identify "large" deviation of the response from its "expected" values under the model. Alternatively they may be plotted as a whole, using QQ-plots or worm plots (discussed in Section 4.7.2), to identify whether the assumed response distribution is adequate for the data. The idea is that if the assumed model distribution is adequate, then the normalized quantile residuals should behave as normally distributed variables. QQ-plots were created to test visually for normality. More precisely, if points in the QQ-plot fall close to the 45° line, it shows that their values are close to what we would expect under the assumption of normality. If this not the case, then the QQ-plot flags possible problems. We use the QQ-plot to this effect in Chapter 1 and specifically in Figures 1.2(b), 1.3(b), and 1.5(b). For discrete or censored responses, the quantile residuals are not unique because of their random component. It is good practice to construct QQ-plots or worm plots for multiple realizations, to avoid faulty decisions based on one possibly unfortunate realization. An example of this with a discrete response variable, using 20 realizations of randomized quantile residuals, is given in Figure 10.3.

One can also define multivariate quantile residuals when dealing with a multivariate response variable y_i. If, for simplicity, we assume a bivariate response $y_i = (y_{i1}, y_{i2})^\top$, one then considers the distribution of y_{i1} (the marginal distribution of the first response element) and the distribution of $y_{i2}|y_{i1}$ (the conditional distribution of the second response element given the first). For each of these distributions, we compute the normalized quantile residuals and if the model fits well, the resulting $\hat{r}_{i1}$ and $\hat{r}_{i2}$ will jointly follow a bivariate normal distribution which can be assessed in a scatterplot. See Hohberg et al. (2021) for details and Section 11.3.2 for an application. An alternative is to consider $\hat{q}_i = \hat{r}_{i1}^2 + \hat{r}_{i2}^2$ which will follow the χ_2^2 distribution if the model fits well.

4.7.2 Worm Plots

A detrended version of the QQ-plot for quantile residuals is also known as the *worm plot*. Worm plots were introduced by van Buuren and Fredriks (2001) for checking growth-curve centile estimation. By detrending the QQ-plots, inadequacies in the fitted distribution are highlighted. For example, Figure 4.5(a) shows the QQ-plot for fitting the BCTo distribution to BMI in the Dutch boys' data (Section 1.5.1). The equivalent worm plot is shown in Figure 4.5(b). The possible area of concern is the tails of the distribution, which is highlighted better in the worm plot than in the QQ-plot. The worm plot points should be close to the horizontal line. The shaded area provides a 95% pointwise confidence interval; if the assumed distribution is adequate for the data, then we expect 95% of the points to lie within the shaded area. There is also a neat way of interpreting the shape of worm plot points:

- if the whole worm plot is shifted up or down from the horizontal line, this indicates that something went wrong with the fitting of the location parameter;

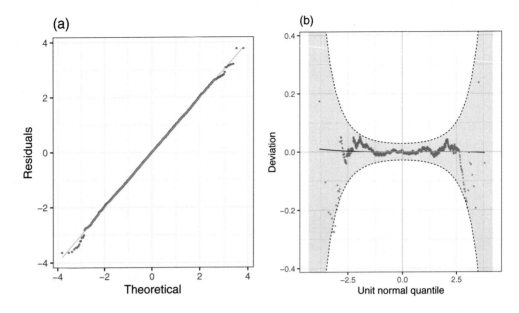

Figure 4.5 (a) QQ-plot from the residuals of the BMI data fitted using a BCTo distribution (identical to Figure 1.5(b)); (b) the equivalent worm plot.

- if the worm plot has a linear shape (lying mostly out of the shaded area) then the scale parameter was fitted badly;

- if the worm plot has a quadratic shape, then the skewness was not taken into account properly;

- if the worm plot has an S-shape then the kurtosis is not fitted properly.

For more details see either the original worm plot paper by van Buuren and Fredriks (2001) or Stasinopoulos et al. (2017, section 12.4).

There are two useful extensions of the simple worm plot: (i) to see how the distribution is fitted at different areas of a predictor, and (ii) to compare different model fits. Feature (i) is discussed in Stasinopoulos et al. (2017, section 12.4.2); (ii) is shown in Figures 8.6 and 8.9.

4.7.3 Bucket Plots

The worm plots provide general information about how well the assumed distribution fits the response variable. The bucket plot, on the other hand, provides more specific information about *skewness* and *kurtosis*, see De Bastiani et al. (2022).

The concepts of skewness and kurtosis are easily understood but their mathematical definition is rather awkward. Rigby et al. (2019, chapters 14 and 15) describe different ways of defining skewness and kurtosis measures. The most well-known population (rather than sample) definitions of skewness and *excess* kurtosis, respectively, are the *moment*-based measures

$$\gamma_1 = \frac{\mu_3}{\mu_2^{1.5}} \tag{4.12}$$

$$\gamma_2 = \frac{\mu_4}{\mu_2^2} - 3, \tag{4.13}$$

where μ_k is the kth central moment of y and where 3 in equation (4.13) corresponds to the normal distribution kurtosis. Both measures can be transformed to a value in the interval $(-1, 1)$ using the transformation:

$$\gamma_{jt} = \frac{\gamma_j}{1 + |\gamma_j|} \qquad \text{for } j = 1, 2. \tag{4.14}$$

The problem with the population moment-based skewness and excess kurtosis measures is that in some long-tailed distributions they do not exist. Equivalent "sample" skewness and excess kurtosis are defined by replacing the population moments by their corresponding sample estimates in equations (4.12) and (4.13); whilst these always exist, they can be very sensitive to outliers.

The "population" centile-based measures described in the following always exist and are robust. The centile measure of skewness is

$$s_p(y) = \frac{(y_p + y_{1-p})/2 - y_{0.5}}{(y_{1-p} - y_p)/2} \tag{4.15}$$

while the centile measure of excess kurtosis is

$$k_p(y) = \frac{y_{1-p} - y_p}{y_{0.75} - y_{0.25}} - 3.449 \tag{4.16}$$

for $0 < p < 0.5$, where $y_p = F^{-1}(p)$, $F^{-1}(\cdot)$ is the inverse cdf, and 3.449 is the centile kurtosis of the normal distribution. Two commonly used measures of skewness are (i) Galton's measure of skewness (also known as *central* measure of skewness), which is defined by setting $p = 0.25$ in equation (4.15); and (ii) the *tail* measure of skewness with $p = 0.01$. Both centile measures can be transformed to values in $(-1, 1)$ using the same transformation as equation (4.14). For sample skewness and excess kurtosis, we replace the population moments with their sample equivalents.

The bucket plot provides a visual tool for checking whether moment- or centile-based skewness and kurtosis are present in the normalized quantile residuals of a fitted model. It is relevant only for continuous response variables. If the presence of skewness and/or kurtosis is detected, this probably reflects the fact that the assumed response distribution is inadequate.

We first used moment bucket plots in Chapter 1, in Figures 1.2(c), 1.3(c) and 1.5(c). Figures 4.6 (a) and (b) show the basic structure of moment and centile bucket plots, respectively. The x axis shows the skewness while the y axis shows the excess kurtosis. The origin $(0, 0)$ represents the normal distribution, which has zero skewness and zero excess kurtosis. We refer to the area within the black line as the *bucket*; it has the

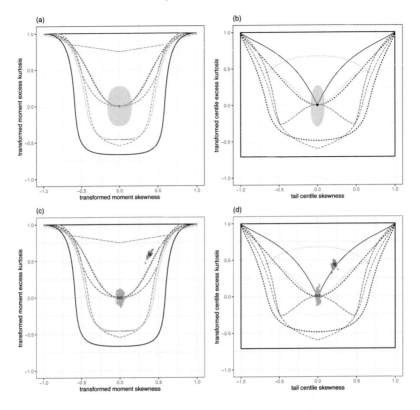

Figure 4.6 (a) A typical example of a moment-based bucket plot, the shaded area in the middle represents a 95% confidence region for the Jarque—Bera test; (b) a typical example of a centile based bucket plot, the shaded area in the middle represents a 95% confidence region for a bootstrap test; (c) moment bucket plot and (d) centile bucket plot, for models fitted to the Dutch boys' BMI data using the `BCTo` and the `IG` distributions.

shape of a bucket for the moment plot and is box-shaped for the centile plot. The bucket represents the possible range of values that the transformed skewness and excess kurtosis measures can take. The curves within the bucket represent different GAMLSS distributions. Those curves are not relevant for the current use of the bucket plot as a diagnostic tool, but are important to demonstrate the flexibility of any theoretical distribution; see Rigby et al. (2019, Chapter 16). Any point lying above the horizontal line passing through zero shows lepto-kurtotic behaviour, while a point under the horizontal line shows platy-kurtotic behaviour. Points to the left of the vertical line passing through zero indicate negative skewness, and points to the right indicate positive skewness.

The shaded area in the moment plot in Figure 4.6(a) represents a 95% confidence region for the Jarque–Bera test (Jarque and Bera, 1987). Any point within this region would not reject the null hypothesis that skewness and excess kurtosis are simultaneously equal to zero. The shaded area in the centile plot in Figure 4.6(b) represents a 95% confidence region from a simulation generated by bootstrapping the original residuals 1000 times. Again any point falling within the region would suggest that there is neither skewness nor excess kurtosis in the residuals. Note that the size of the shaded area for both the Jarque–Bera and bootstrap tests depends on the number of available observations, that is, fewer observations would result in a bigger shaded area. Also note that there are two different measures of centile skewness and centile excess kurtosis, the *tail* and the *central* skewness and excess kurtosis. Figures 4.6(b) and (d) present the tail measure.

Figures 4.6(c) and (d) show the bucket plot in action. Here we compare residuals arising from two models fitted to the Dutch boys' BMI dataset: The first uses the Box–Cox *t* (`BCTo`) response distribution, and the second uses the inverse Gaussian (`IG`). In both cases all distribution parameters (four for `BCTo` and two for `IG`) were fitted using a smooth function for age. The transformed measures of skewness and excess kurtosis for the residuals from the two models are represented by a single point each in the bucket plot and are shown, in Figures 4.6(c) and (d), as `bct` and `ig`, respectively. The transformed measures of skewness and excess kurtosis for the residuals from the `BCTo` distribution model lies in the middle of the figure, close to the point $(0, 0)$ and within the 95% confidence region.(Note that the shaded areas shown in (a) and (b) are present in (c) and (d), but not visible as they are "behind" the 95% confidence regions.) This indicates that the `BCTo` distribution fits adequately, as far as skewness and kurtosis are concerned. The measure of residuals for the `IG` is above and to the right of the middle point, indicating that right skewness and lepto-kurtosis exist in the residuals and therefore a more suitable distribution is needed.

4.8 Visualizing and Interpreting GAMLSS

For a GAMLSS model, the question of "how x affects the response" has a completely different and more complex meaning compared with standard linear, generalized linear and generalized additive models. While these focus exclusively on the mean and therefore involve only a single regression predictor, GAMLSS models feature multiple predictors and the interpretation of the effect that a covariate has on a specific distribution parameter crucially depends on the values assumed for the other predictors. Additionally, each covariate may be included in multiple predictors such that simple ceteris paribus[9] interpretations on the change of one specific distribution parameter regarding the change of the covariate are not available any more. As a consequence, one often resorts to partial effect plots where one (or a small set of) covariate(s) is varied while the others are kept fixed at pre-specified values. This is also closely linked to the notion of effect displays (Fox, 2003). In the following, we introduce different types of partial effect plots, varying according to the visualization of specific aspects of the fitted model.

To formalize the construction of partial effect plots, let x_j denote a single covariate or a small subset of covariates, while x_{-j} denotes the rest of the covariates in the

---

[9] other conditions remain the same, or, literally, "everything else being equal".

model. Let $\mathcal{D}(y)$ denote the conditional distribution of the response variable Y given all the covariates in the analysis. The partial effect that $\boldsymbol{x}_j$ has on some characteristic, say $\omega(\mathcal{D})$, of this conditional distribution then depends in general on the values that the rest of the terms $\boldsymbol{x}_{-j}$ take. These values are defined by subject knowledge or determined as summary statistics derived from the values of $\boldsymbol{x}_{-j}$, which we denote as $g(\boldsymbol{x}_{-j})$ (usually a summary measure such as the mean or median). We refer to the setting under which $g(\boldsymbol{x}_{-j})$ is calculated as the *scenario* $\mathcal{S}$, for example the training data.

The *partial effect* that the term $\boldsymbol{x}_j$ has on the characteristic $\omega(\mathcal{D})$ of the distribution under the scenario $\mathcal{S}$ characterized by $g(\boldsymbol{x}_{-j})$ is then denoted as

$$\mathrm{PE}_{\omega(\mathcal{D})}\left(\boldsymbol{x}_j | g(\boldsymbol{x}_{-j}), \mathcal{S}\right) \ .$$

Some possible characteristics of the distribution of the response can be:

- a specific *predictor* of the model, that is, η_k;

- a specific *parameter* of the distribution of the model, that is, θ_k;

- a specific *moment* of the distribution of the model, for example, $\mathbb{E}(y)$, $\mathbb{V}(y)$, the *skewness* or the *kurtosis*;

- a specific *quantile* of the distribution of the model, for example, y_p for $0 < p < 1$, the centile *skewness* or the centile *kurtosis* measure; or

- the entire distribution $\mathcal{D}(y)$ itself represented, for example, by its density or cdf.

Partial effects are illustrated in detail in Section 8.6, and used in Section 10.4.

5

Penalized Maximum Likelihood Inference

Penalized maximum likelihood is one of the three possible routes for statistical inference in GAMLSS that we discuss in this book. In this chapter, we

- show in detail how the parameters of the GAMLSS model are estimated using penalized maximum likelihood (PML);

- discuss the advantages and disadvantages of using PML; and

- set the basis for the penalized likelihood inference case studies discussed in Part III.

5.1 Introduction

The GAMLSS distributional regression model was introduced in Chapter 1 as

$$y \stackrel{\text{ind}}{\sim} \mathcal{D}(\boldsymbol{\theta}_1, \ldots, \boldsymbol{\theta}_K) \tag{5.1}$$

$$g_k(\boldsymbol{\theta}_k) = \boldsymbol{\eta}_k \tag{5.2}$$

$$\boldsymbol{\eta}_k = \beta_{0k}\mathbf{1}_n + \boldsymbol{B}_{1k}\boldsymbol{\gamma}_{1k} + \cdots + \boldsymbol{B}_{J_k k}\boldsymbol{\gamma}_{J_k k} \tag{5.3}$$

$$\boldsymbol{\gamma}_{jk} \sim \mathcal{N}(\mathbf{0}, \tau_{jk}^2 \boldsymbol{K}_{jk}^-). \tag{5.4}$$

Chapter 2 introduced the distributions $\mathcal{D}(\boldsymbol{\theta}_1, \ldots, \boldsymbol{\theta}_K)$ that GAMLSS can accommodate in equation (5.1). The link function of equation (5.2) connects the predictors $\boldsymbol{\eta}_k$ to the distribution parameters $\boldsymbol{\theta}_k$. Link functions are important for the interpretation of the model, but often are chosen conveniently so that the parameter $\boldsymbol{\theta}_k$ falls in the right range: For example, if σ must be positive we often model $\log(\sigma)$ as a function of the predictors.

Chapter 3 explained how different additive structural terms in equation (5.3) are defined. What the different additive terms have in common is a linear equation of the type $\boldsymbol{B}_{jk}\boldsymbol{\gamma}_{jk}$ and an associated quadratic penalty $\boldsymbol{\gamma}_{ik}^\top \boldsymbol{K}_{ik}(\lambda)\boldsymbol{\gamma}_{ik}$. [1] The beauty of the structural additive terms is that they represent a very general class of statistical models. For example, by using different basis matrices $\boldsymbol{B}_{jk}$ in combination with different quadratic penalties, a variety of types of additive terms can be created: cubic splines, P-splines, random effects, and Gaussian Markov random fields (GMRF) are

---

[1] Note that λ is included in $\boldsymbol{K}$ for possibly vector-valued hyperparameters.

some of the forms that those terms can take. The different penalties $\boldsymbol{K}_{jk}$ put different types of restrictions on the random coefficients $\boldsymbol{\gamma}_{jk}$, which correspond to assuming that the $\boldsymbol{\gamma}_{jk}$ have a normal distribution given by equation (5.4).

This chapter uses three major departures from equations (5.1)–(5.4). The first has to do with the normal assumption of equation (5.4). We will still penalize the random coefficients, that is, $\boldsymbol{\gamma}_{jk}^{\top}\boldsymbol{K}_{jk}(\lambda)\boldsymbol{\gamma}_{jk}$, but we will not use the normality assumption of equation (5.4) explicitly. (The normality assumption is very important for Chapter 6, where Bayesian inference is presented.) The second has to do with equation (5.3). In order to explain the algorithms better, we replace equation (5.3) with

$$g_k(\boldsymbol{\theta}_k) = \boldsymbol{X}_k\boldsymbol{\beta}_k + \boldsymbol{B}_{1k}\boldsymbol{\gamma}_{1k} + \cdots + \boldsymbol{B}_{J_kk}\boldsymbol{\gamma}_{J_kk}, \tag{5.5}$$

where all the linear terms in the model for the parameters $\boldsymbol{\theta}_k$ are bundled together in the design matrix $\boldsymbol{X}_k$. The coefficients $\boldsymbol{\beta}_k$ in equation 5.5 are the *linear* coefficients of the model while the $\boldsymbol{\gamma}$s are the *random effect* coefficients. Finally, we generalize the definition of a GAMLSS model to allow the introduction of terms which do not necessarily have the additive structure of equations (5.3) and (5.4), to blend GAMLSS with other machine learning techniques.

This chapter and Chapters 6 and 7 present different ways of estimating the parameters of a GAMLSS model. The *outer* algorithm (Algorithm 1) in combination with the *inner* algorithm (Algorithm 2) introduced in this chapter, provide a penalized likelihood maximization method for estimation of the models defined by equations (5.3) and (5.4). Chapter 6 shows how the parameters of a GAMLSS model are estimated using Bayesian methodology. The additional structure of the Bayesian MCMC methods allows inference on the fitted parameters, something that can be obtained either with asymptotic theory or bootstrapping simulation within the penalized maximum likelihood approach. Chapter 7 uses a statistical boosting framework to estimate the unknown coefficients while at the same time selecting the informative explanatory variables from a potentially much larger set of potential candidate variables.

5.2 Fitting GAMLSS using Penalized Maximum Likelihood

The maximum likelihood-based approach (ML) for the estimation of distribution parameters in any statistical model has been the dominant method of estimation since the early 1920s, when it was reintroduced by R. A. Fisher. Penalized likelihood inference is a more recent approach:

When the number of parameters is not merely large but infinite, as in nonparametric estimation of a continuous density, ML is certainly inappropriate, but maximum "penalized likelihood" makes good sense, where the penalty depends on "roughness" ... This can be interpreted both in a non-Bayesian manner, and also as maximum posterior density in the shape of density functions (Good, 1976).

The quote above is about density estimation, but the comments apply equally to smooth curve fitting, used in this book. The introduction of quadratic penalties in the likelihood of a regression model allows sophisticated nonlinear smoothing for additive terms, as described in Chapter 3. Those terms can be interpreted as maximum a posteriori (MAP) estimators.

GAMLSS (as most other regression models) assumes that the elements or observations of the response $\boldsymbol{y}$, conditional on the explanatory variables, are independent. Let $f(y_i|\boldsymbol{\theta}_{[i]})$ denote the probability function of the ith observation, for $i = 1, \ldots, n$, where $\boldsymbol{\theta}_{[i]} = (\theta_{i1}, \theta_{i2}, \ldots, \theta_{iK})$ is the vector of distribution parameters for this ith observation. Because of the independence assumption, the likelihood function for the response can be written as the product

$$L(\boldsymbol{\beta}, \boldsymbol{\gamma}) = \prod_{i=1}^{n} f(y_i|\boldsymbol{\theta}_{[i]})$$

and the log-likelihood as:

$$\ell(\boldsymbol{\beta}, \boldsymbol{\gamma}) = \sum_{i=1}^{n} \log f(y_i|\boldsymbol{\theta}_{[i]})$$

$$= \sum_{i=1}^{n} \log f(y_i|\boldsymbol{\beta}, \boldsymbol{\gamma}). \tag{5.6}$$

The $\boldsymbol{\beta} = (\boldsymbol{\beta}_1, \boldsymbol{\beta}_2, \ldots, \boldsymbol{\beta}_k)^\top$ and $\boldsymbol{\gamma} = (\boldsymbol{\gamma}_{11}, \ldots, \boldsymbol{\gamma}_{J_1 1}, \ldots, \boldsymbol{\gamma}_{1k}, \ldots, \boldsymbol{\gamma}_{J_k k})^\top$ are the *linear* and the *random effect coefficients*, respectively, and connect the explanatory terms with the distribution parameters $\boldsymbol{\theta}_k$, for $k = 1, \ldots, K$ through equation (5.5). The addition of the J_k smoothing additive terms, $\boldsymbol{\gamma}_{jk}$, for $j = 1 \ldots, J_k$, for each distribution parameter $\boldsymbol{\theta}_k$, adds the following penalties in the log-likelihood:

$$\ell_p(\boldsymbol{\beta}, \boldsymbol{\gamma}, \boldsymbol{\lambda}) = \ell(\boldsymbol{\beta}, \boldsymbol{\gamma}) - 0.5 \sum_{k=1}^{K} \sum_{j=1}^{J_k} \lambda_{jk} \boldsymbol{\gamma}_{jk}^\top \boldsymbol{K}_{jk} \boldsymbol{\gamma}_{jk}, \tag{5.7}$$

where $\boldsymbol{K}_{jk}$ is the jth penalty (or smoothing) matrix for the jth smoother at the kth parameter and $\boldsymbol{\lambda} = (\boldsymbol{\lambda}_1^\top, \boldsymbol{\lambda}_2^\top, \ldots, \boldsymbol{\lambda}_K^\top)^\top$ is a vector of smoothing coefficients, where each component $\boldsymbol{\lambda}_k = (\lambda_{1k}, \lambda_{2k}, \ldots, \lambda_{J_k k})^\top$ of $\boldsymbol{\lambda}$ represents the smoothing coefficients for each distribution parameter $\boldsymbol{\theta}_k$, for $k = 1, \ldots, K$.

A general GAMLSS model therefore has three different types of coefficients, all of which need to be estimated from the underlying data:

- the linear coefficients $\boldsymbol{\beta}_k$ for $k = 1, \ldots, K$;

- the smoothing coefficients $\boldsymbol{\gamma}_{jk}$, for $k = 1, \ldots, K$ and $j = 1, \ldots, J_k$; and

- the *smoothing coefficients* or *hyperparameters* $\boldsymbol{\lambda}$ (defined earlier), appearing in the penalties in equation (5.7).

The distinction between linear and random effect coefficients plays an important role in the modified backfitting algorithm, and will be discussed later where we consider how those coefficients are estimated within the penalized maximum likelihood framework. At this point we would like to highlight a special case of a GAMLSS model: the *parametric* (linear) GAMLSS model. A pure parametric linear GAMLSS model has only linear effects for all parameters $\boldsymbol{\theta}_k$ and therefore equation (5.5) can be simplified to

$$g_k(\boldsymbol{\theta}_k) = \boldsymbol{\eta}_k = \boldsymbol{B}_{0k}\boldsymbol{\beta}_{0k} = \boldsymbol{X}_k\boldsymbol{\beta}_k. \tag{5.8}$$

$\boldsymbol{B}_{0k} = \boldsymbol{X}_k$ is a linear design matrix modeling the parameter $\boldsymbol{\theta}_k$. In this model, none of the coefficients $\boldsymbol{\beta}_k$ is penalized and therefore the model can be fitted by using straightforward maximum likelihood estimation. The coefficients $\boldsymbol{\beta}_k$ typically represent linear *main* and *interaction* effects or terms fitted using a linear basis, for example polynomials, fractional polynomials, or piecewise polynomials. More details on the use of those terms is given in Stasinopoulos et al. (2017, chapter 8). Note that the first column of $\boldsymbol{X}_k$ is usually a column of ones representing the constant term (intercept) for the parameter. The attractive features of a purely parametric linear model with no smoothers (if it fits the data adequately), are firstly ease and speed of fitting, and secondly, all asymptotic results on the standard errors of the coefficients associated with ML estimation apply; see Rigby et al. (2019, chapter 11).

5.2.1 *Estimating the β and γ Coefficients*

The original GAMLSS model is defined by equations (5.1)–(5.3), while equation (5.4) plays the role of penalizing the $\boldsymbol{\gamma}_{jk}$ coefficients. The redefinition of equation (5.3) to equation (5.5) helps to make explicit that the linear coefficient vector $\boldsymbol{\beta}$ is unpenalized. In this section, we consider the estimation of the $\boldsymbol{\beta}$ and $\boldsymbol{\gamma}$ coefficients (given that the hyperparameters $\boldsymbol{\lambda}$ are fixed), which requires two algorithms: the *outer* algorithm and the *inner* algorithm. One can think of the outer algorithm as a generic algorithm applied to any distributional regression model, while the inner algorithm is more specific to the additive structure of equation (5.5). In order to clarify this point, consider the following generalization of the original GAMLSS definition:

$$\boldsymbol{y} \overset{\text{ind}}{\sim} \mathcal{D}(\boldsymbol{\theta}_1, \dots, \boldsymbol{\theta}_K)$$
$$g_k(\boldsymbol{\theta}_k) = \boldsymbol{\eta}_k = \mathcal{M}_k(\boldsymbol{x}_k) \qquad \text{for } k = 1, \dots, K, \tag{5.9}$$

where the terms $\boldsymbol{x}_k$ are a subset of all available terms $\boldsymbol{x}$ and $\mathcal{M}_k$ stands for different supervised machine learning models, for example linear regression models, neural networks, regression trees, lasso, principal component regression, or structural additive models. Algorithm 1 can be applied to any general GAMLSS model (5.9).

The *working response* (or iterative response) for parameter $\boldsymbol{\theta}_k$ is defined as

$$\tilde{y}_{ik} = \eta_{ik} + v_{ik}/w_{ik} \tag{5.10}$$

or, in compact matrix notation,

$$\tilde{\boldsymbol{y}}_k = \boldsymbol{\eta}_k + \boldsymbol{w}_k^{-1} \circ \boldsymbol{v}_k, \tag{5.11}$$

where $\tilde{\boldsymbol{y}}_k$, $\boldsymbol{\eta}_k$, $\boldsymbol{w}_k$, and $\boldsymbol{v}_k$ are all vectors of length n, for example the weight vector $\boldsymbol{w}_k = (w_{k1}, w_{k2}, \dots, w_{kn})^\top$; $\mathbf{w}_k^{-1} \circ \boldsymbol{v}_k = (w_{k1}^{-1}v_{k1}, w_{k2}^{-1}v_{k2}, \dots, w_{kn}^{-1}v_{kn})^\top$ is the

Algorithm 1 The outer algorithm of GAMLSS

---

initialize $\hat{\boldsymbol{\theta}}_1, \ldots, \hat{\boldsymbol{\theta}}_K$

for $k = 1, \ldots, K$ **do**

  fix all the other parameters $\hat{\boldsymbol{\theta}}_j$ for $j \neq k$ at their current values and calculate the working response $\tilde{\boldsymbol{y}}_k$ and the working weights $\boldsymbol{w}_k$.

  fit model $\mathcal{M}_k(\boldsymbol{x}_k)$ for $\boldsymbol{\theta}_k$ using $\tilde{\boldsymbol{y}}_k$ as response and $\boldsymbol{w}_k$ as prior weights

  calculate the global deviance and exit if there is no change (i.e. if the difference is less than a convergence criterion C_k)

end for

Hadamard element by element product; the vector $\boldsymbol{\eta}_k = g_k(\boldsymbol{\theta}_k)$ is the predictor vector of the kth parameter vector $\boldsymbol{\theta}_k$; and finally the vector of scores

$$\boldsymbol{v}_k = \frac{\partial \ell}{\partial \boldsymbol{\eta}_k} = \left(\frac{\partial \ell}{\partial \boldsymbol{\theta}_k} \right) \circ \left(\frac{\partial \boldsymbol{\theta}_k}{\partial \boldsymbol{\eta}_k} \right)$$

is the first derivative of the log-likelihood with respect to the predictor $\boldsymbol{\eta}_k$, for $k = 1, \ldots, K$. The $\boldsymbol{w}_k$ are the *working weights* (or iterative weights) defined as

$$\boldsymbol{w}_k = \boldsymbol{f}_k \circ \left(\frac{\partial \boldsymbol{\theta}_k}{\partial \boldsymbol{\eta}_k} \right) \circ \left(\frac{\partial \boldsymbol{\theta}_k}{\partial \boldsymbol{\eta}_k} \right), \tag{5.12}$$

where $\boldsymbol{f}_k$ can be either $-\mathbb{E}\left(\partial^2 \ell / \partial \boldsymbol{\theta}_k^2\right)$ for Fisher's scoring, or just the first derivative squared, $-\left(\partial \ell / \partial \boldsymbol{\theta}_k\right)^2$, for quasi-Newton scoring, for $k = 1, \ldots, K$. Note that for most of the response distributions the model can assume, it is rather rare to be able to obtain the Fisher information matrix $-\mathbb{E}\left(\partial^2 \ell / \partial \boldsymbol{\theta}_k^2\right)$. An important point here is the fact that Algorithm 1 can be applied to any regression-type model as long as its implementation allows prior weights. One also should make sure that the fitting of the model does not change radically from one iteration to another, to prevent divergence. Notice that a specific response distribution $\mathcal{D}(\boldsymbol{\theta}_1, \ldots, \boldsymbol{\theta}_K)$ is determined by the construction of $\tilde{\boldsymbol{y}}_k$ and $\boldsymbol{w}_k$.

After this small diversion, let us return to the estimation of the coefficients $\boldsymbol{\beta}_k$ and $\boldsymbol{\gamma}_{jk}$. Rigby and Stasinopoulos (2005) demonstrated that, for fixed hyperparameters $\boldsymbol{\lambda}$, and for the special case where $\mathcal{M}_k(\boldsymbol{x}_k)$ in (5.9) is an additive structural model of the type shown in equation (5.5), maximizing the penalized likelihood in (5.7) with respect to $\boldsymbol{\beta}_k$ and $\boldsymbol{\gamma}_{jk}$ can be achieved by applying the outer Algorithm 1 followed by the inner Algorithm 2. For example the step "fit model $\mathcal{M}_k(\boldsymbol{x}_k)$" of Algorithm 1 should be replaced by "use Algorithm 2."

The important part of Algorithm 2 consists of estimating the coefficients of the smooths $\hat{\boldsymbol{\gamma}}_{jk}$ by fitting different iteratively weighted (penalized) least squares (IWLS)[2]

---

[2] Sometimes also referred to as iteratively reweighted least squares (IRLS) or penalized iteratively reweighted least squares (PIRLS).

Algorithm 2 The inner algorithm of GAMLSS (including modified backfitting)

---

for $\boldsymbol{\theta}_k$ **do** given the working response $\tilde{\boldsymbol{y}}_k$ (equation (5.11)) and working weights $\boldsymbol{w}_k$ (equation (5.12))
    for $j = 0, 1, \ldots, J_k$ **do**
        calculate the partial residuals $\boldsymbol{\epsilon}_{jk}$ (equation (5.15))
        for $r = 1, \ldots$ **do**
            fit a reweighted (penalized) least squares (IWLS) (equation (5.13))
            exit if no change in the vector $\hat{\boldsymbol{\gamma}}_{jk}^{(r)}$ (or in the global deviance)
        end for
        exit if no change in all the vectors of coefficients $\hat{\boldsymbol{\gamma}}_{jk}$ for $j = 1, \ldots, J_k$
    end for
    exit if no change in the global deviance
end for

---

fits of the type:

$$\hat{\boldsymbol{\gamma}}_{jk}^{(r)} = \left(\boldsymbol{B}_{jk}^{\top} \boldsymbol{W}_k^{(r)} \boldsymbol{B}_{jk} + \lambda \boldsymbol{K}_{jk} \right)^{-1} \boldsymbol{B}_{jk}^{\top} \boldsymbol{W}_k^{(r)} \boldsymbol{\epsilon}_{jk}^{(r)}, \tag{5.13}$$

for the different terms $j = 0, 1, \ldots, J_k$ involved in the distribution parameter $\boldsymbol{\theta}_k$. The coefficients $\boldsymbol{\gamma}_{jk}$ and the basis matrices $\boldsymbol{B}_{jk}$ are defined in equation (3.1). The superscript $^{(r)}$ in equation (5.13) denotes the rth iteration of the IWLS algorithm. The diagonal matrix $\boldsymbol{W}_k$ has as elements the working (iterative) weights $\boldsymbol{w}_k$, while $\boldsymbol{\epsilon}_{jk}$ denotes the partial residuals. Note that the first fit of the inner algorithm 2 estimates the coefficients $\boldsymbol{\beta}_k$ and is therefore an unpenalized iteratively weighted least squares:

$$\hat{\boldsymbol{\beta}}_k^{(r)} = (\boldsymbol{X}_k^{\top} \boldsymbol{W}_k^{(r)} \boldsymbol{X}_k)^{-1} \boldsymbol{X}_k^{\top} \boldsymbol{W}_k^{(r)} \boldsymbol{\epsilon}_{0k}^{(r)}. \tag{5.14}$$

The partial residual vector $\boldsymbol{\epsilon}_{jk}$ for the jth term of the distribution parameter $\boldsymbol{\theta}_k$ is defined as:

$$\boldsymbol{\epsilon}_{jk} = \tilde{\boldsymbol{y}}_k - \sum_{\substack{i=0 \\ i \neq j}}^{J_k} \boldsymbol{B}_{ik} \hat{\boldsymbol{\gamma}}_{ik} . \tag{5.15}$$

Both Algorithms 1 and 2 are backfitting algorithms. Algorithm 1 is "simple" backfitting while Algorithm 2 is "modified" backfitting. The backfitting algorithm is a version of the Gauß–Seidel algorithm, in which the parameters are fitted recursively until convergence, and the latest available parameter estimates are used at each iteration. Modified backfitting was introduced for additive smoothing functions by Hastie and Tibshirani (1990) to improve convergence in the presence of *concurvity*, and consists of gathering all main linear effects for the smoother together in the linear matrix $\boldsymbol{X}_k$. Therefore, the linear part of the smoothers is fitted in equation (5.14). This prevents instability in the backfitting algorithm. Concurvity occurs when the linear space which generates the bases for say the jth smoother $\mathcal{M}(\boldsymbol{B}_{jk})$ is close to the linear space generated by all other smoothers $\mathcal{M}(\boldsymbol{B}_{ik})$ for $i \neq j$, where $\mathcal{M}(\cdot)$ denotes the manifold of a linear space spanned by the columns of a matrix. By including all the linear parts of the terms in $\boldsymbol{X}_k$, all possible multicollinearity terms are present and concurvity is less likely to occur during the fitting of the smoothers. The algorithm therefore converges more easily. Note, however, that this approach does not avoid the problem of multicollinearity, which affects the interpretation of the linear coefficients. One could avoid the problem by removing problematic covariate terms.

The unpenalized IWLS algorithm in (5.14) was originally used to fit GLMs, see Francis et al. (1993, p. 266). The modified backfitting algorithm was used by Hastie and Tibshirani (1990, p. 91) to fit a GAM for fixed smoothing parameters where more than one smoother is used within the linear predictor of a GAM. Their methodology is implemented in the **R** package **gam** (originally in Splus; Chambers and Hastie (1992)). Note, however, that the most popular implementation of GAM in **R**, the package **mgcv** (Wood, 2017), does not use backfitting but rather uses a large QR decomposition which has all the explanatory variables together; see Wood (2017, chapter 4). The gam() function estimates all additive smoothing terms simultaneously. The smoothing (hyper) parameters are also estimated efficiently using methods such as GCV or REML.

The main penalized maximum likelihood implementation of GAMLSS in the **R** package **gamlss** is called the RS algorithm and is based on Algorithms 1 and 2. There is an alternative algorithm called CG, a generalization of the algorithm used by Cole and Green (1992), which differs in the way that the inner algorithm works (Stasinopoulos et al., 2017, p. 70). The default is the RS algorithm because it is more reliable and faster in practice, especially if the distribution parameters $\boldsymbol{\theta}_k$ are not correlated; see Section 4.6. The CG method is unstable initially, but converges faster if the parameters are correlated. Sometimes a combination of both algorithms could be beneficial, that is, start with RS and switch to CG in later iterations. Note that the resulting estimated coefficients $\hat{\boldsymbol{\beta}}$ and $\hat{\boldsymbol{\gamma}}$ obtained by using either RS or CG algorithms are MAP estimates; see Stasinopoulos et al. (2017, section 3.2).

5.2.2 Estimating the Hyperparameters $\boldsymbol{\lambda}$

Until now, for both Algorithms 1 and 2 we assumed that the hyperparameters $\boldsymbol{\lambda}$ are fixed, a very unrealistic scenario. The estimation of $\boldsymbol{\lambda}$ can be achieved either *globally* or *locally*, depending on whether the estimation is done outside or within the GAMLSS algorithms, respectively. Both global and local estimation can be achieved by restricted (penalized) maximum likelihood (REML, widely used for random effects models (Pinheiro and Bates, 2000)), GAIC or generalized cross-validation (GCV), and cross-validation itself. Stasinopoulos et al. (2017) provides more details on estimation methods for the $\boldsymbol{\lambda}$s.

Notice, however, that all models fitted in this book using the penalized maximum likelihood paradigm, that is, all models fitted with the algorithm described in Section 5.2.1 which contains smoothing terms, use local estimation methods. These methods estimate a new $\boldsymbol{\lambda}$ at each IWLS fit. The default method implemented in **gamlss** is a form of restricted local maximum likelihood estimation. Rigby and Stasinopoulos (2013) call this method "local maximum likelihood" (LML), while other authors call it the Schall method (Schall, 1991).

Different approaches to estimating the smoothing parameters can be found in Wood et al. (2016) and Marra et al. (2017).

5.3 Inference within Penalized Maximum Likelihood Estimation

A fitted GAMLSS model using penalized maximum likelihood has all its coefficients $\hat{\boldsymbol{\beta}}$, $\hat{\boldsymbol{\gamma}}$, $\hat{\boldsymbol{\lambda}}$ estimated. By substituting the fitted coefficients in equation (5.5), fitted values for the distribution parameters $\hat{\boldsymbol{\theta}}_k$ for $k = 1, \ldots, K$ can be obtained. However, to obtain standard errors, within the penalized maximum likelihood paradigm, some extra computational work is needed for all fitted coefficients, usually involving data reuse techniques. Standard errors for $\hat{\boldsymbol{\beta}}$ are useful for hypothesis testing and for the construction of confidence intervals. Standard errors for the fitted parameters $\hat{\boldsymbol{\theta}}_k$ are useful for prediction. Standard errors for the fitted parameters $\hat{\boldsymbol{\gamma}}_k$ and $\hat{\boldsymbol{\lambda}}_k$ are useful for determining the quality of a smoother. In the special case of a fully parametric (linear) GAMLSS model, when there are no smoothers and the interest is on the coefficients $\hat{\boldsymbol{\beta}}$, *asymptotic* standard errors can be obtained by using classical statistical inference, discussed in Section 5.3.1. Other tools of inference within the penalized maximum likelihood paradigm are profile likelihood (see Section 5.3.2) and *bootstrapping* (discussed in Section 5.3.3). In fact, bootstrapping is the most general method of inference within the penalized likelihood approach.

5.3.1 Inference for the Coefficients β

If the GAMLSS model is fully parametric, then standard maximum likelihood estimation (MLE) properties apply to the fitted model. Strictly speaking, those properties can be divided into two categories depending on the level of the assumptions. The first assumes that the distribution of the response is "correctly" specified. Referring to Figure 4.2, this assumption implies that the "model" line, $f(\boldsymbol{y}|\boldsymbol{\vartheta}(X))$, passes through the "true" population $g(\boldsymbol{y}|\boldsymbol{x})$ and that $\boldsymbol{\vartheta}_c$ is the "true" parameter $\boldsymbol{\vartheta}_T$. This is a rather unrealistic assumption but a convenient one for deriving theoretical results. In the second category, we assume that the distribution of the model is not necessarily correct but rather a "good" approximation of the true population distribution. This is more realistic and, providing that we check the adequacy of the assumption about the distribution using appropriate diagnostic tools, also more practical. The properties of the MLEs under those two different sets of assumptions are described in more detail in Rigby et al. (2019, chapter 11). Here we summarize the results by referring to any parameter as $\boldsymbol{\vartheta}$. Later we describe how this is translated to the fitted parameters $\boldsymbol{\beta}_k$ of a GAMLSS model. There are three basic properties for $\hat{\boldsymbol{\vartheta}}$, the maximum likelihood estimator of $\boldsymbol{\vartheta}$: *invariance*, *consistency*, and *asymptotic normality*.

Invariance has to do with the fact that if $\boldsymbol{\phi} = g(\boldsymbol{\vartheta})$ is a one-to-one transformation of $\boldsymbol{\vartheta}$, then the MLEs $\hat{\boldsymbol{\phi}}$ and $\hat{\boldsymbol{\vartheta}}$ are related by $\hat{\boldsymbol{\phi}} = g(\hat{\boldsymbol{\vartheta}})$. Therefore it does not matter which parametrization is used when fitting the model, since at the point of maximum the likelihood will be the same. Lack of invariance of prior formulations was one of the main arguments held against Bayesian statistical inference, but in practical statistical analyses it seems to be of lesser importance.

Consistency is a more important property and it has to do with the fact that we would like to think that our estimation procedure leads us in the right direction when the sample size increases to infinity. If the population distribution is assumed to be identical to the model distribution, under some regularity conditions the MLE $\hat{\boldsymbol{\vartheta}}$ is a (weakly) consistent estimator of the true parameter $\boldsymbol{\vartheta}_T$, that is, $\hat{\boldsymbol{\vartheta}}$ converges in probability to $\boldsymbol{\vartheta}_T$ as $n \to \infty$. Under the assumption that the two distributions are different, but the model distribution is a good approximation of the population distribution, consistency of the MLE still applies. However, in this case $\hat{\boldsymbol{\vartheta}}$ converges in probability to $\boldsymbol{\vartheta}_C$, that is to the point where the model is closest to the population as, for example, in Figure 4.2(b).

Asymptotic normality is the property which helps us in the calculation of standard errors. We shall start with the more realistic case in which the population and the model distribution differ, and move to the special case when we assume that they are the same. Asymptotically as $n \to \infty$,

$$\hat{\boldsymbol{\vartheta}} \sim \mathcal{N}_K \left(\boldsymbol{\vartheta}_c, n^{-1} \boldsymbol{J}(\boldsymbol{\vartheta}_c)^{-1} \boldsymbol{G}(\boldsymbol{\vartheta}_c) \boldsymbol{J}(\boldsymbol{\vartheta}_c)^{-1} \right) \tag{5.16}$$

where $\boldsymbol{J}(\boldsymbol{\vartheta}_c)$ is the expected information matrix for a single observation Y_i evaluated at $\boldsymbol{\vartheta}_c$:

$$\boldsymbol{J}(\boldsymbol{\vartheta}_c) = - \mathbb{E}_g \left[\frac{\partial^2 \ell_i(\boldsymbol{\vartheta})}{\partial \boldsymbol{\vartheta} \partial \boldsymbol{\vartheta}^\top} \right]_{\boldsymbol{\vartheta}_c} \tag{5.17}$$

and

$$\boldsymbol{G}(\boldsymbol{\vartheta}_c) = \mathbb{V}_g \left[\frac{\partial \ell_i(\boldsymbol{\vartheta})}{\partial \boldsymbol{\vartheta}} \right]_{\boldsymbol{\vartheta}_c} . \tag{5.18}$$

Note that the expectation and variance in equations (5.17) and (5.18) are taken over the true population distribution $g(\boldsymbol{y}|\boldsymbol{x})$. The above quantities are approximated using the corresponding sample estimates:

$$\hat{\boldsymbol{J}}(\boldsymbol{\vartheta}_c) = - \frac{1}{n} \sum_{i=1}^n \left[\frac{\partial^2 \ell_i(\boldsymbol{\vartheta})}{\partial \boldsymbol{\vartheta} \partial \boldsymbol{\vartheta}^\top} \right]_{\boldsymbol{\vartheta}_c} \quad \text{and} \quad \hat{\boldsymbol{G}}(\boldsymbol{\vartheta}_c) = \frac{1}{n-K} \sum_{i=1}^n \left[\frac{\partial \ell_i(\boldsymbol{\vartheta})}{\partial \boldsymbol{\vartheta}} \frac{\partial \ell_i(\boldsymbol{\vartheta})}{\partial \boldsymbol{\vartheta}^\top} \right]_{\boldsymbol{\vartheta}_c} .$$

Of course $\boldsymbol{\vartheta}_c$ is unknown, so is estimated by $\hat{\boldsymbol{\vartheta}}$, giving $\hat{\boldsymbol{J}}(\hat{\boldsymbol{\vartheta}})$ and $\hat{\boldsymbol{G}}(\hat{\boldsymbol{\vartheta}})$. For the first category of assumption in which the model and the population are assumed to be identical, the above asymptotic normality result simplifies to

$$\hat{\boldsymbol{\vartheta}} \sim \mathcal{N}_K \left(\boldsymbol{\vartheta}_c, n^{-1} \boldsymbol{J}(\boldsymbol{\vartheta}_c)^{-1} \right) . \tag{5.19}$$

Note that for a fitted GAMLSS model we have to replace $\frac{\partial \ell_i(\boldsymbol{\vartheta})}{\partial \boldsymbol{\vartheta}}$ in all equations above with

$$\frac{\partial \ell_i(\boldsymbol{\beta})}{\partial \boldsymbol{\beta}} = \frac{\partial \ell_i(\boldsymbol{\theta})}{\partial \boldsymbol{\theta}} \frac{\partial \boldsymbol{\theta}}{\partial \boldsymbol{\eta}_\theta} \frac{\partial \boldsymbol{\eta}_\theta}{\partial \boldsymbol{\beta}}.$$

In practice, it is easier to approximate the expected information matrix $\hat{\boldsymbol{J}}(\hat{\boldsymbol{\beta}})$ using the observed information matrix $\hat{\mathcal{I}}(\hat{\boldsymbol{\beta}})$, which is calculated as the inverse of the Hessian matrix at the point of maximum $\hat{\boldsymbol{\beta}}$. A review of ML theory is given in Rigby et al. (2019, chapter 11).

Standard errors for $\hat{\boldsymbol{\beta}}$ using the asymptotic normality property of MLEs (equation 5.19) can be easily obtained, especially if the expected information matrix is replaced with the observed information matrix. However, asymptotic normality of the MLEs can be rather slow, particularly when the response distribution is skewed, meaning that the t-test (and associated Wald confidence interval) obtained using the asymptotic normality argument may not be accurate (Royston, 2007). The term *robust* standard errors is used when the standard errors are obtained using equation (5.16), and they are usually larger than those obtained using equation 5.19. Alternatively, to obtain standard errors from asymptotic results one may use the *profile likelihood* of the parameters of interest (Venzon and Moolgavkar, 1988); or bootstrapping. Profile likelihood and bootstrapping are described in Sections 5.3.2 and 5.3.3. The fully Bayesian GAMLSS (Chapter 6) solves the problem of obtaining standard errors by adding prior distributional assumptions for the coefficients of the model and thereby obtaining a full posterior distribution of the parameters.

Note that while the asymptotic properties for the penalized likelihood estimates in GAMLSS sound like straightforward extensions of the classical results for maximum likelihood inference in settings with i.i.d. realizations, they require additional assumptions on the data generating process of the covariates. Basically, as the sample size grows, we do not only have to make assumptions on the data generating process of the responses given the covariates, but also on how the increasing number of covariates is generated and whether this generates enough information on the regression effects of interest. The resulting conditions can be quite intricate and technical such that we do not discuss them in detail here, but see Fahrmeir and Kaufmann (1985) for the classical case of generalized linear models. For GAMLSS models the assumptions will in general be even more challenging to derive when including smooth effects, where the number of basis functions as well as the impact of the smoothing parameters have to be taken into account. See, for example, Kauermann et al. (2009).

5.3.2 *Profile Likelihood*

The profile likelihood is an alternative way of checking the information that a GAMLSS model provides for a specific coefficient of interest. It works for both parametric and smoothing GAMLSS models, but extra care is required for the latter case. The method has proved useful when one or two of the coefficients β_{jk} are of main interest while the rest are not of interest. Let us assume that we have a single coefficient of interest, β, and the vector of the coefficients in a model can be split into two sets $\boldsymbol{\vartheta} = (\beta, \boldsymbol{\vartheta}_2^\top)^\top$, where the set $\boldsymbol{\vartheta}_2$ may be considered to be nuisance parameters. The profile likelihood of β is defined as

$$\ell_p(\beta) = \max_{\boldsymbol{\vartheta}_2} \ell(\beta, \boldsymbol{\vartheta}_2).$$

To justify the name "profile", consider that you are looking from a particular direction (at a distance) at a mountain. What you see is the shape or the profile of the mountain. In the same way the the profile likelihood describes the shape of the likelihood from the direction of coefficient β. It is easily calculated, especially for a single coefficient of interest, by defining a grid of values for β and maximizing the likelihood $\ell(\beta, \boldsymbol{\vartheta}_2)$ over $\boldsymbol{\vartheta}_2$ at each point on the grid. The resulting points constitute part of the profile likelihood of β. From standard likelihood theory, given MLE $\hat{\beta}$ and under the null hypothesis $H_0 \colon \beta = \beta_0$, the likelihood ratio statistic

$$\ell_0(\hat{\beta}, \beta_0) = 2[\ell_p(\hat{\beta}) - \ell_p(\beta_0)]$$

has, asymptotically, the χ_1^2 distribution and we would not reject H_0 for $\ell_0(\hat{\beta}, \beta_0) < \chi_{1,\alpha}^2$. Therefore

$$\{\beta_0 \colon \ell_0(\hat{\beta}, \beta_0) < \chi_{1,\alpha}^2\}$$

constitutes a set of points β_0 for which we would not reject $H_0 \colon \beta = \beta_0$ at a $100(1-\alpha)\%$ significance level. This is the *profile likelihood confidence interval*, which is illustrated in Section 8.3 for the ultrasound data analysis. The *profile global deviance* is defined (Rigby et al., 2019, section 11.5.2) as

$$p\text{GDEV}(\beta) = -2\ell_p(\beta) \tag{5.20}$$

and the profile likelihood confidence interval can be expressed as

$$\{\beta_0 \colon p\text{GDEV}(\beta_0) < p\text{GDEV}(\hat{\beta}) + \chi_{1,\alpha}^2\}.$$

When smoothers exist in a GAMLSS model, the profile deviance for a parameter of interest $\boldsymbol{\beta}$ can still be used, but the data analyst has to make sure that the degrees of freedom used for the smoothers remain constant (at their original values) during the maximization over the nuisance parameters, which now include smoothers. Profile likelihood has been criticized from the Bayesian point of view because they do not take into the account the variability of the nuisance parameters $\boldsymbol{\vartheta}_2$ over the range of all possible values, but concentrate only on the maximum $\hat{\boldsymbol{\vartheta}}_2$.

5.3.3 Inference Based on Data Reuse Techniques

In statistical modeling there are usually two ways of utilizing a given dataset in order to obtain more information. One is to *split* the dataset into different subsets and the other is to *reuse* the data several times. Splitting is common for large datasets; the split is usually into a *training* set for fitting the models, a *validation* set for tuning the parameters of the models, and a *test* set used to check prediction accuracy. Figure 5.1 shows that reusing the data can be done in two distinct ways: by rearranging or by reweighting the data. Rearranging is performed by selecting specific observations from the data. This can be achieved by creating an index corresponding to observations in the data, refitting the model using the selected observations only, and then repeating the process B times. Weighting is done similarly but instead of using different observations the observations are weighted differently, using prior defined weights. The model is then refitted B times using the weighted observations. Nonparametric bootstrapping and cross-validation can be seen as rearranging techniques, and Bayesian bootstrapping and boosting as weighted techniques. Cross-validation was explained in Section 4.4 and boosting will be covered in detail in Chapter 7. Here we concentrate on the nonparametric and Bayesian bootstrapping techniques. All of these techniques create B samples which are used to obtain extra information from the data. In this respect, those samples are not very dissimilar to those obtained from an MCMC output using a proper Bayesian analysis, as will be described in Chapter 6. They can be used for inference, model selection, model averaging, or other statistical functions.

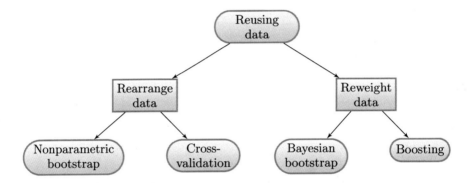

Figure 5.1 Diagrammatic representation of reusing techniques to a single dataset. The data are reused B times. Rearranging the data results in techniques such as *nonparametric bootstrapping* and *cross-validation*. Reweighting the data results in *Bayesian bootstrapping* and *boosting*.

Bootstrapping is a resampling method which provides an easy way to obtain information about the accuracy of an estimate of interest. In bootstrapping the response and covariates are resampled simultaneously. As mentioned above there are two major approaches for bootstrapping from a fitted GAMLSS model: *nonparametric* and *Bayesian* bootstrapping.

Nonparametric bootstrapping was introduced in the 1980s by Bradley Efron; see Efron and Tibshirani (1994) for a comprehensive review of this method. The nonparametric bootstrap draws B samples with replacement from a given dataset of length n, and recalculates the estimate of interest B times. This provides a bootstrap sample of length B of values for the specific estimate of interest, which can then be used to infer information about its behaviour. Note that the nonparametric bootstrap can be performed by creating an index $I = (I_1, I_2, \ldots, I_n)$ with elements a random sample with replacement from $\{1, 2, \ldots, n\}$. The model is then refitted by selecting data using this index.

Algorithm 3 describes the nonparametric bootstrap applied to GAMLSS. Note that by parameter of interest in the algorithm we mean any characteristic of the model which depends on the fitted parameters, for example any of the coefficients, $\boldsymbol{\beta}_k$, $\boldsymbol{\gamma}_k$, and $\boldsymbol{\lambda}_k$, the parameters θ_k, the estimated quantiles of the fitted model, etc.

Algorithm 3 Nonparametric bootstrapping

---

fit a GAMLSS model 'M1' using the training data, 'data'
for $b = 1, \ldots, B$ **do**

   create a **new dataset**, 'data.new', by simultaneously sampling response and covariates with replacement from the original dataset 'data'. In **R** this can be acheived using data.new <- data[I,] where I is an index sampled with replacement from $\{1, 2, \ldots, n\}$.

   refit the model 'M1' using 'data.new', calculate the estimate of interest and store them in the bootstrap dataset 'data.bootstrap'
end for
use 'data.bootstrap' to obtain information about the estimate of interest

---

Bayesian bootstrapping, introduced by Rubin (1981), is a less popular bootstrapping method which, however, is as general and useful as the nonparametric bootstrap. Instead of resampling the data, it refits the data by assigning different prior weights. (In fact, when using integer weights that sum to the sample size n, this exactly resembles the nonparametric bootstrap.) However, the Bayesian bootstrap also allows the consideration of real-valued, noninteger weights and is therefore more general.

More precisely, the Bayesian bootstrap relies on the fact that for parameters that are multinomially distributed and equipped with a noninformative Haldane's distribution prior, the posterior is an $(n-1)$-dimensional Dirichlet distribution. Aitkin (2019) has shown that for any regression type model, Bayesian bootstrapping can be achieved by generating different prior weights $w_1, \ldots, w_n$. In the GAMLSS Bayesian bootstrapping implementation, the randomly generated prior weights sum to the number of observations: $\sum_{i=1}^{n} w_i = n$. Algorithm 4 describes Bayesian bootstrapping for a fitted GAMLSS model. It can be used for generating bootstrap samples for any characteristic of the distribution we are interested in, and those bootstrap samples used for statistical inference. Note that the Bayesian bootstrap generates a posterior probability of the characteristic of interest, while the nonparametric bootstrap generates a sampling distribution for it. In general the results will be different but usually not very dissimilar.

There are various other methods of bootstrapping from a fitted GAMLSS model, for example:

Algorithm 4 Bayesian bootstrapping

---

fit a GAMLSS model `M1` using the training data `data`
for $b = 1, \ldots, B$ **do**
    create new **prior weights**, w_i such as $\sum_{i=1}^{n} w_i = n$
    refit the model `M1` using the new weights w_i, calculate the statistics of interest
and store them in the bootstrap dataset `data.bootstrap`
end for
use `data.bootstrap` to obtain information about the statistics of interest

---

- the *parametric* bootstrap, which simulates new samples of the responses given the covariates from a fitted model,

- the nonparametric bootstrapping of the *residuals*, in which the residuals are re-sampled to restore new values of the responses given the covariates, and

- the asymptotic bootstrap, where samples are drawn from the asymptotic normal distribution of the PML estimate to forward uncertainty in the estimation of the regression coefficients to quantities derived from these.

See Stasinopoulos et al. (2017, section 5.1.4) for more details. The nonparametric and Bayesian bootstraps are illustrated in Section 8.3, in the context of computing a confidence interval for the coefficient of a model term in the ultrasound analysis.

5.4 Covariate Selection via Shrinkage and Regularization

In this section we consider some of the techniques in which the linear design matrix $\boldsymbol{X}_k$ can have more columns than rows, that is, $p_k > n$. Note that for those models the contribution to the predictors will be of the form $\boldsymbol{\eta}_k = \boldsymbol{X}_k \boldsymbol{\beta}_k$, where $\boldsymbol{X}_k$ is suitably standardized, usually with zero mean and standard deviation equal to one. Ridge regression, lasso, elastic net, least angle regression, principal component regression (PCR), and boosting fall in this category of methods, which are often referred to as *dimensionality reduction* techniques. Apart from boosting, these techniques allow the explanatory variables to affect the predictors $\boldsymbol{\eta}_k$ linearly. Often, the fitting algorithms for these techniques do not provide just a single fitted model, but a path of fitted models from the null model (with only constant terms) to the full model where all linear coefficients are fitted unpenalized. Boosting differs by also allowing nonlinear relationships (smoothers) to enter into the model, using smoothers as base-learners.

Ridge regression (Hoerl and Kennard, 1988) was discussed briefly in Section 3.8. It uses a quadratic penalty to penalize the linear coefficients, that is, $\lambda \sum_{l=1}^{p_k} \beta_l^2$.[3] Ridge regression shrinks the coefficients β_k towards zero (something considered to be beneficial for prediction), but does not eliminate them altogether. Lasso, on the other

---

[3] Note we have changed the notation used in Section 3.8 from γ_l to β_l, to align with the notation used in this chapter.

hand, which uses the penalty $\lambda \sum_{l=1}^{p_k} |\beta_l|$, does set some of the coefficients to zero, therefore allowing selection of explanatory variables. Ridge and lasso regressions are covered in James et al. (2013) and Hastie et al. (2009). Elastic net (EN), introduced by Zou and Hastie (2005), has two penalties:

$$\lambda_1 \sum_{l=1}^{p_k} \beta_l^2 + \lambda_2 \sum_{l=1}^{p_k} |\beta_l|,$$

the first a quadratic penalty and the second an absolute values penalty. By letting $\alpha = \lambda_2/(\lambda_1 + \lambda_2)$ and $\lambda = \lambda_1 + \lambda_2$ the penalty can be rewritten as

$$\lambda \left[(1 - \alpha) \sum_{l=1}^{p_k} \beta_l^2 + \alpha \sum_{l=1}^{p_k} |\beta_l| \right]$$

for $0 \leq \alpha \leq 1$. We obtain lasso when we fix $\alpha = 1$ and ridge regression when $\alpha = 0$. Values of α between zero and one produce different EN solutions, and typically $\alpha = 0.5$ produces the halfway solution between ridge and lasso regressions. Note that, while α is fixed, the other smoothing parameter λ needs to be estimated. It is claimed by its authors that EN performs better than lasso with large p_k, and also when "grouped" selection is involved; that is, if a group of highly linearly correlated variables exists, lasso usually picks only one of the terms, ignoring the rest, while EN has the ability to select more than one of them.

Least angle regression (LARS) was created by Efron et al. (2004) as a fast algorithm for fitting lasso and forward stagewise regression. (A forward stagewise regression is very similar to a forward stepwise regression, but the steps taken from the null to the full model are smaller; see, for example, Efron et al. (2004).) The LARS algorithm produces a path for all possible fitted models from the null to the full unpenalized linear model. For example, a lasso model can be written as the solution to the penalized least squares problem:

$$\text{minimize} \sum_{i=1}^{n} (y_i - \mu_i)^2 \text{ subject to } \sum_{j=1}^{p_k} |\beta_j| \leq t,$$

where $\mu_i = \boldsymbol{x}_i^\top \boldsymbol{\beta}$ and $0 < t < \sum_{j=1}^{p_k} |\beta_j|$. For $t = 0$ we have the null model while for $t = \sum_{j=1}^{p_k} |\beta_j|$ we have the full model, that is, least squares solution. Therefore the path of solutions is a function of t, or equivalently the smoothing parameter λ which has an inverse relationship with t. By increasing the smoothing parameter λ from zero (full model) to infinity (null model), different models (and terms) are fitted. Note that when using LARS within the GAMLSS algorithm, one has to fix or estimate the smoothing parameter λ at each iteration. The estimation of λ can be achieved using any of the techniques discussed in Section 4.4, that is, via an information-based criterion or cross-validation.

While lasso, EN, and ridge regression impose a direct penalization to the optimization problem leading to shrinkage and variable selection, similar results can also be achieved via statistical boosting approaches. Boosting for statistical modeling incorporates a more indirect penalization, as the algorithm is typically stopped before it converges (*early stopping*). The resulting coefficients hence do not reflect the optimal solution for the underlying training sample, but are also shrunk towards zero in order to yield more accurate prediction models. As statistical boosting algorithms iteratively select one-by-one updates for the coefficients included in the model, variables that have not been selected for these updates before the algorithm is stopped are effectively excluded from the final model. For details on statistical boosting for variable selection in general, and particularly for GAMLSS, see Chapter 7.

The aim of principal component regression (PCR) is to find a suitable orthogonal linear combination of the columns of the standardized design matrix $\boldsymbol{X}_k$, say $\boldsymbol{T}_k(\lambda)$, which is based on the singular value decomposition of $\boldsymbol{X}_k$. Here, λ is a smoothing parameter taking discrete values $\lambda = 1, \ldots, p_k$ and is the number of columns of $\boldsymbol{T}_k(\lambda)$. The model fits $\boldsymbol{T}_k(\lambda)$ to the parameters of the response distribution of a GAMLSS model. PCR is a shrinking technique since, by fitting only λ components, where $\lambda < p_k$, some of the original coefficients $\hat{\boldsymbol{\beta}}_k$ of the full matrix $\boldsymbol{X}_k$ are shrunk towards zero. The shrinkage depends on the smoothing parameter, and by varying $\lambda = 1, \ldots, p_k$ a path of solutions for $\hat{\boldsymbol{\beta}}_k(\lambda)$ is created. Stasinopoulos et al. (2022) demonstrate two ways of implementing PCR in **gamlss**, depending on whether the singular value decomposition of $\boldsymbol{X}_k$ is performed before or during the IWLS iterations.

We conclude this chapter by highlighting a few important points concerning modeling within GAMLSS. The type of data and the question to answer from the data should determine the path needed to be taken in order to find an adequate GAMLSS model. If there are too many features (explanatory terms), that is, if p_k is close to or greater than the number of observations, then data dimensionality reduction techniques are appropriate. However, using ridge, lasso, EN, or PC regression, the modeling of the relationships between the terms and the parameters is restricted to be linear. Only boosting allows smoothers. Dimensionality reduction techniques are not necessarily appropriate if p_k is small compared with n and if strong nonlinearities exist in the data. Note that by using any dimensionality reduction techniques, the response distribution selection and term selection can be simplified by combining them in a single algorithm. For example, an algorithm which fits different response distributions using a lasso model for all distribution parameters will achieve term reduction and distribution selection in one go. The presence or not of interactions in the model is another possible reason for using dimensionality reduction techniques. First-order linear interactions are much easier to handle than nonlinear interactions. For example, if x_1 and x_2 are both continuous variables, a first-order linear interaction will add one extra linear term $x_1 \times x_2$ in the design matrix, while a nonlinear relationship will add a two-dimensional smoother $s(x_1, x_2)$ (a more difficult proposition altogether). For r main effects there are $r(r-1)/2$ first-order interaction terms, so with r relatively large, dimensionality reduction techniques could be a good solution. Stasinopoulos et al. (2022) use PCR to deal with interactions in the presence of 70 possible main effects terms.

6

Bayesian Inference

In contrast to classical frequentist inference, where a true, fixed parameter $\boldsymbol{\vartheta}$ is assumed and only the data are treated as random, Bayesian inference also assigns probability distributions to the parameters $\boldsymbol{\vartheta}$ of the statistical model and possibly also to further *hyperparameters*. Thereby, Bayesian inference allows the analyst to express prior beliefs about the unknown parameters – before observing the data – via a *prior distribution*. Bayes' theorem then allows for updating these beliefs based on the observed data to obtain a posterior distribution, in order to perform posterior uncertainty assessment about the parameters of interest.

This chapter treats Bayesian inference for GAMLSS and has the following major aims:

- Review the essentials of Bayesian inference and give details of a Bayesian approach to GAMLSS including (hierarchical) prior specifications, posterior sampling, monitoring mixing and convergence and posterior inference (including uncertainty quantification, model fit and model choice).

- Discuss specific advantages of the Bayesian approach to GAMLSS such as routine determination of the smoothing variances based on suitable hyperprior specifications, extensions including effect selection priors or hierarchical model specifications and exact posterior inference on any property of the response distribution and transformations of distribution parameters, without the need to rely on asymptotics or computationally demanding resampling procedures.

The Bayesian case studies discussed in Part III (Chapters 11 and 12) build on this content.

6.1 A Brief Introduction to Bayesian Inference

Consider the case of a statistical model where data $\boldsymbol{y}$ follow a distribution depending on a parameter vector $\boldsymbol{\vartheta}$. The Bayesian model formulation then relies on two modeling assumptions:

(1) The observation model that specifies how the data $\boldsymbol{y}$ are generated for a given parameter vector $\boldsymbol{\vartheta}$. This is formalized via the conditional density function $f(\boldsymbol{y}|\boldsymbol{\vartheta})$

that formally coincides with the likelihood, although it is interpreted as a conditional density in the Bayesian setup.

(2) The prior distribution $f(\boldsymbol{\vartheta})$ that represents prior beliefs about the parameter vector, where high values are assigned to parameter values that are deemed a priori more plausible than parameter values that are assigned small values in the prior density.

Bayesian belief updating results from Bayes' theorem that determines the posterior, that is, the plausibility assessment of different parameter values *after observing the data* as

$$f(\boldsymbol{\vartheta}|\boldsymbol{y}) = \frac{f(\boldsymbol{y}|\boldsymbol{\vartheta})f(\boldsymbol{\vartheta})}{f(\boldsymbol{y})} = \frac{f(\boldsymbol{y}|\boldsymbol{\vartheta})f(\boldsymbol{\vartheta})}{\int f(\boldsymbol{y}|\boldsymbol{\vartheta})f(\boldsymbol{\vartheta})d\boldsymbol{\vartheta}},$$

where $f(\boldsymbol{y})$ is the marginal density of the data without conditioning on a specific parameter value. This marginal density can be determined by marginalizing the joint distribution $f(\boldsymbol{y}, \boldsymbol{\vartheta}) = f(\boldsymbol{y}|\boldsymbol{\vartheta})f(\boldsymbol{\vartheta})$ with respect to $\boldsymbol{\vartheta}$, such that it is in fact completely determined from the observation model and the prior distribution.

The principle of Bayesian belief updating is illustrated in Figure 6.1 for the example of the success probability π in a Bernoulli experiment with $n = 10$ trials and $y = 1$ success. The maximum likelihood estimate, which does not take any potential prior knowledge into account, is in this case given by $\hat{\pi} = 0.1$. As a prior distribution, we choose the beta distribution with parameters $a > 0$ and $b > 0$ with support on the unit interval (making it a suitable candidate for a prior distribution for probabilities). Various shapes can be represented by changing the prior parameters. With $a = b = 1$, we obtain a uniform prior (panel (a)) as a special case which can be considered a representation of prior ignorance; although there is considerable debate about whether this naïve approach to prior ignorance is indeed a sensible approach. Panels (c) and (e) show prior distributions that express a prior preference for small or large success probabilities, respectively. It can be shown that under these specifications (Bernoulli trial, beta prior), the posterior is also a beta distribution with parameters $a + y$ and $b + n - y$; compare panels (b), (d), and (f) of Figure 6.1. For the uniform prior, the posterior mode coincides with the maximum likelihood estimate, while for the other two prior specifications we can see the impact of the prior that shifts the posterior towards either smaller or larger values.

The fact that Bayesian inference treats the parameter vector as a random quantity by assigning a probability distribution to it is often interpreted as the belief that the parameter is indeed a random quantity. This is in fact a misconception: Rather, Bayesian inference relies on the assumption that uncertainty about the parameter vector can be expressed via probability distributions. As a consequence, the question whether one believes in the existence of a true, fixed parameter vector or not is *not* at the heart of the distinction between frequentist and Bayesian inference. This distinction is rather the question of whether prior knowledge about the parameter should be acknowledged at all (which bears the risk of making the outcomes of the analysis subjective due to the subjectivity of prior beliefs) and whether these

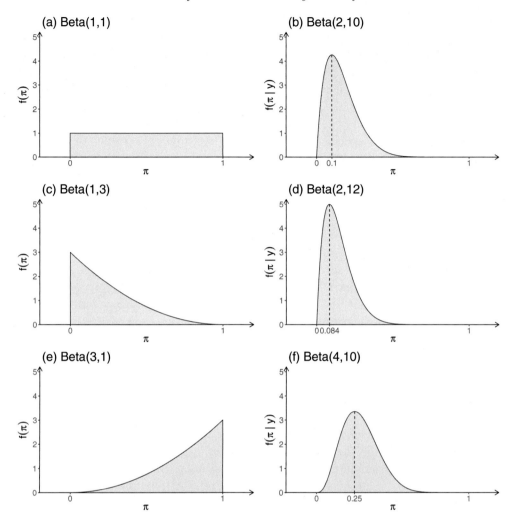

Figure 6.1 Bayesian belief updating for the success probability π in a Bernoulli experiment with $n = 10$ trials and $y = 1$ success. Panel (a), (c), and (e) show different beta priors for the success probability; panels (b), (d), and (f) show the corresponding posteriors. The dashed line represents the posterior mode.

can indeed be represented via a probability distribution. It makes perfect sense to believe in a true, fixed population parameter of interest which, for example, leads to the notion of posterior consistency, where the posterior concentrates around this true value as more and more data accumulate.

Intuitively it may be widely accepted that prior beliefs about parameters exist. For example, in a logistic regression with standardized covariates, there will be wide agreement that covariate effects that are larger in absolute size than three are very unlikely. However, turning these vague beliefs into a prior distribution is not straightforward. Unfortunately, Bayes' theorem does not provide guidance for this task but rather asks the analyst honestly to translate prior beliefs into the prior probability density function $f(\boldsymbol{\vartheta})$. A typical procedure is to start with a parametric assumption for the prior (e.g. a normal distribution) and to determine the prior hyperparameters from belief statements such as

$$\mathbb{P}(c_1 \leq \vartheta \leq c_2) = 1 - \alpha,$$

where c_1 and c_2 are prespecified constants that determine rather small or large values of the (in this case scalar) parameter ϑ, while α determines the confidence of the prior belief. Such choices can of course easily be criticized for (i) reducing the problem to determining the parameters of a parametric class of priors, and (ii) their subjectivity that makes the whole statistical analysis subjective.

An in-depth discussion of constructing sensible priors is beyond the scope of this book and excellent treatments are available in several introductions to Bayesian statistics, for example the introductory discussion in Held and Sabanés Bové (2012) that also provides a nice introduction to maximum likelihood inference; and the more comprehensive treatment in Gelman et al. (2013). In GAMLSS, we will use priors for the regression coefficients that resemble the penalties from penalized likelihood, as discussed in detail in Section 6.2. In addition, we will assign prior distributions to hyperparameters such as the smoothing parameter, to achieve a data-driven amount of smoothing or Bayesian effect selection.

While the posterior provides comprehensive information on the beliefs about the parameter, it is of course typically hard to report and interpret, especially in situations with a high-dimensional parameter vector $\boldsymbol{\vartheta}$ or when $\boldsymbol{\vartheta}$ does not have a direct interpretation for y or the conditional distribution $f(y|\boldsymbol{\vartheta}(\boldsymbol{x}))$. As a consequence, the posterior is often reduced to, for example, point estimates such as the posterior mean, median, or mode of $\boldsymbol{\vartheta}$ (or transformations of $\boldsymbol{\vartheta}$) complemented with uncertainty quantification such as the posterior standard deviation or posterior credible intervals, that is, intervals that achieve a prespecified, high posterior probability.

For a long time, one of the major hurdles in practically applying Bayesian inference has been the determination of the marginal density $f(\boldsymbol{y})$ that is required to normalize the posterior in Bayes' theorem. Analytically, this normalizing constant is available only in very special cases where the observation model and the prior are conjugated, and therefore updating the prior to the posterior corresponds only to a change in the parameters of the prior. However, the advent of Bayesian computing and most notably Markov chain Monte Carlo (MCMC; Hastings, 1970) simulation techniques has boosted the applicability of Bayesian inference to models of considerable complexity.

MCMC simulations rely on the construction of a Markov chain with stationary distribution coinciding with the posterior distribution. As a consequence, iteratively simulating values from this Markov chain leads to samples that are dependent but converge in distribution to the posterior of interest. Importantly, the transition kernel of the Markov chain can be constructed without requiring knowledge of the complete posterior distribution, and in particular the normalizing constant is not required. Moreover, MCMC implements a convenient divide and conquer strategy where complex models with high-dimensional parameter vectors can be broken down into smaller pieces that can be treated conditionally independently. The advent of MCMC made Bayesian inference computationally tractable, as well as providing indirect access to the complete posterior distribution (as discussed in more detail in Section 6.4). Although we focus here on MCMC-based inference, we will also discuss some alternative approaches in Section 6.9.

6.2 Priors for the Regression Coefficients in GAMLSS

While the observation model for a Bayesian GAMLSS is available via the likelihood (see Chapter 5 on penalized likelihood inference), we now have to supplement appropriate prior distributions for all unknown parameters of the model. For the regression coefficients $\boldsymbol{\gamma}_{jk}$ of the jth functional effect in the kth distribution parameter, we will rely on conditionally Gaussian specifications, that is

$$f(\boldsymbol{\gamma}_{jk}|\tau_{jk}^2) \propto \left(\frac{1}{\tau_{jk}^2}\right)^{0.5\,\mathrm{rank}(\boldsymbol{K}_{jk})} \exp\left(-\frac{1}{2\tau_{jk}^2}\boldsymbol{\gamma}_{jk}^\top \boldsymbol{K}_{jk}\boldsymbol{\gamma}_{jk}\right), \qquad (6.1)$$

where τ_{jk}^2 is a hyperparameter that we consider fixed for the moment, while $\boldsymbol{K}_{jk}$ is the prior precision matrix. If $\boldsymbol{K}_{jk}$ has full rank, this is simply the inverse of the prior covariance matrix. However, we will often consider partially improper priors having $\mathrm{rank}(\boldsymbol{K}_{jk}) > 0$ (i.e. the prior is not flat) but $\mathrm{rank}(\boldsymbol{K}_{jk}) < \dim(\boldsymbol{\gamma}_{jk})$ (i.e. the precision matrix does not have full rank but the prior is also not completely flat). In such cases, it is more convenient to express the prior in terms of the precision matrix rather than the covariance matrix, as will become apparent later. As a consequence, the prior is available only upon proportionality and does not integrate to one. Note that the term $(1/\tau_{jk}^2)^{0.5\,\mathrm{rank}(\boldsymbol{K}_{jk})}$ is a normalizing constant when τ_{jk}^2 is a fixed constant. Later we will, however, also perform inference for τ_{jk}^2 and then it is important to include this term in the prior specification.

Interestingly, the kernel of the multivariate Gaussian prior $-1/(2\tau_{jk}^2)\boldsymbol{\gamma}_{jk}^\top \boldsymbol{K}_{jk}\boldsymbol{\gamma}_{jk}$ closely resembles the form of the quadratic penalties considered for penalized likelihood in Chapter 5. This is not a coincidence but rather a dedicated choice that allows us to establish a close link between posterior mode and penalized maximum likelihood estimation in Section 6.3.

When interpreting the implications of the prior (6.1), we have to differentiate between two main components:

- Structural information comes from the term $\boldsymbol{\gamma}_{jk}^\top \boldsymbol{K}_{jk}\boldsymbol{\gamma}_{jk}$ that expresses our prior belief that $\boldsymbol{\gamma}_{jk}^\top \boldsymbol{K}_{jk}\boldsymbol{\gamma}_{jk}$ should not deviate from zero too much. The actual belief then depends on the choice of $\boldsymbol{K}_{jk}$ which, as in penalized likelihood estimation, can express our assumption about desired smoothness or shrinkage of effects. (See Chapter 3.)

- The prior variance τ_{jk}^2 takes a similar role as the smoothing parameter λ_{jk} in the penalized likelihood (with the precise relation detailed in Section 6.3) and determines our prior confidence in the true parameter fulfilling the smoothness or shrinkage properties requested by $\boldsymbol{K}_{jk}$.

Of course, in practice it will be necessary and beneficial to place a hyperprior $f(\tau_{jk}^2)$ on the smoothing variance. This will allow us not only to implement prior beliefs about the required amount of smoothing but also to achieve interesting additional results concerning, for example, effect selection in GAMLSS; see Section 6.7 for details.

Note also that flat priors for $\boldsymbol{\gamma}_{jk}$ can be considered special cases of (6.1) for either $\tau_{jk}^2 \to \infty$ or $\boldsymbol{K}_{jk} \to \boldsymbol{0}$. According to our previous interpretation, both reflect our ignorance about any smoothness or shrinkage properties in $\boldsymbol{\gamma}_{jk}$ such that the estimates for these parameters should be dictated by the data alone.

6.3 Posterior Mode vs. Penalized Likelihood

As previously mentioned, the multivariate Gaussian priors in (6.1) and the quadratic penalties employed in penalized likelihood inference are closely linked. Both effectively incorporate our prior beliefs that the effect to be estimated should not deviate too much from the null space defined by the penalty matrix $\boldsymbol{K}_{jk}$. For example, a polynomial effect of order $r-1$ in the case of a penalized spline with rth-order difference penalty. This similarity can be made more explicit when comparing the posterior and the penalized likelihood criterion, while ignoring the dependence on the hyperparameters τ_{jk}^2 for the moment.

Let $\boldsymbol{\gamma} = (\boldsymbol{\gamma}_1^\top, \ldots, \boldsymbol{\gamma}_K^\top)^\top$, $\boldsymbol{\gamma}_k = (\boldsymbol{\gamma}_{k1}^\top, \ldots, \boldsymbol{\gamma}_{kJ_k}^\top)^\top$ denote, as before, the vector of all regression coefficients in the complete model defined by the regression predictors

$$\boldsymbol{\eta}_k = \boldsymbol{B}_{1k}\boldsymbol{\gamma}_{1k} + \cdots + \boldsymbol{B}_{kJ_k}\boldsymbol{\gamma}_{kJ_k}, \quad k = 1, \ldots, K,$$

where, for simplicity, we have absorbed the intercept into one of the generic regression effects. According to Bayes' theorem, the posterior (given all further hyperparameters) is then given by

$$f(\boldsymbol{\gamma}|\boldsymbol{y}) = \frac{f(\boldsymbol{y}|\boldsymbol{\gamma})f(\boldsymbol{\gamma})}{f(\boldsymbol{y})},$$

where the assumption of prior independence between the regression coefficients of different terms $s_j^{\theta_k}(x)$ and different parameters θ_k allows us to decompose the joint prior into

$$f(\boldsymbol{\gamma}) = \prod_{k=1}^{K} \prod_{j=1}^{J_k} f(\boldsymbol{\gamma}_{jk}),$$

while the individual factors $f(\boldsymbol{\gamma}_{jk})$ are of the generic multivariate normal form (6.1). Since the normalizing constant $f(\boldsymbol{y})$ in Bayes' theorem is just a constant of proportionality, the mode of the posterior can then be obtained by maximizing

$$\log(f(\boldsymbol{\gamma}|\boldsymbol{y})) = \log(f(\boldsymbol{y}|\boldsymbol{\gamma})) + \log(f(\boldsymbol{\gamma}))$$
$$= \ell(\boldsymbol{\gamma}) - \frac{1}{2}\sum_{k=1}^{K}\sum_{j=1}^{J_k}\tau_{jk}^2\boldsymbol{\gamma}_{jk}^{\top}\boldsymbol{K}_{jk}\boldsymbol{\gamma}_{jk},$$

which is exactly the penalized likelihood considered in Chapter 5 (where additive constants have been ignored) with $\lambda_{jk} = 1/(2\tau_{jk}^2)$. Hence, posterior mode estimates and penalized maximum likelihood estimates coincide, if the smoothing parameters λ_{jk} and smoothing variances τ_{jk}^2 have been chosen to match each other.

Interpreting the multivariate Gaussian prior distribution as an equivalent way of enforcing smoothing or shrinkage with quadratic penalties is also closely linked to the mixed model representation of penalized regression models. This link was first identified for smoothing splines (see Wahba, 1978; Green, 1987; Speed, 1991, for early references) and was later transferred to several other components of additive models such as penalized splines (Currie and Durban, 2002; Ruppert et al., 2003) and spatial effects (Kammann and Wand, 2003; Fahrmeir et al., 2004); see Fahrmeir and Kneib (2011, chapter 4) for more details. The basic idea is that the Gaussian prior can also be seen as a random effect specification albeit with a complex dependence structure induced by the penalty matrix $\boldsymbol{K}$. To match with the standard assumption of i.i.d. random effects, the vector of regression coefficients is then often reparametrized to make the individual coefficients independent. While this does not change the model from a theoretical perspective, since the dependence is then contained in the cross-product matrix $\boldsymbol{B}^{\top}\boldsymbol{B}$, it has the advantage that the resulting models can be fit with standard mixed model software, including determination of the smoothing parameter based on restricted maximum likelihood (with well-developed algorithmic solutions for the latter). The main difficulty in the reparametrization arises from the partial impropriety of the Gaussian prior, that is, the fact that $\boldsymbol{K}$ does not have full rank. This basically implies that parts of the regression coefficients $\boldsymbol{\gamma}$ have to be treated as fixed effects (those that correspond to the unpenalized part of $s(\boldsymbol{x})$), while only the penalized components turn into proper random effects. General approaches for handling this issue based on an eigendecomposition of $\boldsymbol{K}$ have been developed; see Fahrmeir et al. (2004) and Fahrmeir and Kneib (2011) for details.

The connection of Gaussian priors with random effects specifications led to extensive statistical research in the late 2000s, where inferential approaches for determining the random effects (co)variance structure were employed for smoothing parameter selection. The most prominent example here is restricted maximum likelihood (REML), as also discussed in Chapter 5. Interestingly, the REML-based determination of smoothing parameters also has a Bayesian interpretation in the case of Gaussian responses with homoscedastic variances, namely for $\boldsymbol{y} \sim \mathcal{N}(\boldsymbol{\eta}, \sigma^2\boldsymbol{I})$. It can then be shown that the restricted likelihood for the smoothing parameters is in fact equivalent to their marginal posterior. More precisely, if $\boldsymbol{\tau}^2$ is the vector consisting of all smoothing variances, the marginal posterior for $\boldsymbol{\tau}^2$ and σ^2 is obtained as

$$f(\boldsymbol{\tau}^2, \sigma^2 | \boldsymbol{y}) = \int f(\boldsymbol{y} | \boldsymbol{\gamma}, \boldsymbol{\tau}^2, \sigma^2) f(\boldsymbol{\gamma} | \boldsymbol{\tau}^2) \, d\boldsymbol{\gamma}$$

and it can be shown that this marginal posterior is proportional to the restricted likelihood $L_{\mathrm{REML}}(\boldsymbol{\tau}^2, \sigma^2)$. As a consequence, penalized likelihood inference in combination with REML estimates for the smoothing variances can be interpreted as an empirical Bayes approach. While this equivalence holds exactly for the case of homoscedastic Gaussian responses, it will still hold approximately when using an approximate restricted likelihood (based, for example, on a Laplace approximation) for more general types of responses where the marginal posterior is not available in closed form. See, for example, Kneib and Fahrmeir (2007) for approximate solutions for Cox-type duration time models.

6.4 The General Idea of Markov Chain Monte Carlo Simulations

The main disadvantage of the posterior mode estimates discussed in Section 6.3 is that they provide us only with point estimates and do not give us access to the complete posterior distribution. The latter is particularly interesting when it comes to uncertainty quantification, which for the posterior mode typically relies on procedures requiring the assumption of an asymptotic normal distribution of the estimates, which usually ignores uncertainty arising from the determination of the smoothing parameter(s).

More detailed Bayesian inference can be achieved by assessing the complete posterior based on MCMC simulations that provide us with samples from the posterior for all model parameters. This also allows for the seamless integration of inference for the smoothing variances by assigning appropriate hyperpriors to them. Moreover, the samples from the posterior will provide us with rich information not only on the regression coefficients and smoothing variances, but also on several interesting quantities related to the distribution of the response, as we will discuss in detail in Section 6.8.

6.4.1 The Basic Setup

The main difficulty in analytically assessing the posterior distribution in any Bayesian model arises from the fact that the marginal density $f(\boldsymbol{y})$ in the denominator of Bayes' formula is not analytically available, with the exception of very few simple settings. Without the normalizing constant, it is still possible to determine the posterior mode and the curvature of the posterior at the mode (as a measure of posterior uncertainty), but many other interesting properties such as the posterior mean, median or quantiles require the availability of the full posterior. While simple numerical integration techniques can be used to approximate the marginal density $f(\boldsymbol{y}) = \int f(\boldsymbol{y} | \boldsymbol{\vartheta}) f(\boldsymbol{\vartheta}) d\boldsymbol{\vartheta}$ in cases where $\boldsymbol{\vartheta}$ is of low dimension, this is clearly not feasible in GAMLSS where the dimension of the parameter space is often of rather large dimensionality.

As a consequence, alternative approaches to assess the posterior distribution are required and MCMC simulations are the most popular way of achieving this. The basic idea is to not study the posterior directly, but to produce a random sample from the posterior where the size of this sample can be specified by the analyst. Once such a sample of large size is available, basically arbitrary properties of the posterior can be determined via their empirical analogues, up to arbitrary precision. For example, the posterior mean can be approximated by the average of the samples, theoretical quantiles can be estimated by empirical quantiles and even marginal distributions of elements of $\boldsymbol{\vartheta}$ can be determined by histograms or kernel density estimates. As a consequence, it would be sufficient to obtain a routine providing us with a sequence of random numbers from the posterior rather than having the posterior itself available.

Of course at first sight it sounds strange to ask for a way of generating random numbers from a distribution one does not know. This is the point where Markov chains come into play. A (discrete time) Markov chain is a specific type of stochastic process where the future of the process is conditionally independent of the past, given the current state of the process. The most prominent examples are first-order random walks or autoregressive processes of first order in time series analysis. One can then show that (under some mathematical regularity conditions) the Markov chain converges in distribution to a limiting, so-called stationary distribution; that is, upon convergence the marginal distribution at any given time point is given by exactly this stationary distribution. Note that this does not imply that the samples of a Markov chain would become independent after convergence. When conditioning on the current state, the future realizations will not follow the stationary distribution. Only the marginal distributions without conditioning on either past or future correspond to the stationary distribution.

To make these ideas a bit more digestible, let us consider a discrete time Markov chain with a discrete state space $\mathcal{S}$ of size $S = |\mathcal{S}|$, such that the transitions from the current state of the Markov chain to future states can be characterized by a transition probability matrix $\boldsymbol{P}$, that is, a matrix of dimension $S \times S$ where each row contains the probabilities of the distribution describing how likely a transition from the state corresponding to the row index to the respective states in the column indices is. One can then show, under some regularity conditions, that

$$\lim_{t \to \infty} \boldsymbol{P}^t = \boldsymbol{P}^\infty;$$

that is, the repeated application of the transition probability matrix converges to a limiting matrix and each row in this matrix has exactly the same entries such that

$$\boldsymbol{P}^\infty = \begin{pmatrix} \boldsymbol{\pi} \\ \vdots \\ \boldsymbol{\pi} \end{pmatrix},$$

where $\boldsymbol{\pi} = (\pi_1, \ldots, \pi_S)$ is the stationary distribution of the Markov chain. The regularity conditions include the absence of structures in $\boldsymbol{P}$ implying strictly periodic visits of the states of the Markov chain, or the decomposition of the state space into mutually unreachable subsets. Interestingly, independently of the starting distribution of the Markov chain, we will always observe convergence to the stationary distribution such that, in the long run, the different states of the Markov chain will be visited with frequencies corresponding to the entries in the stationary distribution.

For Bayesian inference based on MCMC, the goal is now to devise a Markov chain with the posterior distribution as the stationary distribution. However, in contrast to the illustrative example discussed above, this Markov chain will usually have a continuous state space (since the parameter space of most parameters in Bayesian inference is continuous) which, in addition, can be of rather high dimension. Furthermore, the goal is to construct a Markov chain with the prespecified yet unknown posterior distribution as the stationary distribution. The advantage of MCMC, on the other hand, is that the Markov chain can be characterized by its transition distributions, namely the distributions specifying how one moves from the current value of the chain to the next state. We will see in the following how this can be done without requiring knowledge about the complete posterior, and in particular not the normalizing constant $f(\boldsymbol{y})$.

Some issues arising from producing samples from the posterior as the stationary distribution of a Markov chain deserve further attention:

- Mathematical theory ensures that appropriately designed Markov chains will converge to the their stationary distribution as the number of steps taken by the chain goes to infinity. However, the speed of this convergence varies considerably such that the analyst has to determine from the observed Markov chain whether it is close enough to the stationary distribution. Unfortunately, in Bayesian inference we do not know the stationary distribution and it is difficult to compare the actual observed chain to the limit distribution. One common way out is visually to assess one or multiple chains to detect signs of nonconvergence. We discuss this in more detail in Section 6.4.3.

- The convergence behavior also depends on the initial state considered for the Markov chain (i.e. the starting values). If these are generated from the stationary distribution, one can show that there is no convergence necessary; all following realizations immediately stem from the stationary distribution. Unfortunately, again we cannot use this result since we do not know the posterior distribution. As a consequence, one usually picks some good guesses for the parameters of interest (e.g. the posterior mode estimate determined via penalized likelihood) to initialize the Markov chain. From the generated samples, one then discards the so-called burn-in phase to remove dependence of the samples on the initialization.

- Naturally the samples from a Markov chain exhibit dependence since the current status of the chain forms the basis for generating the next sample. While at any point in time the samples have marginal distributions coinciding with the stationary distribution, we would like to use the complete set of samples as a basis for determining properties of the posterior. While some results for the asymptotic behavior of the arithmetic mean for dependent samples are available, this is not the case for many other empirical estimates such as variances or kernel density estimates. Moreover, the presence of dependence usually implies that one needs a larger sample to achieve the same accuracy as with an independent sample. As a consequence, it is common practice to perform thinning on the generated Markov chain where only every lth sample is stored, and l is determined to achieve approximate independence between successive samples.

Some of these aspects are illustrated in Figure 6.2. Panel (a) shows kernel density estimates for the posterior distribution $f(\boldsymbol{\vartheta}|\boldsymbol{x})$ after different numbers of iterations for the complete chain including burn-in and without any thinning. While convergence (in distribution) is clearly visible as the number of iterations increases, one can also see the impact of starting the Markov chain in the wrong place, as indicated by the density estimate being centered around a value of 1 after 10 iterations. The impact of this remains visible after 50 and 100 iterations but then diminishes. After applying thinning and removing burn-in (panel (b)), there is no longer any impact of the starting values and convergence to the posterior is considerably faster (although of course the kernel density estimate is still unstable for a small number of samples).

Panels (c) and (d) in Figure 6.2 show convergence of some properties of the posterior including the posterior expectation, posterior variance and a specific posterior quantile. For these quantities, we see convergence towards a deterministic value corresponding to the true values from the posterior with increasing samples from the posterior, as suggested by the law of large numbers. Again, the impact of starting with a value far out in the tails of the posterior is clearly visible in panel (c) where the burn-in phase has not been removed, while convergence is much quicker after removing burn-in and applying thinning (panel (d)).

We will return to these issues after introducing the basic recipe for MCMC in Bayesian inference.

6.4.2 A Generic MCMC Algorithm

While it is possible to treat all parameters $\boldsymbol{\vartheta}$ simultaneously in the construction of an MCMC algorithm, this is often inconvenient because of the size of $\boldsymbol{\vartheta}$ and the different nature of the parameters included in $\boldsymbol{\vartheta}$. We therefore split the complete vector of unknowns into S blocks $\boldsymbol{\vartheta}_1, \ldots, \boldsymbol{\vartheta}_s, \ldots, \boldsymbol{\vartheta}_S$, where each of the subblocks $\boldsymbol{\vartheta}_s$ can be either scalar-valued or a vector. Often the blocks reflect the structure of the model where, for example, the regression coefficients $\boldsymbol{\gamma}_{jk}$ for the jth effect on the kth distribution parameter form one block while the associated smoothing variance τ_{jk}^2 forms another block.

We then proceed as follows to construct the Markov chain: Given appropriate starting values $\boldsymbol{\vartheta}_s^{[0]}$, $s = 1, \ldots, S$ for all parameter blocks, we iteratively loop over the MCMC iterations $t = 1, \ldots, T$ and do the following:

- Generate *proposals* for a new value of $\boldsymbol{\vartheta}_s$ from a so-called proposal density that depends on the data and the current state of all parameters. Ideally, this pro-

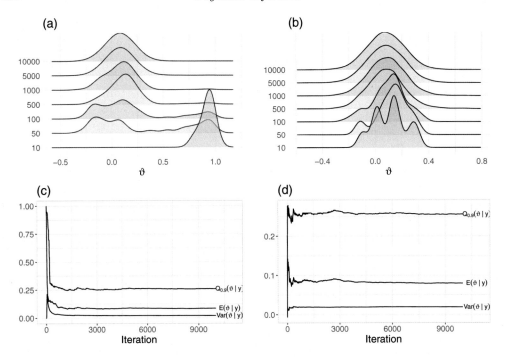

Figure 6.2 Convergence behavior of MCMC samples. Panels (a) and (b) show kernel density estimates obtained for the posterior distribution $f(\boldsymbol{\vartheta}|\boldsymbol{y})$ after different numbers of iterations for the complete chain including burn-in and without thinning (a) and after removing samples from the burn-in phase plus thinning the Markov chain (b). Panels (c) and (d) show the convergence of the posterior expectation $\mathbb{E}(\boldsymbol{\vartheta}|\boldsymbol{y})$, the posterior variance $\mathbb{V}(\boldsymbol{\vartheta}|\boldsymbol{y})$, and the 90% quantile of the posterior distribution $Q_{0.9}(\boldsymbol{\vartheta}|\boldsymbol{y})$, again with/without burn-in and thinning.

posal would come from the posterior distribution of $\boldsymbol{\vartheta}_s$ but since this is generally unknown, proposals can be made from distributions that deviate from this ideal situation. This is corrected for by the acceptance step.

- Accept the proposal only with a certain probability that depends on the posterior as well as proposal density. If the proposal is not accepted, the parameter remains in its current state. When taking the right acceptance probability, one can then ensure that – after a certain burn in period – the realizations from the resulting Markov chain can be treated as realizations from the posterior.

Mathematically more precisely, we do the following:

- Propose a new value for $\boldsymbol{\vartheta}_s$ from a proposal density

$$q_s\left(\boldsymbol{\vartheta}_s^*|\boldsymbol{\vartheta}_1^{[t]},\ldots,\boldsymbol{\vartheta}_{s-1}^{[t]},\boldsymbol{\vartheta}_s^{[t-1]},\ldots,\boldsymbol{\vartheta}_S^{[t-1]}\right)$$

that is allowed to depend on the data $\boldsymbol{y}$, all current values of the other parameter blocks, and the previous value $\boldsymbol{\vartheta}_s^{[t-1]}$ of $\boldsymbol{\vartheta}_s$ itself. Note that by iteratively proceeding through the sequence of parameters, $\boldsymbol{\vartheta}_1$ to $\boldsymbol{\vartheta}_{s-1}$ are already updated to iteration t when constructing the proposal for $\boldsymbol{\vartheta}_s$ while all the further parameters $\boldsymbol{\vartheta}_{s+1}$ to $\boldsymbol{\vartheta}_S$ are still in state $t-1$. We will discuss ways of constructing sensible proposal densities later.

- Accept the proposed new value $\boldsymbol{\vartheta}_s^*$ with probability

$$
\alpha\left(\boldsymbol{\vartheta}_s^*|\boldsymbol{\vartheta}_s^{[t-1]}\right) = \frac{f\left(\boldsymbol{\vartheta}_s^*|\boldsymbol{\vartheta}_{-s}^{[t-1]},\boldsymbol{y}\right) q_s\left(\boldsymbol{\vartheta}_s^{[t-1]}|\boldsymbol{\vartheta}_1^{[t]},\ldots,\boldsymbol{\vartheta}_{s-1}^{[t]},\boldsymbol{\vartheta}_s^*,\ldots,\boldsymbol{\vartheta}_S^{[t-1]}\right)}{f\left(\boldsymbol{\vartheta}_s^{[t-1]}|\boldsymbol{\vartheta}_{-s}^{[t-1]},\boldsymbol{y}\right) q_s\left(\boldsymbol{\vartheta}_s^*|\boldsymbol{\vartheta}_1^{[t]},\ldots,\boldsymbol{\vartheta}_{s-1}^{[t]},\boldsymbol{\vartheta}_s^{[t-1]},\ldots,\boldsymbol{\vartheta}_S^{[t-1]}\right)}
$$

(truncated to 1 in case the value exceeds 1), where the so-called full conditional distribution of $\boldsymbol{\vartheta}_s$

$$
f\left(\boldsymbol{\vartheta}_s|\boldsymbol{\vartheta}_{-s}^{[t-1]},\boldsymbol{y}\right) = f\left(\boldsymbol{\vartheta}_s|\boldsymbol{\vartheta}_1^{[t]},\ldots,\boldsymbol{\vartheta}_{s-1}^{[t]},\boldsymbol{\vartheta}_{s+1}^{[t-1]},\ldots,\boldsymbol{\vartheta}_S^{[t-1]},\boldsymbol{y}\right)
$$

corresponds to the conditional distribution of $\boldsymbol{\vartheta}_s$ given the data and the current values of all the other parameters. If the proposal is not accepted, we set $\boldsymbol{\vartheta}_s^{[t]} = \boldsymbol{\vartheta}_s^{[t-1]}$ such that the Markov chain remains in its current state. The min-operator is simply applied to ensure that the acceptance probability is at most one and can therefore be interpreted as a proper probability.

Note that the full conditional distribution can be obtained from the posterior distribution by considering only factors that involve the parameter block $\boldsymbol{\vartheta}_s$ of interest since

$$
f\left(\boldsymbol{\vartheta}_s|\boldsymbol{\vartheta}_{-s}^{[t-1]},\boldsymbol{y}\right) \propto f(\boldsymbol{\vartheta}|\boldsymbol{y}).
$$

Importantly, the ratio of full conditionals in the acceptance probability does not require the normalizing constant $f(\boldsymbol{y})$ of the posterior since this will cancel in the ratio. This is where the magic trick of MCMC, that enables sampling from a distribution while only knowing the density of this distribution upon proportionality, happens. Note that evaluating the acceptance probability should in practice be done on the log-scale (i.e. working with the log-full conditional and the log-proposal density) to ensure numerical stability. Deriving the ratio of two densities which are both close to zero is numerically ill-conditioned and can easily induce large rounding errors or even overflow errors.

Some intuition for the acceptance step of the MCMC algorithm outlined above can be gained by comparing the two situations depicted in Figure 6.3, where two proposals are considered that lead the Markov chain either towards the center (a) or the tail (b) of the full conditional. For the move towards the center, the first factor in the acceptance probability, that takes the ratio of the full conditionals, is considerably larger than for the move towards the tail. As a consequence, moves towards the center are accepted with a much larger probability than moves towards the tail. The second factor in the acceptance probability depends on the likelihood of either obtaining the proposal (denominator) or reversing the observed transition (numerator). This ratio is required to obtain a reversibility condition that is needed to ensure that the Markov chain indeed converges to the posterior as the stationary distribution.

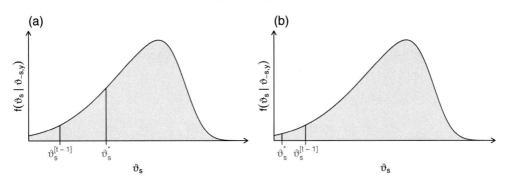

Figure 6.3 Two potential proposals leading either towards the center (a) or the tail (b) of the full conditional distribution.

From a theoretical perspective, the order in which the parameter blocks are treated does not matter, and could even be changed over the course of the MCMC iterations, but some orders result in better mixing and convergence properties than others. Often the order reflects levels in the prior specification such that one starts with a parameter "close" to the data and then iteratively moves away from the data along the prior hierarchy. Similarly, the proposal distribution can be chosen arbitrarily, subject to it respecting the domain of the parameter space, but again some choices will lead to better mixing and convergence than others.

This becomes more obvious when looking more closely at the acceptance probability. First of all, we should avoid the Markov chain not moving due to the non-acceptance of the proposal. This will lead to a high persistence which results in large auto-correlation, and therefore will require large thinning of the Markov chain (see the top row in Figure 6.4). We should therefore aim at acceptance probabilities close to one to achieve enough movement of the Markov chain. However, it is important to determine this in the appropriate way. One obvious, but not desirable, strategy would be to enforce the proposal always being very close to the current value. As a consequence, $\boldsymbol{\vartheta}_s^* \approx \boldsymbol{\vartheta}_s^{[t-1]}$ and therefore both the ratio of the full conditionals and the ratio of the proposal densities will be close to one. However, this will result in large autocorrelation again and will also lead to a Markov chain that moves only very slowly through the parameter space. Such an effect can be observed with the popular approach of random walk proposals where

$$\boldsymbol{\vartheta}_s^* = \boldsymbol{\vartheta}_s^{[t-1]} + \boldsymbol{u}_t$$

with some disturbance $\boldsymbol{u}_t$ added to the current value of the parameter. If, for example, $\boldsymbol{u}_t \sim \mathcal{N}(\boldsymbol{0}, \tau_u^2 \boldsymbol{I})$ and τ_u^2 is small, then the random walk will take only small steps, resulting in the behavior described above (see the middle row of Figure 6.4). On the other hand, setting τ_u^2 large will usually lead to small acceptance probabilities since one often proposes values outside the range of parameters supported by the posterior (see again the top row of Figure 6.4).

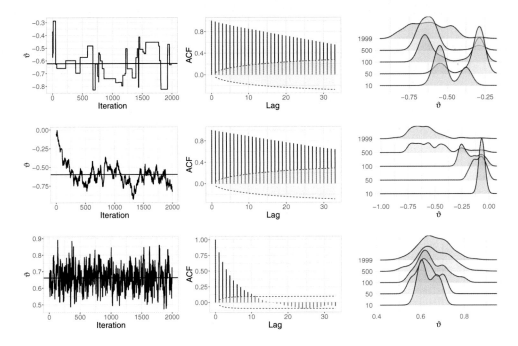

Figure 6.4 Sampling paths (left-hand column), autocorrelation functions (middle column), and posterior kernel density estimates (right-hand column) for three different samplers. Top row: random walk proposals with large variance; middle row: random walk proposal with small variance; bottom row: local quadratic approximation of the log-full conditional.

What would be desirable is a proposal density $q_s\left(\boldsymbol{\vartheta}_s|\boldsymbol{\vartheta}_1^{[t]},\dots,\boldsymbol{\vartheta}_{s-1}^{[t]},\boldsymbol{\vartheta}_s^{[t-1]},\dots,\boldsymbol{\vartheta}_S^{[t-1]}\right)$ that is very close to the full conditional $f\left(\boldsymbol{\vartheta}_s|\boldsymbol{\vartheta}_{-s}^{[t-1]},\boldsymbol{y}\right)$. This again leads to cancellation in the acceptance probability, but now not within but across the ratios of full conditionals and proposal densities. We will discuss a particular example of such a proposal for the regression coefficients in GAMLSS in Section 6.5.1, where we approximate the full conditional with a multivariate normal distribution fitting the mode and the curvature at the mode of the full conditional. See the bottom row in Figure 6.4 for a graphical illustration of the resulting mixing behavior.

In the design of the proposal density, it is important to include information on the current state of the Markov chain when updating $\boldsymbol{\vartheta}_s$. This includes the data as well as the state t information on $\boldsymbol{\vartheta}_1,\dots,\boldsymbol{\vartheta}_{s-1}$ and the state $t-1$ information on $\boldsymbol{\vartheta}_{s+1},\dots,\boldsymbol{\vartheta}_S$. No information on the previous behavior of the chain can be included, since we would then lose the Markov property of the chain. An exception is to use information from the burn-in phase, for example to choose hyperparameters of the proposal density such as the variance τ_u^2 of a random walk proposal, since the samples from the burn-in phase are not taken into account in the final analysis. Similarly, one can conduct trial runs of the MCMC algorithm to determine such hyperparameters.

In some specific cases, the full conditional may in fact turn out to be a known type of distribution from which one can directly generate samples, for example, if the full conditional happens to be a normal distribution. In this case, the acceptance probability $\alpha\left(\boldsymbol{\vartheta}_s^* | \boldsymbol{\vartheta}_s^{[t-1]}\right)$ is always equal to one such that the acceptance step can be discarded from the MCMC algorithm for this specific parameter. This situation is referred to as a Gibbs update; an MCMC algorithm consisting only of Gibbs updates is also called a Gibbs sampler.

The main virtues of MCMC-based inference can now be summarized as follows:

- It provides access to the complete posterior distribution and therefore enables inference without requiring asymptotic considerations.

- The divide and conquer approach resulting from updating blocks of parameters separately allows the breakdown of very complex models having hundreds or thousands of parameters into smaller pieces that can be handled much more easily.

- In addition, the resulting modularity allows the replacement of certain parts of the model without affecting the updates of the other model components. For example, when replacing the prior of one of the smoothing variances τ_{jk}^2, this changes only the update of τ_{jk}^2 itself while all other components and their updates remain completely unaffected.

- From the samples of the model parameters, we can determine not only inferences about these parameters themselves, but also inference for complex functionals of these parameters. We will demonstrate this in more detail in Section 6.8.

On the downside, MCMC requires the design of appropriate proposal densities for a potentially large set of parameter blocks and careful monitoring of mixing and convergence of the Markov chain.

6.4.3 Monitoring Mixing and Convergence

Theoretical results for MCMC-simulations indicate that, regardless of the choice of the exact sampling approach, the definition and order of updates for parameter blocks, or the proposal distribution, we will always achieve convergence of the underlying Markov chain to its stationary distribution. Unfortunately, this does not by any means imply that this result carries over to the actual application of MCMC-based inference, since we will never run the sample for an infinite number of iterations. To detect potential difficulties with the current setup, it is of considerable importance to monitor mixing and convergence of the realized Markov chains to detect

- whether the Markov chain has indeed already converged to the posterior distribution of interest, that is, whether the chosen burn-in period has been sufficient,

- high autocorrelations of the samples that would imply that additional thinning has to be applied, and

- whether the Markov chain moves sufficiently well through the posterior support of the parameter of interest.

While adjusting the number of iterations, the size of the burn-in period, and the thinning parameter are natural options to counterbalance potential deviations from ideal mixing and convergence, severe issues should also be taken as a sign to modify the general setup, including the choice of starting values for the sample, the blocking strategy applied to decompose the parameter vector, the order of the updates, and the chosen proposal distribution.

For example, for Bayesian inference in GAMLSS sensible starting values for the regression coefficients can be obtained by first performing penalized maximum likelihood estimation where, for simplicity, the smoothing variances can be fixed at some reasonable starting value. In this way, one can ensure that the Markov chain does not start at a point without any support for the data, and the burn-in period should be relatively short.

In addition, it makes sense to employ the natural blocking of the parameter vector induced by the model construction where the regression coefficients relating to the different effects in the additive predictor form the subblocks of the complete parameter vector. Compared with a single-site update where each scalar parameter forms one block, this has the distinct advantage of respecting autocorrelation induced by the correlated normal prior and/or the data.

Finally, a sensible strategy is usually to try to match the full conditional as closely as possible with the proposal distribution. This then implies acceptance probabilities close to one and good mixing over the support of the full conditional distribution. We will discuss this in more detail in Section 6.5 for the regression coefficients in a GAMLSS.

A first obvious way of graphically checking mixing and convergence is simply to visualize the realized Markov chains for all model parameters to identify

- trends in the Markov chain indicating possible nonconvergence (especially for the initial part of the sample) or bad mixing, where the Markov chain moves slowly resulting in the impression of a trend,

- long periods of nonaccepted proposals that will induce high autocorrelation and indicate a mismatch between the proposal distribution and the full conditional, or

- slowly moving Markov chains that insufficiently explore the posterior space, again indicating bad mixing.

It also makes sense to initialize multiple Markov chains (potentially with different starting values) and to compare their sampling paths to detect potential nonconvergence.

A simple summary measure for potential dependence in the realized Markov chains is to compute empirical autocorrelation and partial autocorrelation functions. After applying thinning, ideally no detectable autocorrelation should remain, such that large sample results such as the law of large numbers for independent random variables can be applied. From the empirical autocorrelations, one can also compute the effective sample size, that is, the number of independent samples that would carry the same information about the parameter of interest as the dependent data generated with MCMC; see Robert and Casella (2010, chapter 8) for details.

While both plots of the sampling paths and (partial) autocorrelation plots provide useful and easily digestible information, their application only works for models with a small to moderate number of parameters. Graphically assessing hundreds or even thousands of sampling paths is cumbersome and it is then desirable to supplement the manual inspection by the determination of appropriate summary measures. These often rely on comparing the behavior of multiple chains initialized at either the same or different starting values to compare the differences between the behavior of these chains.

In the following, we informally discuss a number of measures that are considered in the literature and that are available in many software packages for Bayesian inference or the postprocessing of samples provided by such packages; see, for example, Robert and Casella (2010, chapter 8) for more details.

- Gelman–Rubin: Initialize multiple chains with different, dispersed starting values to determine whether all chains converge to the same stationary distribution. If there are large differences, this could for example indicate multimodality of the posterior, which makes it difficult to construct well-mixing Markov chains since the chains may then converge to and stay in different modes; or an insufficient length of the burn-in period.

- Geweke: Rather than considering multiple chains, a single Markov chain is split into multiple parts and is then treated in a similar way as with the Gelman–Rubin statistic. This detects trends in the chain that may reflect an insufficient length of the burn-in period or bad mixing. Multimodality may be easily missed by this approach, unless the proposal distribution enables switching of the chain between the different modes.

- Heidelberger–Welch Stationarity Test: Perform a stationarity test to detect whether the Markov chain behaves like a weakly stationary process. Deviations from stationarity may again indicate the need for a longer burn-in period or insufficient mixing.

- Heidelberger–Welch Half-Width Test: Determine whether the sample size of the realized chain is large enough relative to the variability of the chain to achieve a certain target precision on estimating the mean of the samples.

- Raftery-Lewis: Similar to the Heidelberger–Welch Half-Width Test, but focuses on some prespecified quantiles of the posterior distribution rather than the mean.

- Effective sample size: Determines the effective number of independent samples in the Markov chain based on the autocorrelation function. The effective sample size should be close to the actual number of samples; large deviations indicate the need for additional thinning or improved mixing of the Markov chain.

When applying any of these approaches on samples for a model with many parameters, one should always take the resulting multiple testing problem into account. When checking hundreds or thousands of parameters, a small number of deviations from an ideal setting has to be expected and should not necessarily be seen as an indication of the need for improving the sampling process.

6.5 MCMC for GAMLSS

For a Bayesian treatment of GAMLSS, the main difficulty is to construct a suitable proposal distribution for the regression coefficients $\boldsymbol{\gamma}_{jk}$. Owing to the general nonconjugacy between the complex likelihoods employed in GAMLSS and the multivariate normal distribution, there is in general no possibility of establishing Gibbs sampling strategies (but see Section 6.5.3 for some counterexamples). Furthermore, the fact that the elements of $\boldsymbol{\gamma}_{jk}$ are usually strongly correlated both a priori (owing to a nondiagonal structure of $\boldsymbol{K}_{jk}$) and a posteriori (owing to a nondiagonal structure of the cross-product matrix $\boldsymbol{B}_{jk}^{\top}\boldsymbol{B}_{jk}$), it is not advisable to perform single updates for the elements of $\boldsymbol{\gamma}_{jk}$. In summary, we are looking for a multivariate proposal density that adequately reflects the dependence structure in $\boldsymbol{\gamma}_{jk}$.

6.5.1 Iteratively Weighted Least Squares Proposals

General asymptotic theory for Bayesian inference suggests that the joint posterior tends to be Gaussian for large sample sizes and therefore also all full conditionals should asymptotically be close to Gaussian distributions. While it is often questionable in applications whether approximate normality is a good enough approximation on which to base statistical inference, it can be used to construct a proposal density that matches the shape of the full conditional well enough to ensure good mixing and convergence. Essentially, the acceptance step in the MCMC algorithm will be responsible for adjusting for any differences between the asymptotic normal distribution and the actual shape of the full conditional, such that the proposal can also be used in situations with small- to medium-sized samples without any difficulties.

To find the best-fitting normal approximation to the full conditional, we establish a local quadratic approximation of the log-full conditional. This is illustrated in Figure 6.5, which shows the log-full conditional $f(\boldsymbol{\gamma}_{jk}|\cdot)$ (solid line) and the local quadratic approximation (dashed line). The local quadratic approximation of the log-full conditional then corresponds to a Gaussian approximation of the full conditional, where the mode and the curvature are determined by the root of the first derivative on the one hand and the negative second derivative evaluated at the mode on the other hand. Note that both quantities can be determined without requiring the normalizing constant of the full conditional, since all normalizing constants will cancel when taking derivatives on the log-scale. Since the steps involved in the local quadratic approximation resemble those from iteratively weighted least squares

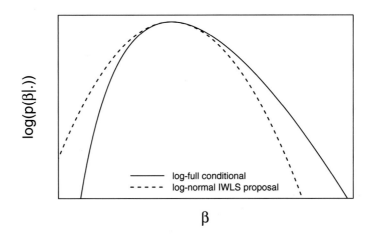

Figure 6.5 Principle of IWLS proposals: The log-full conditional $f(\boldsymbol{\gamma}_{jk}|\cdot)$ (solid line) is approximated quadratically (dashed line) which leads to a Gaussian proposal for $\boldsymbol{\gamma}_{jk}$.

estimation in penalized likelihood inference, the proposal mechanism arising from local quadratic approximation is called the iteratively weighted least squares proposal (IWLS; Gamerman, 1997).

More precisely, the log-full conditional for $\boldsymbol{\gamma}_{jk}$ is (up to additive constants) given by

$$\log(f(\boldsymbol{\gamma}_{jk}|\cdot)) = \ell(\boldsymbol{\eta}_k) - \frac{1}{2\tau_{jk}^2}\boldsymbol{\gamma}_{jk}^\top \boldsymbol{K}_{jk}\boldsymbol{\gamma}_{jk},$$

where $\ell(\boldsymbol{\eta}_k)$ denotes the part of the log-likelihood that depends on the predictor $\boldsymbol{\eta}_k$ which includes $\boldsymbol{\gamma}_{jk}$. A quadratic approximation to this log-full conditional can then be motivated from a normal approximation of the working observations

$$\tilde{\boldsymbol{y}}_k = \boldsymbol{\eta}_k + (\boldsymbol{W}_k)^{-1}\boldsymbol{v}_k,$$

where

$$\boldsymbol{v}_k = \frac{\partial \ell(\boldsymbol{\eta}_k)}{\partial \boldsymbol{\eta}_k}$$

is the vector of first derivatives of the log-likelihood with respect to the predictor (i.e. the score vector) and

$$\begin{aligned}\boldsymbol{W}_k &= \mathrm{diag}(\{w_{ik}, i = 1, \ldots, n\}) \\ &= \mathrm{diag}\left(\left\{\mathbb{E}\left(-\frac{\partial^2 \ell_i(\eta_{ik})}{\partial \eta_{ik}^2}\right), i = 1, \ldots, n\right\}\right)\end{aligned}$$

is the diagonal matrix of working weights corresponding to the individual contributions to the expected Fisher information. The resulting approximate normal distribution for the working observations then implies

$$\tilde{\boldsymbol{y}}_k \sim \mathcal{N}\left(\boldsymbol{\eta}_k, \boldsymbol{W}_k^{-1}\right).$$

Since both the working observations $\tilde{\boldsymbol{y}}_k$ and the working weights $\boldsymbol{W}_k$ are also included in the iterative maximization of the penalized log-likelihood corresponding to a GAMLSS, the approximation is also referred to as an iteratively weighted least squares (IWLS) approximation.

The IWLS proposal density for $\boldsymbol{\gamma}_{jk}$ is then constructed from the resulting working model for $\tilde{\boldsymbol{y}}_k$ as $\boldsymbol{\gamma}_{jk} \sim \mathcal{N}(\boldsymbol{\mu}_{jk}, \boldsymbol{P}_{jk}^{-1})$ with expectation

$$\boldsymbol{\mu}_{jk} = \boldsymbol{P}_{jk}^{-1} \boldsymbol{B}_{jk}^{\top} \boldsymbol{W}_k (\tilde{\boldsymbol{y}}_k - \boldsymbol{\eta}_{-jk})$$

and precision matrix

$$\boldsymbol{P}_{jk} = \boldsymbol{B}_{jk}^{\top} \boldsymbol{W}_k \boldsymbol{B}_{jk} + \frac{1}{\tau_{jk}^2} \boldsymbol{K}_{jk},$$

where $\boldsymbol{\eta}_{-jk} = \boldsymbol{\eta}_k - \boldsymbol{B}_{jk} \boldsymbol{\gamma}_{jk}$ is the predictor without the jth component. More precisely, a proposal $\boldsymbol{\gamma}_j^*$ is determined from the density $q(\boldsymbol{\gamma}_{jk}^* | \boldsymbol{\gamma}_{jk}^c)$ of the normal distribution

$$\mathcal{N}\left(\boldsymbol{\mu}_{jk}^c, (\boldsymbol{P}_{jk}^c)^{-1}\right),$$

where $\boldsymbol{\gamma}_{jk}^c$ denotes the current value of the parameter vector $\boldsymbol{\gamma}_{jk}$ and

$$\boldsymbol{\mu}_{jk}^c = \boldsymbol{\mu}_{jk}(\boldsymbol{\gamma}_{jk}^c) = \boldsymbol{P}_{jk}^{-1}(\boldsymbol{\gamma}_{jk}^c) \boldsymbol{B}_{jk}^{\top} \boldsymbol{W}_k(\boldsymbol{\gamma}_{jk}^c) (\tilde{\boldsymbol{y}}_k(\boldsymbol{\gamma}_{jk}^c) - \boldsymbol{\eta}_{-jk})$$

and

$$\boldsymbol{P}_{jk}^c = \boldsymbol{P}_{jk}(\boldsymbol{\gamma}_{jk}^c) = \boldsymbol{B}_{jk}^{\top} \boldsymbol{W}_k(\boldsymbol{\gamma}_{jk}^c) \boldsymbol{B}_{jk} + \frac{1}{\tau_{jk}^2} \boldsymbol{K}_{jk}$$

correspond to the mean and precision matrix with the matrix of working weights $\boldsymbol{W}_k(\boldsymbol{\gamma}_{jk}^c)$ and the vector of working observations $\tilde{\boldsymbol{y}}_k(\boldsymbol{\gamma}_{jk}^c)$ evaluated at this current value. The proposal $\boldsymbol{\gamma}_{jk}^*$ is then accepted with probability

$$\alpha\left(\boldsymbol{\gamma}_{jk}^* | \boldsymbol{\gamma}_{jk}^c\right) = \min\left\{\frac{f(\boldsymbol{\gamma}_{jk}^* | \cdot) q(\boldsymbol{\gamma}_{jk}^c | \boldsymbol{\gamma}_{jk}^*)}{f(\boldsymbol{\gamma}_{jk}^c | \cdot) q(\boldsymbol{\gamma}_{jk}^* | \boldsymbol{\gamma}_{jk}^c)}, 1\right\}.$$

The major advantages of IWLS proposals are that they

- automatically adapt to the shape of the full conditional and therefore ensure high acceptance probabilities as well as good mixing properties,

- do not require any type of manual tuning by the analyst (unlike alternatives such as random walk proposals), and

- are of a multivariate nature that takes the dependence between elements in the vector of regression coefficients into account and therefore typically yield very good mixing properties.

Furthermore, IWLS updates can be derived quite generically for different types of distributions since the only ingredients that have to be derived for a given type of distribution are the entries of the score vectors $\boldsymbol{v}_k$ and the working weights $\boldsymbol{W}_k$, that is, the first and expected (negative) second derivatives of the log-likelihood with respect to the predictors of the different distribution parameters. For the working weights, it may be tempting to consider the negative second derivative without taking the expectation, to simplify the derivation. This would also be supported by the asymptotic equivalence of expected and observed Fisher information, but taking the expectation is often beneficial for a number of reasons:

- Often the expressions for the expectation are considerably simpler than the ones without the expectation since some terms cancel on average. This allows for easier computations.

- For many distributions the expected second derivative can be shown to be strictly positive, which also ensures that the precision matrix $\boldsymbol{P}_{jk}$ is positive definite such that no additional checks or modifications are required.

- Empirically, it has been observed that taking the expectation yields numerically more stable algorithms.

Still, it has to be kept in mind that the IWLS proposal density is only an approximation to the log-full conditional such that the proposal will still work well even when simplifying the expressions for the score vector and the working weights. Such simplifications can be particularly attractive if some components are hard to evaluate numerically such that there is a trade-off between the goodness of the quadratic approximation and computational speed.

In general, the naïve implementation of an IWLS proposal scheme can suffer from slow computations since densities of high-dimensional normal distributions have to be evaluated, and random draws from their normal distributions have to be obtained. Fortunately, in many cases sparse matrix structures can be employed to perform both tasks rather efficiently. More precisely, the precision matrix $\boldsymbol{K}_{jk}$ as well as the cross-product $\boldsymbol{B}_{jk}^{\top}\boldsymbol{W}_k(\boldsymbol{\gamma}_{jk}^c)\boldsymbol{B}_{jk}$ are often band matrices or more generally sparse matrices with a large fraction of zeros, which can be exploited in various computational aspects. More details on this can be found in Rue and Held (2005) for general Gaussian Markov random fields; and Lang et al. (2014) for structured additive regression models, including penalized splines and spatial Markov random fields as the most important special cases.

In some specific settings, the IWLS proposal density in fact matches the full conditional such that a Gibbs sampling step is obtained and acceptance probabilities are equal to one. This is, for example, the case for all regression coefficients in the mean equation for normal responses, even if a regression specification has been added for the variance. Similar situations arise for the log-normal distribution or when the model involves a probit specification that enables augmentation of a latent Gaussian response (e.g. in zero-adjusted models or in partially binary multivariate models; see Section 6.5.3 below).

Of course other strategies for designing proposal distributions and/or MCMC simulation schemes in general are available and can also be employed as building blocks for Bayesian inference in GAMLSS. This includes components such as adaptive rejection sampling (Gilks and Wild, 1992), slice sampling (Neal, 2003), and different variants of Hamiltonian MCMC with the no-U-turn sampler of Homan and Gelman (2014) as the most prominent example.

6.5.2 Sampling Subject to Linear Constraints

In many GAMLSS specifications (and in fact also in the corresponding simplified GAMs), appropriate constraints have to be incorporated to ensure identifiability of the regression specification. This is, for example, the case when combining multiple nonlinear effects of continuous covariates based on penalized splines, that is, when

$$\eta_{ik} = \cdots + s_{jk}(x_{ij}) + s_{j+1,k}(x_{i,j+1}) + \cdots$$

since shifting one function upwards by a constant while subtracting the same constant from the other function yields the same predictor. Such identifiability issues can typically be resolved by adding a linear constraint

$$\boldsymbol{A}_{jk}\boldsymbol{\gamma}_{jk} = \boldsymbol{0}$$

to the prior distribution of $\boldsymbol{\gamma}_{jk}$ leading to

$$f(\boldsymbol{\gamma}_{jk}|\tau_{jk}^2) \propto \left(\frac{1}{\tau_{jk}^2}\right)^{0.5\,\mathrm{rank}(\boldsymbol{K}_{jk})} \exp\left(-\frac{1}{2\tau_{jk}^2}\boldsymbol{\gamma}_{jk}^\top\boldsymbol{K}_{jk}\boldsymbol{\gamma}_{jk}\right) \mathbb{1}\left(\boldsymbol{A}_{jk}\boldsymbol{\gamma}_{jk} = \boldsymbol{0}\right).$$

The $(a_{jk} \times L_{jk})$-dimensional constraint matrix $\boldsymbol{A}_{jk}$ is assumed to be of full rank $a_{jk} = \mathrm{rank}(\boldsymbol{A}_{jk})$ and chosen according to the problem of interest. For example, to achieve the sum to zero constraint

$$\sum_{i=1}^n s_{jk}(x_{ij}) = 0$$

on all function evaluations $s_{jk}(x_j)$, the constraint matrix is of rank $a_{jk} = 1$ with entries given by

$$\boldsymbol{A}_{jk}[l, 1] = \sum_{i=1}^n B_l(x_{ij}),$$

that is, $\boldsymbol{A}_{jk} = \boldsymbol{1}_n\boldsymbol{B}_{jk}$ where $\boldsymbol{1}_n$ is an n-dimensional vector of ones.

Adding a linear constraint to the prior can also be used as a means of removing the partial impropriety induced by the rank deficiency in the precision matrix $\boldsymbol{K}_{jk}$. To achieve this, we constrain $\boldsymbol{\gamma}_{jk}$ to contain only deviations from the null space of the precision matrix $\boldsymbol{K}_{jk}$. Therefore, let the rank deficiency of $\boldsymbol{K}_{jk}$ be given by a_{jk}, namely, $a_{jk} = L_{jk} - \mathrm{rank}(\boldsymbol{K}_{jk})$. Then the eigendecomposition of $\boldsymbol{K}_{jk}$ will comprise a total of a_{jk} zero eigenvalues while all other eigenvalues are positive, due to the assumption that $\boldsymbol{K}_{jk}$ is positive semidefinite. Setting $\boldsymbol{A}_{jk} = \boldsymbol{\Gamma}_{jk}^\top$, where

$\boldsymbol{\Gamma}_{jk}$ contains the eigenvectors corresponding to the zero eigenvalues of $\boldsymbol{K}_{jk}$, then effectively removes the improper part of the prior distribution. For example, in the case of a penalized spline estimate with rth-order random walk prior, this would remove a polynomial of degree $r - 1$ from the spline effect.

Applying the linear constraint based on the eigendecomposition of $\boldsymbol{K}_{jk}$ does not only remove the impropriety of $\boldsymbol{K}_{jk}$ by constraining the effective vector of regression coefficient to a subspace of dimension $\text{rank}(\boldsymbol{K}_{jk})$, but also enables a decomposition of the total effect $s_{jk}(x_j)$ into a penalized and an unpenalized component. For example, in the case of a penalized spline, we can formulate the model

$$\eta_{ik} = \beta_0 + x_{ij}\beta_1 + \cdots + x_{ij}^{r-1}\beta_{r-1} + \tilde{s}_{jk}(x_{ij}) + \cdots,$$

where $\tilde{s}_{jk}(x_{ij})$ corresponds to the nonlinear effect of x_j with the polynomial of degree $r - 1$ removed. Such a decomposition can, for example, be useful in the context of effect selection priors (see Section 6.7 for details), where one can separately decide whether a polynomial is enough to represent the effect of x_j (i.e. deciding whether $\tilde{s}_{jk}(x_{ij}) \approx 0$) or whether the polynomial effect can be removed as well.

In fact, applying the constraint based on the eigendecomposition yields a similar reparametrization of the model as the mixed model representation of penalized regression. However, in contrast to the explicit reparametrization, the constraint leaves the model parametrization untouched. This has the advantage that sparse matrix structures in $\boldsymbol{K}_{jk}$ or $\boldsymbol{B}_{jk}$ can still be exploited.

An alternative, more flexible way of constructing a linear constraint is to work with a formulation based on function spaces. To discuss this in more detail, we drop the indices j and k such that we are considering a generic effect $s(\boldsymbol{x})$. Let $\mathcal{S}$ be the space of all functions $s(\boldsymbol{x})$ that can be generated from the basis expansion

$$s(\boldsymbol{x}) = \sum_{l=1}^{L} \gamma_l B_l(\boldsymbol{x}).$$

If we are interested in removing all effects from the function space $\mathcal{H} \subset \mathcal{S}$ generated as

$$\mathcal{H} = \left\{ g(\boldsymbol{x}) : g(\boldsymbol{x}) = \sum_{a=1}^{A} \delta_a H_a(\boldsymbol{x}) \right\},$$

this implies a set of A linear constraints on the vector of regression coefficients $\boldsymbol{\gamma}$ derived from orthogonality of the basis functions of $s(\boldsymbol{x})$ to the basis functions of $\mathcal{H}$, that is,

$$\int B_l(\boldsymbol{x})H_a(\boldsymbol{x})d\boldsymbol{x} = 0, \quad a = 1, \ldots, A, \; l = 1, \ldots, L.$$

This then leads to elements of the constraint matrix given by

$$\boldsymbol{A}[a, l] = \int B_l(\boldsymbol{x})H_a(\boldsymbol{x})d\boldsymbol{x}, \quad a = 1, \ldots, A, \; l = 1, \ldots, L.$$

As a major advantage, this approach gives us considerably more freedom in deciding which portions of the effect to remove via the constraint. For example, we could also remove quadratic effects from a penalized spline with second-order random walk penalty, although the null space in this case comprises only linear effects. However, when removing portions of $s(\boldsymbol{x})$ that are not contained in the null space of the precision matrix, we also have to adjust the prior distribution to

$$f(\boldsymbol{\gamma}|\tau^2) \propto \left(\frac{1}{\tau^2}\right)^{\frac{\text{effdim}(\boldsymbol{\gamma})}{2}} \exp\left(-\frac{1}{2\tau^2}\boldsymbol{\gamma}^\top \boldsymbol{K}\boldsymbol{\gamma}\right) \mathbb{1}\left(\boldsymbol{A}\boldsymbol{\gamma} = \boldsymbol{0}\right),$$

where

$$\text{effdim}(\boldsymbol{\gamma}) = L - \tilde{L} \tag{6.2}$$

represents the effective dimension of $\boldsymbol{\gamma}$ after applying the constraint and

$$\tilde{L} = \dim\left(\text{span}(\ker(\boldsymbol{K})) \cup \text{span}\left(\boldsymbol{A}^\top\right)\right)$$

measures the amount of overlap between the nullspace of $\boldsymbol{K}$ and the dimension of the space spanned by the constraint matrix. This complex definition of the effective dimension evaluates the overlap between the impropriety of the prior as characterized by the null space (i.e. the kernel) of the precision matrix $\boldsymbol{K}_{jk}$ and the space of functions removed by the constraint.

While the approach based on basis functions appears far more complicated than that based on the null space, it opens up considerably more flexible ways of separating regression effects $s(\boldsymbol{x})$ into components. For example, in the case of a tensor product interaction spline $s(x_1, x_2)$ of two continuous covariates, one can decompose the effect as

$$s(x_1, x_2) = \beta_0 + \beta_1 x_1 + \beta_2 x_2 + s_1^c(x_1) + s_2^c(x_2) + s_{1,2}^c(x_1, x_2),$$

where $s_{1,2}^c(x_1, x_2)$ represents the interaction surface with all main effects removed; $s_1^c(x_1)$ and $s_2^c(x_2)$ are nonlinear main effects of x_1 and x_2, respectively, with the linear main effects removed; and $\beta_1 x_1$, $\beta_2 x_2$ represent those linear effects. A more complete discussion of constraints from function spaces for models involving more complex interaction terms such as spatio-temporal regression models can be found in Kneib et al. (2019).

In any case, the constraint has to be taken into account when sampling from the proposal distribution. This can be achieved by applying the following algorithm for sampling from a multivariate normal distribution $\mathcal{N}(\boldsymbol{\mu}_{jk}, \boldsymbol{P}_{jk}^{-1})$ with mean vector $\boldsymbol{\mu}_{jk}$ and (full rank) precision matrix $\boldsymbol{P}_{jk}$ subject to the constraint $\boldsymbol{A}_{jk}\boldsymbol{\gamma}_{jk} = \boldsymbol{0}$ (Rue and Held, 2005, algorithm 2.6):

- Compute the $L_{jk} \times a_{jk}$ matrix $\boldsymbol{V}_{jk} = \boldsymbol{P}_{jk}^{-1}\boldsymbol{A}_{jk}^\top$ by solving the system of equations $\boldsymbol{P}_{jk}\boldsymbol{V}_{jk} = \boldsymbol{A}_{jk}^\top$ for each of the columns of $\boldsymbol{V}_{jk}$.

- Compute the $a_{jk} \times a_{jk}$ matrix $\boldsymbol{R}_{jk} = \boldsymbol{A}_{jk}\boldsymbol{V}_{jk}$.

- Compute the $a_{jk} \times J_{jk}$ matrix $\boldsymbol{U}_{jk} = \boldsymbol{R}_{jk}^{-1}\boldsymbol{V}_{jk}^\top$ by solving the system of equations $\boldsymbol{R}_{jk}\boldsymbol{U}_{jk} = \boldsymbol{V}_{jk}^\top$ for each of the columns of $\boldsymbol{U}_{jk}$.

- Compute the constrained sample $\boldsymbol{\gamma}_{jk}^* = \boldsymbol{\gamma}_{jk} - \boldsymbol{U}_{jk}^\top \boldsymbol{A}_{jk} \boldsymbol{\gamma}_{jk}$, where $\boldsymbol{\gamma}_{jk}$ is an unconstrained sample from $\mathcal{N}(\boldsymbol{\mu}_{jk}, \boldsymbol{P}_{jk}^{-1})$.

In each of the steps, methods for dealing with sparse matrices can be employed to achieve numerical efficiency.

6.5.3 Auxiliary Variable Approaches

Even though IWLS proposals can be implemented quite generically and efficiently when exploiting sparse matrix structures, avoiding the acceptance step via Gibbs sampling is still very attractive. On the one hand, questions of mixing and convergence are often of much smaller relevance in Gibbs samples. On the other hand, the evaluation of the proposal density and the posterior at the current and the proposed value can be avoided. As a consequence, formulating model variants that allow for Gibbs sampling is of considerable interest in Bayesian inference. One possibility to achieve this is via augmenting auxiliary variables in the prior hierarchy.

To illustrate this point, consider a probit model with linear predictor $\boldsymbol{x}_i^\top \boldsymbol{\gamma}$ where

$$y_i \sim \mathrm{B}(\pi_i), \quad \text{and} \quad \pi_i = \Phi(\boldsymbol{x}_i^\top \boldsymbol{\gamma}).$$

In this formulation, a Metropolis–Hastings sampler is required to implement MCMC-based inference for $\boldsymbol{\gamma}$. However, resorting to a latent variable representation allows for the construction of a Gibbs sampler. This is based on the well-known representation of the probit model as

$$y_i = \mathbb{1}\left(\tilde{y}_i > 0\right), \quad \text{and} \quad \tilde{y}_i \sim \mathcal{N}(\boldsymbol{x}_i^\top \boldsymbol{\gamma}, 1).$$

If we were to know the latent variables $\tilde{y}_i$, a Gibbs sampler could be constructed based on the normal model assumption for these.

The Bayesian paradigm now enables us to treat the latent variables as additional unknowns that are included in the hierarchical model specification and are determined along with the regression coefficients in an MCMC approach. More precisely, one iteratively updates the latent variables and the regression coefficients from their full conditionals. For the latent variables, we obtain a truncated normal distribution where

$$\tilde{y}_i | y_i, \boldsymbol{\gamma} \sim \begin{cases} \mathrm{TN}_{(0,\infty)}\left(\boldsymbol{x}_i^\top \boldsymbol{\gamma}, 1\right) & y_i = 1, \\ \mathrm{TN}_{(-\infty,0)}\left(\boldsymbol{x}_i^\top \boldsymbol{\gamma}, 1\right) & y_i = 0 \end{cases}$$

and $\mathrm{TN}_{(a,b)}(\mu, \sigma^2)$ denotes the normal distribution with mean μ and variance σ^2 truncated to the interval (a, b). The full conditional for $\boldsymbol{\gamma}$ (under a flat prior $f(\boldsymbol{\gamma}) \propto$ const) is then given by

$$\boldsymbol{\gamma} | \tilde{\boldsymbol{y}}, \boldsymbol{y} \sim \mathcal{N}((\boldsymbol{X}^\top \boldsymbol{X})^{-1} \boldsymbol{X}^\top \tilde{\boldsymbol{y}}, (\boldsymbol{X}^\top \boldsymbol{X})^{-1}).$$

Hence, at the cost of imputing the latent variables, the requirement of a Metropolis–Hastings sampler can be circumvented. Since sampling from univariate truncated normal distributions is quite efficient, it usually pays off to invest the additional effort of imputing the latent variables.

Similar strategies can be developed for other model specifications such as logistic regression (Holmes and Held, 2006) or via approximating the desired distribution by mixtures of normals (Frühwirth-Schnatter et al., 2009).

6.6 Hierarchical Predictor Specifications

Bayesian formulations of GAMLSS are particularly attractive for datasets that comprise multiple hierarchical levels. The most obvious example of such a structure is a dataset with hierarchical clustering as encountered, for example, in data that are collected on multiple aggregation levels of administrative regions. For example, Lang et al. (2014) analyze data on real estate evaluation with individual house prices in Austria as the response variable of interest. To explain house prices, they consider several types of covariates characterizing either the house itself or its surrounding. For the latter type of covariates, information is available on various spatial levels corresponding to municipalities nested within districts, which are themselves nested in counties. For hierarchical clustering in general, a common example is pupils clustered in classes, which are then clustered within schools at a higher level of the hierarchy. When analyzing individual grades, information on the classes and the schools then constitutes covariate information at different aggregation levels.

For such types of data structures, it is helpful to structure the model specification in a similar manner. In a generic approach, this can be achieved by replacing the multivariate Gaussian prior (6.1) assumed for the vector of regression coefficients $\boldsymbol{\gamma}$ by a compound prior specification where another additive predictor is assigned to $\boldsymbol{\gamma}$, that is,

$$\boldsymbol{\gamma} = \boldsymbol{B}_1\boldsymbol{\gamma}_1 + \cdots + \boldsymbol{B}_J\boldsymbol{\gamma}_J + \boldsymbol{\varepsilon}, \qquad \boldsymbol{\varepsilon} \sim \mathcal{N}\left(\mathbf{0}, \tau^2\boldsymbol{I}_L\right). \tag{6.3}$$

For example, let $\boldsymbol{\gamma}$ be the vector of spatial regression coefficients of a Markov random field term such that the elements in $\boldsymbol{\gamma}$ represent the spatial effect in different spatial regions. We then treat these spatial effects as responses in a regression specification based on covariates that are only varying at the level of these regions and that determine the different effects $\boldsymbol{B}_j\boldsymbol{\gamma}_j$. The error terms $\boldsymbol{\varepsilon}$ comprise the deviations from these covariate effects and represent a spatially unstructured random effect. The prior for $\boldsymbol{\gamma}$ then explicitly reads

$$\boldsymbol{\gamma}|\boldsymbol{\gamma}_1, \ldots, \boldsymbol{\gamma}_J, \tau^2 \sim \mathcal{N}\left(\boldsymbol{\mu}, \tau^2\boldsymbol{I}_L\right),$$

where $\boldsymbol{\mu} = \boldsymbol{B}_1\boldsymbol{\gamma}_1 + \cdots + \boldsymbol{B}_J\boldsymbol{\gamma}_J$. For the individual effect components $\boldsymbol{\gamma}_1, \ldots, \boldsymbol{\gamma}_J$ defining the prior mean, one can then either assume another hierarchical level of priors such as (6.3) to introduce multiple, hierarchical scales, or assign the standard prior (6.1).

In principle, it is straightforward to derive a reduced form equation from the hierarchical specification by plugging equation (6.3) into $\boldsymbol{B}\boldsymbol{\gamma}$, such that the total effect decomposes to

$$\boldsymbol{B}\boldsymbol{\gamma} = \boldsymbol{B}\boldsymbol{B}_1\boldsymbol{\gamma}_1 + \cdots + \boldsymbol{B}\boldsymbol{B}_J\boldsymbol{\gamma}_J + \boldsymbol{B}\boldsymbol{\varepsilon}.$$

Plugging this effect into the original model equation, one can then use standard samplers to estimate the parameters $\boldsymbol{\gamma}_1, \ldots, \boldsymbol{\gamma}_J$ and $\boldsymbol{\varepsilon}$ when using $\boldsymbol{B}\boldsymbol{B}_j$ as the design matrix. However, it is algorithmically advantageous to sample parameters within the hierarchical framework:

- The compound prior treats the effects in $\boldsymbol{\gamma}$ as the response of a regression specification, which therefore consists of only L observations. This considerably speeds up the computations compared to the updates in the reduced form equation which has n observations.

- Regardless of the type of response distribution, the regression specification for $\boldsymbol{\gamma}$ relies on a normal distribution such that Gibbs updates are available for all parameters $\boldsymbol{\gamma}_j$. This does not only leave aside the question of how large the acceptance probability is in practice, but also leads to considerable computational savings since the acceptance probability does not have to be computed.

Full details on such an implementation can be found in Lang et al. (2014).

6.7 Hyperpriors for the Smoothing Variances

The Bayesian specification of a GAMLSS model is completed by assigning appropriate hyperpriors to the smoothing variances τ_{jk}^2 controlling the amount of prior regularization for the corresponding term in the predictor for the kth distribution parameter. Assigning such hyperpriors allows for the identification of data-driven amounts of regularization simultaneously with estimating the regression effects of interest, avoiding the need to resort to numerically demanding approaches such as cross-validation.

6.7.1 Inverse-Gamma-Type Priors

The conjugate (and therefore most frequently used) choice as a prior for the variance component τ_{jk}^2 of a multivariate normal prior are inverse gamma priors

$$\tau_{jk}^2 \sim \mathrm{IG}(a, b)$$

with parameters $a > 0$ and $b > 0$, density

$$f(\tau_{jk}^2) \propto \frac{1}{(\tau_{jk}^2)^{a+1}} \exp\left(-\frac{b}{\tau_{jk}^2}\right), \tag{6.4}$$

expectation

$$\mathbb{E}(\tau_{jk}^2) = \frac{b}{a-1}, \quad a > 1,$$

and variance

$$\mathbb{V}(\tau_{jk}^2) = \frac{b^2}{(a-1)^2(a-2)}, \quad a > 2.$$

Due to the conjugacy between the inverse gamma and the multivariate normal distribution, the resulting full conditional for τ_{jk}^2 is always of the inverse gamma type, independently of the distributional assumption for the response. More precisely, the full conditional distribution is given by $\tau_{jk}^2|\cdot \sim \mathrm{IG}(\tilde{a}, \tilde{b})$ with updated parameters

$$\tilde{a} = a + \frac{1}{2}\mathrm{effdim}(\boldsymbol{\gamma}_{jk}), \qquad \tilde{b} = b + \frac{1}{2}\boldsymbol{\gamma}_{jk}^{\top}\boldsymbol{K}_{jk}\boldsymbol{\gamma}_{jk},$$

where the effective dimension of $\boldsymbol{\gamma}_{jk}$ was defined in (6.2), which in many cases reduces to $\mathrm{effdim}(\boldsymbol{\gamma}_{jk}) = \mathrm{rank}(\boldsymbol{K}_{jk})$.

Of course the specification of an inverse gamma prior for τ_{jk}^2 again involves unknown hyperparameters a and b that have to be chosen by the user. One possibility to do so is to come up with sensible values for the mean and variance of the inverse gamma prior and to invert the relations from above to find corresponding values for a and b. However, this is usually complicated by the fact that τ_{jk}^2 is not the marginal variance of the regression coefficients $\boldsymbol{\gamma}_{jk}$ since $\mathrm{Cov}(\boldsymbol{\gamma}_{jk}) = \tau_{jk}^2\boldsymbol{K}^-$ involves the generalized inverse of the precision matrix $\boldsymbol{K}$. Hence, appropriate choices for the expectation and variance of τ_{jk}^2 have to take the structure and values in $\boldsymbol{K}$ into account. For example, changing the number of knots employed for a penalized spline effect leads to very different covariance matrices for $\boldsymbol{\gamma}_{jk}$ and therefore these differences should also be reflected in the hyperparameters a and b (which is not the case when resorting to fixed default values); see Sørbye and Rue (2014).

Still, many applications of Bayesian inference in semiparametric regression models rely on default choices for a and b such as the following:

- $a = b = \varepsilon$ with ε small to obtain a proper approximation to Jeffreys' (improper) prior

$$f(\tau_{jk}^2) \propto \frac{1}{\tau_{jk}^2},$$

  which therefore avoids the risk of improper posteriors resulting from the limiting case $\varepsilon \to 0$. Equivalently, this choice can be derived as an approximation to the flat prior for $\log(\tau_{jk}^2)$.

- $a = 1$, b small, which approximates a flat prior for the precision $1/\tau_{jk}^2$ while again avoiding an improper specification resulting for the limiting case when b tends to zero.

Two further choices arise when allowing for improper variants of the density (6.4) resulting from $a \leq 0$ or $b \leq 0$ implying that (6.4) cannot be normalized to yield a proper density. Still, the corresponding full conditionals will be of the form $\tau_{jk}^2|\cdot \sim \mathrm{IG}(\tilde{a}, \tilde{b})$, with parameters specified above if the posterior actually exists. Two common specifications for such improper choices are

- $a = -1$, $b = 0$ corresponding to an (improper) flat prior for the variance τ_{jk}^2, and

- $a = -0.5$, $b = 0$ corresponding to an (improper) flat prior for the standard deviation τ_{jk}.

In both cases, additional assumptions are required to ensure that the joint posterior is proper, see for example Fahrmeir and Kneib (2009); Klein et al. (2015a).

6.7.2 Scale-Dependent Priors

Although being mathematically and algorithmically convenient, the inverse gamma prior has been criticized for several reasons. One point concerns the choice of default values for the hyperparameters that do not adapt to the structure of the corresponding prior for the regression coefficients. For example, Sørbye and Rue (2014) show that the scaling of the random walk prior commonly assumed for the basis coefficients of Bayesian penalized splines changes considerably when changing the number of basis functions. As a consequence, the hyperparameters themselves should also be adapted to this number, or the prior for the regression coefficients should itself be scaled.

Furthermore, it can be shown that the inverse gamma prior a priori favors complex models with $\tau_{jk}^2 > 0$ over the simplifying models obtained with $\tau_{jk}^2 = 0$, even if there is no evidence in the data for $\tau_{jk}^2 > 0$. This is often not a plausible prior assumption. For example, in the case of a Bayesian penalized spline with second-order random walk prior, $\tau_{jk}^2 = 0$ corresponds to a purely linear effect of the covariate of interest. Usually one would favor such a linear specification over a nonlinear one, unless the data contain convincing evidence in favor of the latter.

To overcome this drawback, Klein and Kneib (2016a) introduce the concept of scale-dependent hyperpriors to GAMLSS models, following the principle of penalized complexity priors by Simpson et al. (2017). Scale-dependent priors rely on four basic principles:

- Occam's razor implies that one should favor simpler models over complex ones unless the data provide convincing evidence for the more complex alternative. For example, in the absence of evidence for nonlinear models, favor the linear specification.

- The discrepancy between the simple model and the more complex alternative can be measured in terms of the Kullback–Leibler discrepancy.

- For the distance scale, an exponential decay prior provides a sensible prior assumption.

- The rate of the exponential decay can be determined based on an a priori assumption for the scaling of the corresponding effect, that is,

$$\mathbb{P}(|s_{jk}(x)| \leq c) \geq 1 - \alpha \tag{6.5}$$

with prespecified values $c > 0$ and $\alpha \in (0, 1)$. Note that the probability statement in (6.5) relates to the marginal distribution of $s_{jk}(x)$ marginalized over the hyperprior for τ_{jk}^2.

One can then show that these principles imply a Weibull prior

$$p(\tau^2) = \frac{1}{2} \left(\frac{\tau^2}{b_{\alpha,c}} \right)^{-1/2} \exp \left(- \left(\frac{\tau^2}{b_{\alpha,c}} \right)^{1/2} \right)$$

with shape parameter $a = 0.5$ and scale parameter $b_{\alpha,c}$ determined by the scaling criterion (6.5). Since the Weibull prior is not conjugate to the multivariate normal prior for the regression coefficients, the implementation of scale-dependent priors requires a Metropolis–Hastings update. Klein and Kneib (2016a) derive a generic Metropolis–Hastings proposal scheme based on local quadratic approximation for the log-variances that allows the convenient sampling of proposals from a normal distribution.

6.7.3 Further Prior Choices for the Smoothing Variances

Several alternative prior structures for the variance parameters have been suggested in the literature, in particular in the context of random effects models which are typically more sensitive to hyperprior settings than penalized spline or spatial smoothing terms in additive regression models. Popular examples advocated in Gelman (2006) include

- the half-normal prior $\tau^2 \sim \mathcal{N}^+(0, b^2)$ with density

$$f\left(\tau^2\right) \propto \left(\tau^2\right)^{1/2-1} \exp\left(-\tau^2/\left(2b^2\right)\right),$$

  that is, a normal distribution truncated to the positive half-axis with the variance parameter b^2 as a hyperparameter,

- the half-Cauchy prior $\tau^2 \sim \mathcal{C}^+(0, b^2)$ with density

$$f\left(\tau^2\right) \propto \left(1 + \tau^2/b^2\right)^{-1} \left(\tau^2/b^2\right)^{-1/2},$$

  that is, a Cauchy distribution truncated to the positive half-axis with the scale parameter b^2 as a hyperparameter, and

- the uniform prior $\tau^2 \sim \mathcal{U}(0, b)$ with the upper limit b as a hyperparameter.

As discussed in Klein and Kneib (2016a), the choice of the hyperparameters for all these priors is crucial to adapt to the scale of the effects under scrutiny. While a criterion such as (6.5) can be employed to guide the choice of the hyperparameter, the scale-dependent prior discussed in Section 6.7.2 seems to be more robust to different challenges such as weakly identified regression parameters or smoothing variances close to zero.

All three suggestions lead to nonconjugate prior hierarchies, but the Metropolis–Hastings proposal developed in Klein and Kneib (2016a) remains readily applicable and only requires changes with respect to derivatives of the log-priors.

6.7.4 Effect Selection Priors

The specification of hyperpriors for the smoothing variances can also be utilized to achieve effect selection, that is, to facilitate the decision as to which covariates should be included in which predictor of a GAMLSS specification. An automatic procedure for effect selection is very desirable since, in addition to choosing the right distributional specification for the response, the selection of the relevant covariate effects in multiple regression specifications is one of the most challenging aspects when applying GAMLSS in practice. In fact, many effect selection questions in Bayesian GAMLSS can be traced back to the question of whether $\tau_{jk}^2 = 0$ or $\tau_{jk}^2 > 0$, where the former implies a restrictive, simple model, while the latter corresponds to the more complex alternative. The basic idea of effect selection priors is then to assign a prior structure that specifically includes the probability of $\tau_{jk}^2 > 0$ as a parameter.

In the simplest case, one could resort to a mixed discrete–continuous prior where, instead of a continuous prior density $g(\tau_{jk}^2)$, we assume a hierarchical prior where, with probability $1 - \omega_{jk}$ we have $\tau_{jk}^2 = 0$ and with probability ω_{jk} we have that τ_{jk}^2 comes from the continuous prior. This leads to the mixed discrete–continuous prior density

$$f\left(\tau_{jk}^2\right) = (1 - \omega_{jk})\mathbb{1}\left(\tau_{jk}^2 = 0\right) + \omega_{jk}g\left(\tau_{jk}^2\right).$$

Conceptually, we can now include the parameter ω_{jk} into the Bayesian inferential scheme, which gives us access to the posterior probability

$$\mathbb{P}\left(\tau_{jk}^2 > 0 \middle| \boldsymbol{y}\right)$$

which can, for example, be estimated by the posterior mean of ω_{jk}. Unfortunately, the direct implementation of this approach is numerically challenging since the mixed discrete–continuous prior for τ_{jk}^2 often leads to difficulties in achieving satisfactory mixing of the MCMC algorithm, except for special cases where marginalization over parts of the parameter space makes the problem tractable.

As a consequence, one often resorts to spike and slab prior structures where, instead of a discrete point mass at zero, one considers a continuous approximation of this point mass. For example, a simple version of such a prior can be constructed by assuming $\tau_{jk}^2 = \delta_{jk}v_{jk}^2$, where $v_{jk}^2 \sim \text{IG}(a, b)$ and

$$f(\delta_{jk}) = (1 - \omega_{jk})\mathbb{1}\left(\delta_{jk} = v_0\right) + \omega_{jk}\mathbb{1}\left(\delta_{jk} = v_1\right)$$

with constants $0 \leq v_1 \ll v_1$. As a consequence, the smoothing variance τ_{jk}^2 follows a scaled inverse gamma prior that is either scaled by a small number v_0 (such that the prior for τ_{jk}^2 effectively concentrates closely to zero) or by a larger value v_1 (typically assumed to be equal to one for simplicity). The prior probabilities for these two cases are again given by $(1 - \omega_{jk})$ and ω_{jk}, respectively.

The main advantage of spike and slab priors is the avoidance of mixed discrete–continuous priors, which usually facilitates mixing in particular in non-Gaussian models. Still, the exact design of spike and slab priors requires considerable care, for example to specify sensible prior specifications for the prior inclusion probability ω_{jk} or for choosing the even larger number of hyperparameters involved in the prior. Scheipl et al. (2012) develop corresponding approaches for generalized additive models while Klein et al. (2021) propose a specification that is tailored towards GAMLSS, including details on the resulting MCMC algorithm.

6.8 Posterior Inference in GAMLSS

When conducting Bayesian inference based on MCMC simulations, the resulting sampling paths for the coefficients provide us with rich and universally accessible information on the posterior distribution. In particular, based on the sample $\boldsymbol{\vartheta}^{[1]}, \ldots, \boldsymbol{\vartheta}^{[T]}$ from the posterior of a generic parameter vector $\boldsymbol{\vartheta}$, we can determine moments, quantiles, or other shape characteristics of the marginal posteriors by their empirical analogues. We can even estimate the complete density by, for example, utilizing a kernel density estimate. In the following, we discuss some specific aspects of Bayesian posterior inference based on MCMC samples, such as the construction of credible intervals and bands, sample-based inference beyond the original set of parameters, model choice, and posterior predictive checks.

6.8.1 Pointwise Credible Intervals and Simultaneous Credible Bands

A Bayesian credible interval $[\vartheta_{s,\text{low}}, \vartheta_{s,\text{upp}}]$ for a scalar parameter ϑ_s is characterized by the posterior coverage probability

$$\mathbb{P}(\vartheta_{s,\text{low}} \leq \vartheta_s \leq \vartheta_{s,\text{upp}} | \boldsymbol{y}) \geq 1 - \alpha,$$

where $1 - \alpha$ denotes the desired coverage level. In contrast to the frequentist interpretation of the coverage probability, the posterior credibility statement is conditional on the data such that the "active" role in the probability is attributed to the parameter ϑ_s and not to the boundaries of the credible interval. The latter are only random insofar as the data $\boldsymbol{y}$ that determine the posterior distribution are random.

There are multiple ways of choosing the boundaries of the credible interval. The easiest case is to consider the $\alpha/2$ and $1 - \alpha/2$ quantiles of the posterior distribution, which can also be consistently estimated via the empirical $\alpha/2$ and $1 - \alpha/2$ quantiles from the samples $\vartheta_s^{[1]}, \ldots, \vartheta_s^{[T]}$ (see panel (a) in Figure 6.6 for an illustration). An alternative is highest posterior density intervals, where $\vartheta_{s,\text{low}}$ and $\vartheta_{s,\text{upp}}$ are chosen such that (i) the posterior density for any value inside the credible interval is at least as high as the posterior density for any value outside the credible interval, and (ii) the desired coverage probability is reached (see pane (b) in Figure 6.6). The quantile-based credible interval is obviously in general not a highest posterior density interval, unless the posterior is symmetric around a central modal point. While highest posterior density intervals appear more attractive from a conceptual

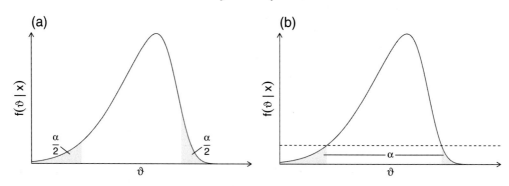

Figure 6.6 Illustration of (a) symmetric and (b) highest posterior density (right) credibility intervals.

perspective, their determination is more demanding and usually requires numerical integration of a kernel density estimate for the posterior distribution.

Credible intervals can easily be extended from scalar to vector-valued parameters when considering a pointwise credibility approach, namely when the desired coverage probability is to be achieved for each element in the parameter vector separately. However, in this case, the overall coverage probability will typically be far smaller than desired, because of the resulting multiple testing problem. The Bonferroni adjustment is one obvious possibility to address this issue, where the probability α of noncoverage is divided by the dimension of the parameter vector. However, this approach tends to be quite conservative, in particular if the elements of the parameter vector tend to be correlated. This is typically the case when the parameter vector consists of elements relating to the same regression effect $s(\boldsymbol{x})$ or when looking at the complete curve estimate for $s(\cdot)$ over a prespecified range of covariate values. In such cases, dedicated approaches for simultaneous credibility regions are preferable.

For the specific case when $s(\cdot)$ relates to a nonlinear effect of a continuous covariate x, simultaneous confidence bands in a frequentist sense can be constructed by studying the asymptotic behavior of the estimate $\hat{s}(\cdot)$. Under suitable conditions, the normalized estimate then behaves like a Gaussian process such that the volume-of-tube formula can be applied (see Krivobokova et al., 2010, for details). In a sample-based Bayesian approach, a simpler possibility (that also extends to general vector-valued parameters) is also proposed by Krivobokova et al. (2010). Here the basic idea is to take a pointwise credibility assessment for a grid of covariate values as a starting point and to scale the resulting pointwise credible interval by a constant factor until a prespecified fraction of sampled curves is completely covered by the credible band. This approach is easily implemented based on MCMC-samples and has been found to be a strong competitor to previous suggestions for Bayesian simultaneous confidence bands (see again Krivobokova et al., 2010).

To discuss this approach in somewhat more detail, let $\boldsymbol{x}_1, \ldots, \boldsymbol{x}_p$ be a set of design points for the effect $s(\boldsymbol{x})$ (e.g. all distinct observed covariate values in the dataset or an equidistant grid of covariate values) and let $\boldsymbol{s} = (s(\boldsymbol{x}_1), \ldots, s(\boldsymbol{x}_p))^\top$ be the corresponding vector of function evaluations. MCMC then provides us with samples $\boldsymbol{s}^{[1]}, \ldots, \boldsymbol{s}^{[T]}$ from which a set of pointwise credible intervals for the elements in $\boldsymbol{s}$ can be obtained by the corresponding sample quantiles. The resulting collection of pointwise credible intervals is then taken as representing the relative uncertainty for the different design points. The pointwise intervals are then scaled up with a common factor, until the desired fraction of complete sampled vectors $\boldsymbol{s}^{[1]}, \ldots, \boldsymbol{s}^{[T]}$ is simultaneously contained in the resulting credible band. As a major advantage, this construction principle can not only be applied to the function estimates but also to any other (covariate-dependent) quantity of interest derived from the model.

6.8.2 MCMC-Based Inference for Parameter Transformations

In maximum likelihood inference, it is well known that the maximum likelihood estimate is invariant under one-to-one transformations, i.e. if $\hat{\boldsymbol{\vartheta}}$ is the maximum likelihood estimate for $\boldsymbol{\vartheta}$ and $\boldsymbol{\xi} = g(\boldsymbol{\vartheta})$ is a one-to-one transformation, then $g(\hat{\boldsymbol{\vartheta}})$ is the maximum likelihood estimate for $\boldsymbol{\xi}$. Similarly, the delta rule allows us to establish asymptotic normality for the maximum likelihood estimate for $\boldsymbol{\xi}$ from the asymptotic normal distribution of $\hat{\boldsymbol{\vartheta}}$.

In sample-based Bayesian inference, access to the posterior distribution of transformations of the original parameter vector is even easier. It is neither restricted to one-to-one transformations, nor does it rely on asymptotic normality. If $\boldsymbol{\vartheta}^{[1]}, \ldots, \boldsymbol{\vartheta}^{[T]}$ is a sample from the posterior of $\boldsymbol{\vartheta}$, transformations $g(\cdot)$ of arbitrary complexity can be applied to the individual samples, and $g(\boldsymbol{\vartheta}^{[1]}), \ldots, g(\boldsymbol{\vartheta}^{[T]})$ will be a sample from the posterior of the transformed parameter such that properties of this posterior can be derived based on empirical analogues.

To make the idea concrete, consider the case of Bayesian inference for the mean μ and the variance σ^2 based on an i.i.d. sample of normally distributed observations such that $\boldsymbol{\vartheta} = (\mu, \sigma^2)^\top$. If an MCMC sampler has been implemented to obtain samples for $\boldsymbol{\vartheta}$, one can directly obtain samples for quantities such as the standard deviation σ, the coefficient of variation σ/μ, or other transformations of the original parametrization. In fact, one can easily go further and derive samples for quantities such as tail probabilities, the interquartile range, etc., by applying corresponding formulae for the normal distribution with the parameter samples plugged in. It is even possible to get a sample of densities or cumulative distribution functions such that the posterior distribution of such functionals is accessible for further analysis.

As an example, Klein et al. (2015c) investigate the spatial variation in income inequality in Germany based on GAMLSS specifications, assuming either a log-normal, gamma, inverse Gaussian, or Dagum distribution for the conditional income distribution (given characteristics of the individuals under study such as labor market experience or age). Instead of studying the regression effects directly on the distribution parameters (which will be hardly comparable across the four different distributions owing to very different parametrizations), they conduct posterior inference on measures such as the expected income, the income standard deviation, the income Gini coefficient, and the probability of falling below certain income thresholds. For all these quantities, exact posterior inference is directly provided via applying transformations to the sampled parameters, while avoiding the need for asymptotic normality and complex considerations as in the delta rule.

6.8.3 Model Choice Based on Information Criteria

The classical approach to Bayesian model choice is to compare competing models through their posterior probabilities. If there are L competing models $M_1, \ldots, M_L$ with associated parameters $\boldsymbol{\vartheta}_1, \ldots, \boldsymbol{\vartheta}_L$, the posterior for $\boldsymbol{\vartheta}_l$ given the model M_l is given by

$$f(\boldsymbol{\vartheta}_l | \boldsymbol{y}, M_l) = \frac{f(\boldsymbol{y} | \boldsymbol{\vartheta}_l, M_l) f(\boldsymbol{\vartheta}_l | M_l)}{f(\boldsymbol{y} | M_l)},$$

where $f(\boldsymbol{y} | \boldsymbol{\vartheta}_l, M_l)$ and $f(\boldsymbol{\vartheta}_l | M_l)$ are the likelihood and the prior of $\boldsymbol{\vartheta}_l$ under model M_l, respectively, and

$$f(\boldsymbol{y} | M_l) = \int f(\boldsymbol{y} | \boldsymbol{\vartheta}_l, M_l) f(\boldsymbol{\vartheta}_l | M_l) d\boldsymbol{\vartheta}_l$$

is the marginal likelihood of model M_l.

For model selection, we then also have to assign prior distributions to the competing models, $f(M_l)$, and compare models through their marginal posteriors

$$f(M_l | \boldsymbol{y}) = \frac{f(\boldsymbol{y} | M_l) f(M_l)}{f(\boldsymbol{y})}, \qquad l = 1, \ldots, L,$$

where $f(\boldsymbol{y}) = \sum_{l=1}^{L} f(\boldsymbol{y} | M_l) f(M_l)$. The marginal posterior can be interpreted as the average ability to explain the observed data where the average is taken subject to weights introduced by the prior distribution.

We would now prefer model M_l against a model M_s, $s \neq l$ if $f(M_l | \boldsymbol{y}) > f(M_s | \boldsymbol{y})$ or, in other words, if

$$\frac{f(M_l | \boldsymbol{y})}{f(M_s | \boldsymbol{y})} = \frac{f(M_l) f(\boldsymbol{y} | M_l)}{f(M_s) f(\boldsymbol{y} | M_s)} > 1.$$

Assuming the same prior probability for all models, namely $f(M_1) = \cdots = f(M_L)$, this ratio of marginal posteriors simplifies to the Bayes factor

$$\text{BF}_{ls} = \frac{f(\boldsymbol{y} | M_l)}{f(\boldsymbol{y} | M_s)}.$$

In many problems the exact computation of Bayes factors is, however, difficult because determining the marginal likelihoods $f(\boldsymbol{y} | M_l)$ is infeasible. In principle, it can be estimated as the arithmetic mean

$$\hat{f}(\boldsymbol{y}) = \frac{1}{T} \sum_{t=1}^{T} f\left(\boldsymbol{y} | \boldsymbol{\vartheta}^{[t]}\right)$$

with samples $\boldsymbol{\vartheta}^{[t]}$, $t = 1, \ldots, T$ from the prior distribution, or the harmonic mean

$$\hat{f}(\boldsymbol{y}) = \left(\frac{1}{T} \sum_{t=1}^{T} \frac{1}{f\left(\boldsymbol{y}|\boldsymbol{\vartheta}^{[t]}\right)} \right)^{-1}$$

with samples $\boldsymbol{\vartheta}^{[t]}$, $t = 1, \ldots, T$ from the posterior distribution. Both estimates suffer from numerical instability as well as conceptual difficulties. Samples from the posterior are readily available in MCMC inference, but the variance of the inverse likelihood terms $1/f(\boldsymbol{y}|\boldsymbol{\vartheta}^{[t]})$ is usually much larger if not infinite, making the harmonic mean estimate very inefficient.

Another possibility to conduct Bayesian model comparison is based on Laplace approximations given by

$$-2f(\boldsymbol{y}|M_l) \approx -2 \log \left(f\left(\boldsymbol{y}|\hat{\boldsymbol{\vartheta}}_l, M_l\right) \right) + \log(n)p_l,$$

where p_l is the dimension of $\boldsymbol{\vartheta}_l$ and $\hat{\boldsymbol{\vartheta}}_l$ denotes the posterior mode. This approximation leads to the Bayesian information criterion (BIC)

$$\mathrm{BIC}(M_l) = -2l\left(\hat{\boldsymbol{\vartheta}}_l\right) + \log(n)p_l.$$

Although being in principle Bayesian, the BIC is hardly used in Bayesian analyses. The main reasons are that the Laplace approximation is rather poor in many high-dimensional or complex problems and the fact that for models with nonindependent observations n should be replaced by an effective sample size that would have to be estimated as well. As a consequence, model choice in Bayesian GAMLSS often resorts to analogues to Akaike's information criterion (AIC) such as the deviance information criterion (DIC) and the widely applicable information criterion (WAIC), which are both easily calculated from MCMC samples.

Let $\boldsymbol{\vartheta}^{[1]}, \ldots, \boldsymbol{\vartheta}^{[T]}$ denote the MCMC sample for the complete parameter vector $\boldsymbol{\vartheta}$ comprising all unknowns of the model to be estimated. Then the deviance information criterion (DIC; Spiegelhalter et al., 2002) is defined as

$$\begin{aligned}
\mathrm{DIC} &= \overline{D(\boldsymbol{\vartheta})} + p_{\mathrm{DIC}} \\
&= 2\overline{D(\boldsymbol{\vartheta})} - D(\overline{\boldsymbol{\vartheta}}) \\
&= \frac{2}{T} \sum_{t=1}^{T} D\left(\boldsymbol{\vartheta}^{[t]}\right) - D\left(\frac{1}{T} \sum_{t=1}^{T} \boldsymbol{\vartheta}^{[t]}\right),
\end{aligned}$$

where $D(\boldsymbol{\vartheta}) = -2 \log(f(\boldsymbol{y}|\boldsymbol{\vartheta}))$ denotes the model deviance (including all normalizing constants), $\overline{D(\boldsymbol{\vartheta})}$ is the estimated posterior expected model deviance, and $p_{\mathrm{DIC}} = \overline{D(\boldsymbol{\vartheta})} - D(\overline{\boldsymbol{\vartheta}})$ provides an estimate for the effective parameter count.

The basic interpretation of the DIC is similar to that of the AIC, that is, $\overline{D(\boldsymbol{\vartheta})}$ provides a measure for the fit of the model while p_{DIC} measures the model complexity such that minimizing the DIC yields a compromise between fit and complexity (see also Section 4.4.1). While it is easily calculated based on MCMC samples, the DIC has a number of drawbacks:

- The value of the DIC obtained for a given model depends on the model parametrization. Hence, different values for the DIC can be obtained when, for example, parameterizing a model in terms of standard deviations instead of variances, even if the two models are exactly equivalent to each other.

- The degrees of freedom can assume negative values, which does not fit well with the intuition of a measure for model complexity.

Both problems result from the fact that, implicitly, the DIC relies on asymptotic normality of the posterior such that the posterior mean can be considered a good point estimate representing the central tendency. Especially in complex models or for small samples, the DIC may therefore be a problematic choice.

To overcome the drawbacks of the DIC, Watanabe (2010) developed the widely applicable information criterion (WAIC), where

$$\mathrm{WAIC} = 2\left(D_{\mathrm{WAIC}} + p_{\mathrm{WAIC}}\right)$$

with

$$D_{\mathrm{WAIC}} = -\sum_{i=1}^{n}\left(\log\left(\frac{1}{T}\sum_{t=1}^{T} p\left(y_i|\boldsymbol{\vartheta}^{[t]}\right)\right)\right)$$

as the measure of model fit,

$$p_{\mathrm{WAIC}} = \sum_{i=1}^{n}\hat{\mathbb{V}}\left(\log\left(p\left(y_i|\boldsymbol{\vartheta}\right)\right)\right)$$

as the measure of model complexity, and the empirical variance

$$\hat{\mathbb{V}}\left(a\right) = \frac{1}{T-1}\sum_{t=1}^{T}\left(a_t - \bar{a}\right)^2.$$

While the WAIC is somewhat more complicated to calculate, it is indeed independent of the chosen parametrization and the model complexity is always positive. As a consequence, the WAIC is increasingly popular in Bayesian inference.

6.8.4 Posterior Predictive Checks

While model choice criteria allow the assessment of the fit on the observed data, posterior predictive checks focus on the predictive abilities of a model specification. In general, predictive distributions of the form

$$f(\tilde{y}|\boldsymbol{y}) = \int f(\tilde{y}|\boldsymbol{\vartheta})f(\boldsymbol{\vartheta}|\boldsymbol{y})\,d\boldsymbol{\vartheta}$$

are commonly used in Bayesian statistics to evaluate the ability of a model with parameters $\boldsymbol{\vartheta}$ estimated from data $\boldsymbol{y}$ in predicting a new observation $\tilde{y}$.

The posterior predictive ordinate (PPO) utilizes this principle for the individual observations such that $\tilde{y} = y_i$ and therefore

$$\text{PPO}_i = f(y_i|\boldsymbol{y}) = \int f(y_i|\boldsymbol{\vartheta})f(\boldsymbol{\vartheta}|\boldsymbol{y})\,d\boldsymbol{\vartheta}$$

which can conveniently be estimated from MCMC samples as

$$\widehat{\text{PPO}}_i = \frac{1}{T}\sum_{t=1}^{T} f\left(y_i|\boldsymbol{\vartheta}^{[t]}\right).$$

However, the posterior predictive ordinate is conceptually somewhat inconsistent since it conditions on all data including the data point y_i when evaluating the ability to predict a future observation y_i.

To rectify this, the conditional predictive ordinate (CPO) considers the leave-one-out cross-validation (LOO-CV) predictive density

$$\text{CPO}_i = f(y_i|\boldsymbol{y}_{-i}) = \left[\int \frac{1}{f(y_i|\boldsymbol{\vartheta})}f(\boldsymbol{\vartheta}|\boldsymbol{y})d\boldsymbol{\vartheta}\right]^{-1},$$

where $\boldsymbol{y}_{-i}$ denotes the observed data with the ith data point removed. Estimation can in principle be conducted via

$$\widehat{\text{CPO}}_i = \left[\frac{1}{T}\sum_{t=1}^{T}\frac{1}{f\left(y_i|\boldsymbol{\vartheta}^{[t]}\right)}\right]^{-1}$$

but this estimate tends to be unstable owing to the inversion of densities that easily leads to numerical overflow if $f(y_i|\boldsymbol{\vartheta}^{[t]})$ is close to zero.

As a consequence, one often resorts to cross-validation approaches based on k-fold cross-validation of the data. This evaluates the predictive ability on the hold-out sample based on proper scoring rules such as the log-score (which is then conceptually close to the conditional predictive ordinate but based on log-densities), the Brier score, or the continuous ranked probability score.

6.9 Other Bayesian Approaches

While Bayesian inference based on MCMC simulations yields rich and accurate information on the posterior for all model parameters and quantities of interest, it may be computationally demanding especially for complex models with a large number of parameters. In such cases, approximate solutions can be of interest. The easiest (yet also not very accurate) option is to rely on the relation to penalized likelihood sketched out in Section 6.3 and to rely on empirical Bayes approaches for estimating any hyperparameters in the model (e.g. restricted maximum likelihood estimation). Relying on asymptotic normal theory then allows the determination of not only point estimates but also corresponding credible intervals and other inferential statements.

In this way, the penalized likelihood approaches discussed extensively in Chapter 5

can be seen as an approximate empirical Bayes counterpart to the results derived
in this chapter. This also holds true for other variants of penalized likelihood approaches, presented for example in Yee and Wild (1996), Marra and Wood (2012)
and Wood et al. (2016). To circumvent the complexities of transferring the asymptotic normal results obtained for the regression coefficients to derived quantities,
one may resort to an asymptotic bootstrap procedure, that is, simulating from the
asymptotic normal distribution followed by applying the transformation of interest.

Refined approximations to the posterior that avoid the necessity to rely on asymptotic
normality have been suggested, for example, based on integrated nested Laplace
approximations (INLA; Rue et al., 2009) or utilizing variational inference (Waldmann
et al., 2013). However, these are mostly still restricted to generalized additive models
featuring only a single regression predictor. Conceptually, they can, however, also be
applied to GAMLSS-type models and we expect to see corresponding developments
in the future.

7

Statistical Boosting for GAMLSS

This chapter introduces an alternative estimation scheme for GAMLSS and distributional regression based on statistical learning. The core of the approach is a component-wise gradient boosting algorithm with statistical models as base-learners, which we refer to as *statistical boosting*. It incorporates variable selection, effect selection, and also shrinkage of effect estimates in the fitting of GAMLSS models, but does not change the underlying model formulation. Also the main interpretation of the resulting models remains the same as for GAMLSS models.

This chapter hence has the following major aims:

- highlight the main idea behind boosting algorithms in machine learning,

- discuss how this concept can be used to fit statistical models,

- highlight potential advantages of boosting in the context of statistical modeling, and

- introduce the application of statistical boosting algorithms for distributional regression.

The structure follows the aims. In Section 7.1 the original concept of boosting for machine learning is presented in a non-technical manner. In Section 7.2 we introduce how the concept can be used for classical statistical models. In Section 7.3 we further extend this statistical boosting approach to GAMLSS models.

The main advantage of applying statistical boosting is its ability to incorporate effect selection in the fitting process: The algorithm allows to identify the most informative predictors, and their corresponding effects, automatically from a much larger candidate model. It hence presents a potential solution to the classical trade-off between variance and bias in statistical modeling and variable selection for GAMLSS. See also Section 4.2 for an in-depth discussion on the relationship between model and the underlying truth, as well as Section 4.5 for model selection strategies.

This variable selection and model fitting via boosting even works in settings where the number of candidate effects p is larger than the number n of observations ($p > n$). These $p > n$ situations are often referred to as *high-dimensional* data problems and are typically not identifiable for commonly used statistical inference methods without incorporating additional constraints or penalizations. Note that in the case of

GAMLSS, this problematic situation of needing to estimate more covariate effects than observations may appear far more often than in classical regression: As the covariates may be included in the additive predictors for multiple distribution parameters, in fact when the sum of these dimensions is larger than the number of observations ($p = \sum_{k=1}^{K} p_k > n$), classical approaches become infeasible.

Furthermore, the effect selection process for GAMLSS models becomes even more challenging than in classical statistical modeling, as the variables for the different regression predictors of the distribution parameters need to be selected simultaneously. These two circumstances make the need for an inference scheme suitable for data analyses with larger number of potential explanatory variables even more pressing.

7.1 Boosting in Machine Learning

Boosting emerged from the field of supervised machine learning, where it was introduced as a way to increase the performance of weak classifiers $h(x)$. The task of the classifier is to predict the labels of $y_i \in \{-1, 1\}$ for the underlying sample $\boldsymbol{y} = (y_1, \ldots, y_n)^\top$. These so-called *base-learners* $h(x)$ perform only slightly better than random guessing for this task. In supervised learning, these base-learners are boosted in order to get a nearly perfect classification. This is achieved by applying them repeatedly on reweighted observations, and aggregating this sequential ensemble of solutions to a more powerful classifier. We will now first highlight the general concept of boosting for classification tasks in machine learning. The approach is later extended to statistical modeling and distributional regression.

Base-learners in the context of machine learning are often simple decision trees or stumps (trees with only one single split). While a single tree might lead to relatively good performance on the underlying dataset, it typically performs poorly on new observations. That is because of the high variability in the estimation process of a tree. To overcome this, *bagging* uses bootstrap-aggregated trees. The algorithm samples B bootstrap versions (sampling with replacement) of the underlying data and combines the different solutions to a more stable classification rule. The concept of *boosting* follows a similar aim, but uses an entirely different concept.

Instead of applying the tree on various resampled datasets and forming a parallel ensemble as in bagging, boosting sequentially applies the base-learner on the same observations, but reweights the data in every iteration $m = 1, \ldots, m_{\text{stop}}$ according to its classification result in the previous iteration $(m - 1)$. It hence builds a *sequential* ensemble (see Figure 7.1). The concept of boosting, however, is not really to manipulate the base-learner itself to increase its performance, but to manipulate the underlying training data by iteratively reweighting the observations with weights $\boldsymbol{w}^{[m]} = (w_1^{[m]}, \ldots, w_n^{[m]})^\top$. As a result, the base-learner at every iteration will find a new solution from the data. These weights $\boldsymbol{w}$ now play a crucial role: Observations that were misclassified in the previous iteration get higher weights, while observations

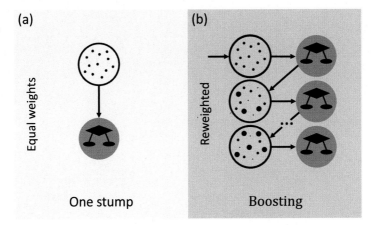

Figure 7.1 Illustration of the boosting concept with respect to enhancing the accuracy of simple trees or stumps, adapted from a popular blog-entry[1]. Panel (a) depicts a single stump. In the case of boosting (b), the algorithm sequentially uses reweighted observations to fit different trees.

that were already identified correctly get lower weights. As a result, the algorithm is forced to concentrate on data points that are not yet predicted well by the model.

This sequential ensemble algorithm hence mimics a rather human behavior in learning: We (hopefully) learn from our mistakes and try to do better the next time. Similarly, boosting techniques try to learn from data by learning from the errors made by the previous base-learner.

This concept was first introduced by Schapire (1990) and Freund (1995), who later characterized this idea as *garnering wisdom from a council of fools* (Schapire and Freund, 2012). The fools in this case are the base-learners – they need to perform only slightly better than flipping a coin. A simple base-learner (e.g. a stump) is by no means a practical classification rule. However, even for performing only slightly better than random guessing, the simple base-learner must contain some valid information about the underlying structure of the problem. The task of a boosting algorithm is hence to learn from the iterative application of one or multiple weak learners and to combine this information into an accurate classification. Garnering wisdom is performed by trial and error: The base-learners try to classify, but their following base-learner will attempt to learn from their mistakes.

The accuracy of boosting is hence achieved by *increasing* the importance of observations that are problematic to classify (where the previous learner made errors). During the iteration cycle, the focus is shifted towards observations that were misclassified up to the current iteration (see Figure 7.2).

Afterwards, the final aggregation of the sequential ensemble is generated by weighted

[1] Adapted from:
https://quantdare.com/what-is-the-difference-between-bagging-and-boosting/.

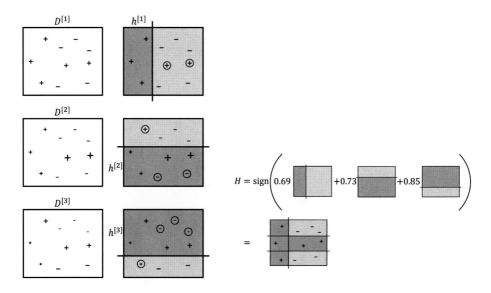

Figure 7.2 Boosting toy example (adapted from Schapire and Freund (2012)): Illustration of a sequential ensemble of three iterations $\left(\text{base-learners } h^{[1]}, h^{[2]}, h^{[3]}\right)$ on reweighted observations $\left(D^{[1]}, D^{[2]}, D^{[3]}\right)$ that is boosted to a perfect classification. In the left part of the plot, each row refers to a new iteration. The symbols represent the observations $y_i \in \{-1, 1\}$. In the first iteration (top row) all observations have equal weights (size of the symbols), in the later iterations observations that have been misclassified (symbols that are circled) get larger weights. The final aggregation is performed via weighted majority weighting (right part of the plot) to a combined solution. By reweighting the observations (left) and weighting the solutions (right), in this toy example, boosting is able to perform a perfect classification after three iterations.

majority voting: Every base-learner solution $h^{[1]}(x), \ldots, h^{[m_{\text{stop}}]}(x)$ enters in the final aggregation $H(x)$, but each solution is weighted by an iteration specific factor $\boldsymbol{\alpha}$ representing the base-learner's success $\boldsymbol{\alpha} = (\alpha_1, \ldots, \alpha_{m_{\text{stop}}})^{\top}$:

$$H(x) = \text{sign}\left(\sum_{m=1}^{m_{\text{stop}}} \alpha_m h^{[m]}(x)\right).$$

The concept of boosting can hence be explained by two different weighting schemes: Each observation is associated in each iteration with weight $w_i^{[m]}$ for $i = 1, \ldots, n$ and $m = 1, \ldots, m_{\text{stop}}$, where $\sum_{i=1}^{n} w_i^{[m]} = 1$. Observations that were previously misclassified receive higher weights, while observations that were correctly classified get downweighted (see the left-hand part of Figure 7.2). On the other hand, in the final aggregation, each base-learner solution is weighted again with α_m. Here, solutions with good accuracy receive higher weights, while unsuccessful iterations

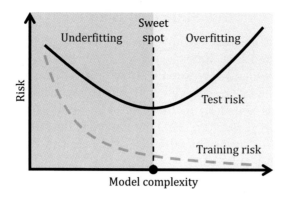

Figure 7.3 The relationship between model complexity, underfitting and overfitting (adapted from Belkin et al., 2019). In the case of boosting, the complexity or size of the model depends on the number of iterations m_{stop} that is carried out. This general trade-off between overfitting and underfitting also represents the trade-off between variance and bias.

are downweighted (see the right-hand part of Figure 7.2). This system of weights $\boldsymbol{w}^{[m]}$ and α_m that work in different directions is the key to the success of boosting.

An important discussion in the context of boosting algorithms in machine learning is their overfitting behavior. *Overfitting* describes the common phenomenon that when a prediction rule concentrates too much on peculiarities (*noise*) of the specific sample of training observations it was optimized on, it will often perform poorly on a new dataset. To avoid overfitting, the task for the algorithm therefore should not be to find the best possible classifier for the underlying training sample, but rather to find the best prediction rule for a set of new observations. The main control instrument to avoid overfitting in boosting algorithms is the number of boosting iterations m_{stop} that are carried out. Very late stopping may favor overfitting, as the complexity of the aggregated solution increases. On the other hand, stopping the algorithm too early not only inevitably leads to higher error on the training data, but can also result in poorer prediction on new data (*underfitting*). The aim for the data analyst is hence to find the so-called *sweet spot* between under- and overfitting, where the model is as complex as necessary to learn the necessary information from the training data, but also simple enough to still perform well on external validation data (Figure 7.3).

The model complexity is determined by both the number of boosting iterations and the complexity of base-learners (e.g. number or depth of trees). A recent discussion in this context also focuses on achieving a *double-descent* by further increasing the complexity beyond the limit of interpolating the training data. Under specific circumstances, this can lead to a second descent of the risk on the validation data (test risk) which at first glance seems to contradict the classical bias–variance trade-off. Investigations for such a behavior typically focus on complex neural networks, but there are also examples constructed with boosting algorithms using multiple trees (forests) as base-learners (Belkin et al., 2019).

7.2 Statistical Boosting

Although developed in machine learning, the general idea of boosting was soon transferred to statistical modeling. Here the main concept is adapted to estimate the unknown quantities in general regression settings (e.g. GLMs, GAMs) via univariate regression models as base-learners. This concept of component-wise gradient boosting with statistical models as underlying base-learners is often referred to as model-based boosting (Bühlmann and Hothorn, 2007) or statistical boosting (Mayr et al., 2014a). The latter also takes similar likelihood-based approaches into account and is the term we use throughout this chapter.

Statistical boosting is methodologically situated between the worlds of statistics and computer science. The underlying concept bridges the gap between two rather different points of view on how to gather information from data (Breiman, 2001): On the one hand there is the classical statistical modeling view which focuses on models *describing* and *explaining* the outcome in order to somehow find an approximation to the underlying stochastic data generating process. The machine learning approach, on the other hand, focuses primarily on algorithmic models *predicting* the outcome while treating the nature of the underlying process as unknown.

The advantage of machine learning is its flexibility, as the structure of the prediction rule is extremely flexible and not limited to additive effects. The main disadvantage of this prediction-focused approach is its limited interpretation. Although nowadays various techniques exist to assess the *importance* of single features or variables (e.g. variable importance measures), this is not the same as relying on an interpretable effect of a variable. This interpretability is the main advantage of statistical modeling approaches: Owing to the typical additive structure of statistical models, one can not only rank the variables in their importance but actually describe the effect of a single variable independently of the others (ceteris paribus).

Statistical boosting adapts an advanced machine learning algorithm to perform inference for classical statistical models, thereby combining the two views on data modeling described above. The aim is hence an inference scheme that incorporates the interpretability (or *descriptive accuracy*; Murdoch et al., 2019) of statistical modeling with the prediction accuracy of machine learning.

7.2.1 Base-learners for Statistical Boosting

While the general idea of boosting is to build a sequential ensemble of different weak solutions for supervised learning (see Figure 7.2), in the case of statistical boosting these weak solutions $h(x)$ most often refer to simple regression functions that are aggregated in an additive manner to build an additive predictor η for a statistical regression model:

$$\eta(\boldsymbol{x}) = \eta = \beta_0 + \sum_{j=1}^{p} h_j(x_j).$$

In this case, β_0 corresponds to a global intercept and the functions $h_1(x_1), \ldots, h_p(x_p)$ refer to base-learners representing different types of effects of the explanatory variables $x_1, \ldots, x_p$. In the context of statistical boosting, the base-learner functions $h_j(x_j)$ hence reflect the terms of a structured additive regression predictor. The choice of base-learners is crucial for the application of statistical boosting, as they define the type(s) of effect(s) that covariates will have on the predictors of the model (cf. Chapter 3).

In most situations, each explanatory variable x_j is included in one base-learner $h_j(x_j)$. However, there are also base-learners incorporating two or more explanatory variables, for example for spatial effects or interactions. For the sake of simplicity we drop these particular cases w.r.t. notation. These base-learners, which in classical boosting are most often tree-based classifiers, in the case of statistical boosting are typically univariate regression models. Often-used base-learners include simple linear models $h_j(x) = \beta_j x_j$ for linear effects, and penalized regression splines for nonlinear smooth effects $h_j(x_j) = s_j(x_j)$ (cf. Section 3.1). Spatial or other bivariate effects can be incorporated by setting up a bivariate tensor product extension of P-splines for two continuous variables (cf. Section 3.3). In some settings it also makes sense that a covariate will enter in more than one base-learner, representing different types of effects of the same variable (e.g. a decomposition of smooth and linear effects). The advantage is that the algorithm can decide in a data-driven way which base-learner is more suitable. For approaches to model choice via such a decomposition, see Kneib et al. (2009) and Hofner et al. (2011).

Another way to include spatial effects is the adaptation of Markov random fields for modeling a neighborhood structure or radial basis functions (cf. Sections 3.4 and 3.5). Constrained effects such as monotonic or cyclic effects can be specified as well. Random effects can be taken into account by using ridge-penalized base-learners for fitting categorical grouping variables such as random intercepts or slopes. One has to keep in mind, though, that the general idea of boosting algorithms is that the base-learners are *weak*. They will be updated iteratively, and no base-learner should hence lead to a reasonable fit in only one iteration.

7.2.2 Loss Functions for Statistical Boosting

While the base-learners define the type of effect of a single covariate, the loss function $\rho(y_i, \eta_i)$ defines the type of regression setting. The loss function is responsible for describing the discrepancy between observations y_i and model fit η_i as well as for defining the actual optimization problem (cf. Sections 4.3 and 1.6.1). Typical examples are the L_2 loss

$$\rho(y_i, \eta_i) = (y_i - \eta_i)^2 \ ,$$

and the L_1 loss

$$\rho(y_i, \eta_i) = |y_i - \eta_i| \ ,$$

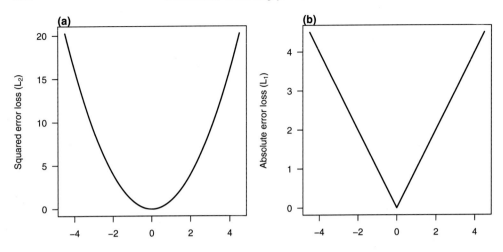

Figure 7.4 Comparing the L_2 loss (a) which leads to mean regression and the L_1 loss (b) leading to median regression.

leading to classical mean regression and median regression, respectively (see Figure 7.4). In case of GAMs or GAMLSS, the typical loss function is the negative (log) likelihood of the response distribution.

The optimization problem for statistical boosting is now to minimize the expected loss (often denoted as risk) with respect to the unknown quantities $\boldsymbol{\theta}$. In practice, this means that the algorithm minimizes the empirical risk

$$r = \sum_{i=1}^{n} \rho(y_i, \eta_i) \,, \tag{7.1}$$

where $\eta_1, \ldots, \eta_n$ depend on $\boldsymbol{\theta}$. (The empirical risk is discussed in Section 4.3.) The concept of statistical boosting is to iteratively minimize r by *gradient descent in function space*.

7.2.3 The Algorithm

In case of the classical L_2 loss, optimizing the empirical risk r is equivalent to minimizing the mean squared error

$$\frac{1}{n} \sum_{i=1}^{n} (y_i - \eta_i)^2 \,,$$

where η_i refers to the structured additive model fit (additive predictor) of observation i.

While there exist different approaches for statistical boosting (e.g. likelihood-based boosting), in this book we will focus on component-wise gradient boosting. The main concept is to apply simple regression-type functions as base-learners and fit them one-by-one (component-wise), not on the observations $y_1, \ldots, y_n$ but on the negative gradient vector (first derivative) $u_1, \ldots, u_n$ of the loss

$$
\boldsymbol{u}^{[m]} = \left(u_i^{[m]} \right)_{i=1,\ldots,n} = \left(-\frac{\partial}{\partial \eta} \rho(y_i, \eta_i) \Big|_{\eta_i = \hat{\eta}^{[m-1]}(x_i)} \right)_{i=1,\ldots,n} .
$$

In every iteration m, the gradient is recomputed based on the current model fit $\hat{\eta}^{[m-1]}$. In the case of L_2 loss, this means that the base-learners in iteration m are actually fitted on the residuals from iteration $m-1$.

The algorithm decreases the empirical risk step-by-step; this behavior has been described as *gradient descent in function space*. Comparing this fitting scheme with the classical boosting approaches in machine learning, the analogy becomes clear: While in classical boosting algorithms for supervised learning, the weights of the observations $\boldsymbol{w}^{[m]} = (w_1^{[m]}, \ldots, w_n^{[m]})^\top$ refer to their correct classification or misclassification at iteration $m-1$, in the case of statistical boosting this role is played by the residuals. Observations that are somehow "hard" to predict will have larger residuals and, as the base-learners are directly fitting them, will play a more important role in the next iteration.

Before the first boosting iteration is carried out, a set of base-learners $h_1(x_1), \ldots, h_J(x_J)$ has to be specified (see Section 7.2.1). Instead of fitting the original data points for iterations $m = 1, \ldots, m_{\text{stop}}$, the boosting algorithm iteratively fits the negative gradient vector of the loss function $\boldsymbol{u}$ separately to each of the base-learners:

$$
\boldsymbol{u}^{[m]} \xrightarrow{\text{base-learner}} \hat{h}_j^{[m]}(x_j) \quad \text{for } j = 1, \ldots, J.
$$

In every iteration, only the best-performing base-learner j^*

$$
j^* = \underset{1 \leq j \leq J}{\operatorname{argmin}} \sum_{i=1}^{n} \left(u_i^{[m]} - \hat{h}_j^{[m]}(x_j) \right)^2,
$$

and, hence, the best-performing covariate is selected to be included in the update step. During the fitting process the algorithm iteratively selects variables, in order to get the most accurate prediction: It searches the function space, which is spanned by the base-learners, for the steepest descent. In every update step, the current version of η is additively updated only by a small step-length (denoted "sl"), in order to approximate a minimum:

$$
\hat{\eta}^{[m]} = \hat{\eta}^{[m-1]} + \text{sl} \cdot \hat{h}_{j^*}^{[m]}(x_{j^*}).
$$

It is noteworthy that since the base-learner fit refers to the negative gradient, in this step of the algorithm an estimate of the true negative gradient of the empirical risk (7.1) is added to the current estimate of the additive predictor η. As a result, the presented component-wise boosting algorithm descends along the gradient of the empirical risk.

Owing to the additive updates, the final model $\hat{\eta}^{[m_{\text{stop}}]}$ follows an additive structure

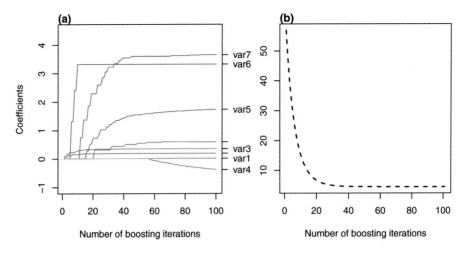

Figure 7.5 Example of a coefficient path for statistical boosting with linear base-learners (a). In each iteration, either a new base-learner and corresponding covariate is included in the model, or an already incorporated effect is updated. In panel (b), the resulting empirical risk $r(\boldsymbol{\theta})$ is displayed.

and is thus interpretable as every common GLM or GAM. The type of the covariate effect $f_j(x_j)$ inside η refers to the effect represented by the corresponding base-learner $h_j(x_j)$.

Note that the base-learner $h_j(\cdot)$ can be selected and updated various times; the partial effect of variable x_j is the sum of all corresponding base-learners that have been selected:

$$\hat{f}_j(x_j) = \sum_m \mathrm{sl} \cdot \hat{h}_j^{[m]}(x_j) \mathbb{1}\left(j = j^*\right),$$

where $\mathbb{1}\left(\cdot\right)$ represents the indicator function. In the case of a linear base-learner, $\hat{h}_j(x_j) = x_j^\top \beta_j$, the final coefficient β_j simply boils down to the sum of $\hat{\beta}_j^{[m]}$ from the iterations where it was the best-performing base-learner ($j = j^*$):

$$\hat{\beta}_j = \sum_m \mathrm{sl} \cdot \hat{\beta}_j^{[m]} \mathbb{1}\left(j = j^*\right).$$

This iterative procedure then leads to coefficient paths similar to those resulting from L_1 penalization with regularized regression approaches such as the lasso (e.g. Figure 7.5). In fact it can be shown that statistical boosting with linear base-learners and very small step-lengths, under some restriction, lead to the same coefficient paths as the lasso (Hepp et al., 2016).

7.2.4 Tuning of the Algorithm

The main control instrument to steer the variable selection and shrinkage properties of statistical boosting is the stopping iteration m_{stop}. Larger numbers of boosting iterations lead to larger models containing more selected variables with bigger effect estimates. Smaller numbers of boosting iterations, on the other hand, lead to smaller, sparser models that are easier to interpret, and shrunken effect estimates.

The selection of m_{stop} hence represents the classical trade-off in penalized regression and statistical modeling: One has to choose between a very accurate model for the underlying training data (larger model with small bias) or a smaller, more strictly regularized model with larger bias but smaller variance (and hence better prediction accuracy). The main idea in selecting m_{stop} is hence *not* to find the best value to model the underlying (training) sample. The tuning parameter should be selected in order to find a model that can be generalized: This is reflected in good prediction accuracy,that is, by the fit on *new* observations (test data). In practice, the selection of m_{stop} is typically guided by optimizing the predictive performance of the model (evaluated by the corresponding loss function) via cross-validation or bootstrapping techniques.

The selection of m_{stop} in the case of statistical boosting, however, not only steers the variable selection and the induced shrinkage; in the case of nonlinear effects (splines) it also controls the level of smoothness for single predictor variables. As a P-spline base-learner, for example, is typically incorporated with a fixed strong penalty (and hence fixed large amount of smoothness) on a fixed number of equidistant knots, the coefficients for the basis-functions are also added for each boosting iteration where the corresponding base-learner had been selected to be updated. The more often this base-learner is selected, the rougher the nonlinear effect gets, until it finally simply overfits by following every observation of the training data. This is illustrated in Figure 7.6.

7.2.5 Properties of Statistical Boosting in Practice and Current Developments

One of the main advantages of the boosting approach is of course the intrinsic variable selection. By stopping the algorithm before it converges, variables that have never been selected to be updated via their corresponding base-learner are effectively excluded from the final model. Another advantage is that statistical boosting also works for high-dimensional data. As the algorithm simply cycles through the different base-learners and fits them one-by-one, the dimensionality problem (more covariates than observations, $p > n$) is avoided during the fitting. Together with the intrinsic variable selection, this makes the concept particularly suitable for high-dimensional prediction problems.

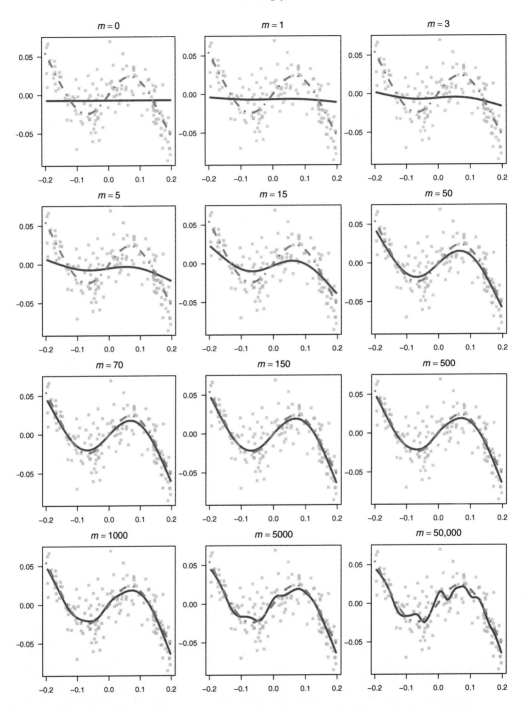

Figure 7.6 Impact of the number of iterations m on the smoothness of a nonlinear effect in a boosting model (see also Mayr and Hofner, 2018). The gray points are the observations. The dashed blue line represents the true underlying relationship between explanatory variable and outcome. The red line is the aggregated partial effect in the final model, shown from the initialization with a horizontal line at $m = 0$ until clear overfitting at $m = 50,000$. The optimal stopping iteration determined via bootstrapping in this case was at $m_{stop} = 70$. One can clearly observe how the stopping iteration influences the shape of the effect; it is also apparent that the algorithm shows a rather slow overfitting behavior.

The design of the algorithm makes it rather easy to extend: In general, any base-learner $h_j(x_j)$ can be combined with any convex loss function $\rho(y, f)$. Hence boosting algorithms can also be advantageous in low-dimensional settings, without any need for variable selection in order to combine non standard effect types (e.g. spatial effects) with non standard loss functions (e.g. robust loss functions, quantile regression) for which no implementation may currently be available. Over the past several years, various research articles have focused on the extension of statistical boosting approaches to new regression settings and model families. (See Mayr et al., 2014b, 2017, for overviews.)

Another aspect of boosting is the possibility of performing *model choice*, for example, the automated selection of the type of effect for an explanatory variable: The algorithm is adapted to use different types of base-learners for particular covariates. For a continuous variable, this could be a linear and a nonlinear base-learner. For these covariates, both types of base-learners are included in the list of potential updates. To achieve a fair comparison, these base-learners are specified so as to ensure a similar level of complexity. Then the algorithm can select in every iteration which base-learner, and hence which type of effect, is updated. Therefore, not only the selection of whether a covariate is included at all in the final additive predictor, but also via which base-learner and which type of effect, is based on the boosting procedure (Hofner et al., 2011; Kneib et al., 2009).

In many practical data situations, particularly in cases with large sample size and a moderate number of variables ($n > p$), boosting algorithms tend to yield models with a relatively large number of selected variables. The reason for this behavior can be found in the sequential updating and both the indirect and implicit regularization. While direct regularization techniques such as the lasso optimize the corresponding model for each given shrinkage parameter, in the case of boosting the regularization is an *indirect* effect of early stopping. The stopping iteration m_{stop} is selected optimizing the predictive risk (i.e. the loss evaluated on test data) via cross-validation or resampling techniques. However, if a variable and its corresponding base-learner have been selected once to be updated in one of the earlier iterations, this update cannot be reversed. As this component will enter with only a small contribution (also because of the small learning rate that is typically set to 0.1), this will not lead to overfitting and does not have a large effect on prediction accuracy. However, if the main goal of the analysis and the application of boosting was to select the most influential predictors, rather than fit a prediction model, including variables with a negligible contribution is problematic.

To overcome this issue, several extensions to enhance the sparsity of statistical boosting models are currently discussed in the literature. If the focus of the analysis is mainly the selection of variables, then one could use the *stability selection* approach (Meinshausen and Bühlmann, 2010; Shah and Samworth, 2013). Stability selection is a general technique to enhance variable selection procedures via subsampling: Boosting is hence applied on various subsamples and only variables that pass a threshold with respect to their selection rates on the subsamples will be included in the final model (Hofner et al., 2015; Mayr et al., 2016). For more details on this approach see Section 7.3.4 and Chapter 13 for an illustrative gene expression example.

Another more direct way to enhance the sparsity of statistical boosting is via *probing*. Instead of applying boosting multiple times, the model is fitted only once but considering also additional shadow variables (or probes). These shadow variables are constructed as permuted versions of actual candidate variables, and are hence surely noninformative. The boosting algorithm is stopped once the first probe is falsely selected (Thomas et al., 2017). This way, not only is the time-consuming tuning of m_{stop} via resampling avoided, but also the focus of the early stopping procedure is shifted from traditional prediction accuracy towards variable selection properties. This procedure is not only simple, but also very fast. Actually, the model has to be fitted only once; there is no additional tuning necessary. Experience suggests that probing leads to relatively sparse models which might not be always optimal with respect to prediction accuracy. But that is basically just a consequence of focusing on variable selection rather than prediction accuracy to determine m_{stop}. For high-dimensional data, a drawback of probing is that one might run into computational memory problems: Owing to the p additional probe variables, the size of the dataset (and the memory demand) doubles. For an example of probing see Chapter 13.

Another way to ensure earlier stopping is the one standard error rule (1se-rule). It goes back to a proposal of Breiman et al. (2017) in the context of random forests. The idea is not to determine the optimal tuning parameter via resampling or cross-validation, but to select the smallest model that provides a test risk in the range of $\pm$ one standard error of the risk. Nowadays, one of the most popular implementations of the lasso, the **R** package **glmnet**, incorporates it as the default criterion to determine the shrinkage parameter λ. In the context of boosting, this means that we stop the algorithm not at the optimal m_{stop} but earlier: We select the smallest m_{stop} where the test risk is only one standard error larger that the minimal one. This procedure typically leads to smaller and less complex models, but might also influence the prediction accuracy as the optimal m_{stop} is no longer used. Another disadvantage of this approach is that the margin of one standard error is actually rather arbitrary, as it depends on the size and variability of the data as well as the resampling scheme used (see also Ellenbach et al., 2021).

A more recent development focuses on actively deselecting base-learners that have only a minor impact on the total risk reduction of the final model (Strömer et al., 2022). The concept of deselection is considerably newer and was particularly developed for statistical boosting (Strömer et al., 2022). Instead of stopping the algorithm earlier (as in probing or via the 1se-rule), which inevitably leads to more shrinkage of effect sizes, the m_{stop} is chosen as usual. Afterwards, selected variables are deselected again, if their contribution to the model fit is only negligible. To identify variables that were selected but are of minor importance for the prediction, the idea is to evaluate the overall risk reduction of the model. If only a small proportion of the risk reduction can be attributed to updates from the corresponding base-learner of this variable (the authors propose a threshold of 0.01, i.e. 1%) the variable is *deselected*.

Afterwards, the model is refitted with the same m_{stop} but only with the initially selected variables which passed the threshold. This pragmatic procedure works fairly well in practice, and leads typically to smaller models with a very similar prediction accuracy

These approaches for enhanced variable selection are illustrated in the application on variable selection for gene expression data (Chapter 13).

7.2.6 Implementation in R

The most important implementation of statistical boosting in the statistical programming environment **R** is the **mboost** package. For a tutorial article on the usage of **mboost** see Hofner et al. (2014). The main function is `glmboost()` for fitting linear models (similar to the basic `glm()`). As in the standard function, the actual loss (e.g. the corresponding distribution) is specified via a `family` argument. If this argument is omitted, a classical L_2 loss is used.

```
library("mboost")
m1 <- glmboost(y ~ x1 + x2 + x3, data = data, family = Poisson())
```

By default, the linear Poisson model is fitted here with 100 iterations. The tuning of the algorithm (e.g. the selection of the stopping iteration m_{stop}) can be done via `cvrisk()`. The selected grid of potential stopping iterations can be specified; the default is to search for the best solution among the already fitted iterations (e.g. up to $m_{\text{stop}} = 500$).

```
cvr <- cvrisk(m1, grid = 1:500)
```

To set the model now to the resulting optimal number of iterations, one can specify the following.

```
mstop(m1) <- mstop(cvr)
```

To fit additive models (including not only linear effects), the function `gamboost()` allows the specification of different types of base-learners that can be combined with the same set of loss functions as in the linear case (also via `family`).

```
m2 <- gamboost(y ~ bols(x1) + bbs(x2) + bbs(x3), data = data,
               family = Poisson())
```

The Poisson regression model `m2` now incorporates a linear ordinary least squares base-learner (`bols()`) for variable `x1` and separate nonlinear penalized B-spline base-learners (`bbs()`) for variables `x2` and `x3`.

Another very flexible implementation of boosting is the **xgboost** framework (extreme gradient boosting) which is particularly suitable for large datasets (Chen et al., 2023). It typically uses tree-based base-learners, but there is also an implementation for linear models via `booster = "gblinear"`. Note that **xgboost** does not in general perform component-wise boosting or variable selection, but incorporates all candidate variables. However, specifying `updater = "coord_descent"`, `feature_selector = "greedy"` and `top_k = 1` leads to models very similar to the classical `glmboost()` function from **mboost**.

Example: Birthweight Prediction Based on Ultrasound Data

To analyze and illustrate the properties of statistical boosting with respect to variable selection, shrinkage of effect estimates, and prediction accuracy, we use the fetal ultrasound dataset, which will be described and treated in more detail in Chapter 8. While ultrasound examinations of the fetus during pregnancy are typically performed in order to detect abnormalities and potential risk factors for standard delivery, the particular research question in our context is to predict the birthweight via the biometric parameters on the fetus available from ultrasound. The predicted birthweight is an important clinical indicator in delivery management and is often the basis for decisions on how to proceed.

Regarding the properties of the boosting approach, the aim of this analysis is to illustrate whether statistical boosting is able to (i) carry out variable selection and (ii) to outperform classical inference approaches regarding the resulting prediction accuracy. We carry out a 10-fold cross-validation (see Section 4.4.3) on the complete dataset with $n = 1038$ deliveries, in order to separate between training and test observations and get realistic estimates for the prediction accuracy. We will therefore compare different potential models (represented by sets of base-learners) with varying inference schemes.

- **Set 1**: Linear effects (base-learners) for the seven potential explanatory variables: Fetal biometric measurements abdominal circumference (`AC`), biparietal diameter (`BPD`), head circumference (`HC`), femur length (`FL`), as well as maternal factors `age`, `parity` (number of previous births), and the interval between ultrasound scan and delivery (days before delivery, `DBD`).

- **Set 2**: Linear effects (base-learners) for the seven potential explanatory variables from set 1, but additionally also all pairwise interactions leading to $7 + \binom{7}{2} = 28$ potential base-learners.

- **Set 3**: Linear effects (base-learners) for the seven potential explanatory variables from set 1, together with all interactions, leading to $7 + \binom{7}{2} + \binom{7}{3} + \binom{7}{4} + \binom{7}{5} + \binom{7}{6} = 126$ potential base-learners.

- **Set 4**: Nonlinear effects via smooth base-learners for the seven explanatory variables of set 1.

For each of the four models, we applied statistical boosting for variable selection and model fitting, as well as classical inference based on (penalized) maximum likelihood approaches. For statistical boosting, we used the **R** package **mboost** with default linear base-learners in `glmboost()` to fit models for sets 1–3 and penalized splines in `gamboost()` for set 4. As loss function, we considered the classical L_2 loss, which is equivalent to the optimization problem of classical mean regression. The stopping iteration m_{stop} was optimized with an additional bootstrap approach (using $B = 25$ bootstrap samples) via the pre implemented `cvrisk()` inside the training samples.

For the classical approaches we used the ordinary least squares estimator `lm()`, while for fitting a model on set 4 we applied the `gam()` function in the **mgcv** package. Note that the boosting models perform variable selection and tuning within each fold – so the models vary between the folds. For the classical approaches, only the coefficients of the models are different for the different folds.

As an additional comparison, we also applied boosting with the L_1 loss function, which effectively leads to a median regression model. (For a comparison of the L_2 loss to the L_1 loss, see Figure 7.4.) Quantile regression (discussed in Section 1.6.1), and therefore also median regression, is more robust to outliers than classical mean regression and hence might also be favorable in some settings, leading to higher prediction accuracy. Furthermore, quantile regression (similar to GAMLSS) goes beyond the classical mean regression, as many different aspects of the conditional distribution can be modeled. The main advantage of quantile regression over distributional regression approaches such as GAMLSS is its absence of distributional assumptions. Due to the modular nature of statistical boosting, the L_1 loss can easily be included in the algorithm and combined with any type of base-learner.

We evaluate the prediction accuracy regarding different criteria, where BW_i is the *true* birthweight of observation i and $\widehat{\text{BW}}_i$ is the estimated birthweight based on the corresponding model. These criteria are quite common in the analysis of prediction formulas for fetal weight (e.g. Faschingbauer et al., 2015). The first criterion is the median absolute percentage error

$$\text{MAPE} = \text{med} \left(\left| 100 \cdot \frac{\widehat{\text{BW}}_i - \text{BW}_i}{\text{BW}_i} \right|_{i=1,\ldots,n} \right), \tag{7.2}$$

whose values should be as small as possible. The second criterion is the mean percentage error

$$\text{MPE} = \frac{1}{n} \sum_{i=1}^{n} \left(100 \cdot \frac{\widehat{\text{BW}}_i - \text{BW}_i}{\text{BW}_i} \right), \tag{7.3}$$

which can highlight a systematic overestimation (MPE > 0) or underestimation (MPE < 0). A well-fitting model should lead to an MPE close to zero on training data.

We also evaluate the standard deviation of the percentage error, which is often referred to as *random error* (RE). It measures the variability of the prediction error and should also be as small as possible.

Table 7.1 *Results of the ultrasound application, comparing the performance of statistical boosting with classical inference schemes based on four different sets of explanatory variables. MAPE = median absolute percentage error, MPE = mean percentage error, RE = random error (sd of percentage error), selected = average number of selected variables. For MAPE and RE smaller values indicate a higher prediction accuracy, while the MPE should be as close as possible to 0. Note that as the classical inference approaches do not aim to select variables, the corresponding numbers refer to the size of the candidate model.*

		MAPE	MPE	RE	Selected
Model set 1	Classical	5.92	−0.45	9.54	7
	Boosting	5.90	−0.45	9.55	7
Model set 2	Classical	5.63	−0.46	9.22	28
	Boosting	5.66	−0.44	9.16	17
Model set 3	Classical	5.87	−1.34	19.94	126
	Boosting	5.43	−0.49	9.31	19.4
Model set 4	Classical	5.80	−0.44	9.32	7
	Boosting	5.79	−0.46	9.43	7

Results from 10-Fold Cross-Validation

To provide a fair evaluation of prediction accuracy, we fitted all candidate models with classical and boosting approaches in a 10-fold cross-validation scheme. The complete sample is hence randomly partitioned into 10 different folds, where 9 of them are used to train and fit the models (*training data*) while the remaining observations are used to test the accuracy (*test data*).

Comparing the results of the different inference schemes regarding prediction accuracy (Table 7.1), it is clear that most of the benefit of applying boosting in this application can be observed only for the large candidate model where variable selection is really beneficial (set 3). For the smallest model (set 1) with only the main effects of the seven potential explanatory variables, the application of boosting here leads basically to the same results as classical inference via a least squares estimator (e.g. MAPE = 5.90% vs. MAPE = 5.92%). For the model corresponding to set 2, boosting even leads to a slightly worse MAPE than the classical estimation.

Only for the larger and more complex models with set 3, including all potential interactions, do we observe a noteworthy increase in prediction accuracy. Following our analysis, the best model performance is observed for the boosted version with this set 3 (with a MAPE = 5.43% compared with MAPE = 5.87% with a classical fit). This is in line with the literature on birthweight prediction, where many of the standard formulas contain interactions between the biometric parameters (Anderson et al., 2007). The variable selection by boosting led to an average of 19.4 included variables out of the 126 candidate variables.

Noteworthy also is the performance regarding the nonlinear model with set 4. Here, the boosting approach also showed basically the same accuracy regarding the MAPE than classical inference. The inclusion of nonlinear effects only slightly increased the prediction performance (compared to set 1), but is less beneficial than the inclusion of interactions.

Table 7.2 *Results of the ultrasound application, comparing the performance of statistical boosting for median regression (L_1 loss) with mean regression (L_2 loss) based on four different sets of explanatory variables. MAPE = median absolute percentage error, MPE = mean percentage error, RE = random error (sd of percentage error).*

Model		MAPE	MPE	RE
Boosting model set 1	L_2	5.90	−0.45	9.55
	L_1	5.95	−0.15	9.5
Boosting model set 2	L_2	5.66	−0.44	9.16
	L_1	5.65	−0.34	9.14
Boosting model set 3	L_2	5.43	−0.49	9.31
	L_1	5.58	−0.42	9.43
Boosting model set 4	L_2	5.79	−0.46	9.44
	L_1	5.86	−0.38	9.60

Regarding the comparison of the prediction performance between applying statistical boosting with the L_2 loss (mean regression) and the L_1 loss (median regression), the loss functions lead to relatively similar accuracy (Table 7.2). The classical mean regression shows a slightly better performance for three of the four different models (regarding the MAPE). Regarding the MPE and RE, the lowest errors in this example were observed for the median regression models. A reason for this might be the better robustness with respect to outliers of models optimized via the L_1 loss.

7.3 Boosting for Distributional Regression

Statistical boosting algorithms have been adapted to fit distributional regression and GAMLSS. While there exist various sophisticated and stable inference approaches, allowing very flexible different modeling options (see Chapters 5 and 6), a remaining challenge in the context of GAMLSS is the selection of the most informative variables for the different components of a distributional regression model in the context of high-dimensional data with many potential predictor variables.

While the task of variable selection is already a widely discussed issue in classical regression approaches, the issue obviously becomes even more pressing in the context of GAMLSS, where one typically has to select variables for modeling not only the conditional mean, but also for modeling all other distribution parameters.

Typically, one can distinguish between several general aspects of variable selection, which are related to the aim of the analysis. If, for example, the objective is to estimate reference growth charts for children, it is obvious that age should be included for all relevant distribution parameters. In this example, the aim is hence to estimate and describe the effect of a known predictor (age) on the conditional distribution (e.g. height, weight, BMI). The selection of the variable(s) is in this case already motivated by the subject matter information on the problem.

The situation changes if the general aim is to estimate the conditional distribution of the outcome without relying on predefined covariates. This could be the case, if the aim is prediction as in the ultrasound birthweight example (Section 7.2.6 and

Chapter 8). There are various fetal biometric measurements and their interactions available to estimate infant birthweight; however, it is unclear which of these measurements and which interactions should be included. Here, a data-driven variable selection process might be advantageous. Boosting is one way to perform data-driven variable selection in the context of distributional regression, particularly suitable for a larger number of potential candidate variables. For a more classical stepwise selection procedure see Section 4.5.2.

However, besides the selection of the most informative variables, the great flexibility of GAMLSS leads also to the burden of many other modeling decisions such as the selection of the most suitable response distribution, and the specific type of effect (e.g. linear or nonlinear effect) for each included covariate. Similar to boosting for classical regression models, the multi parameter boosting algorithm to fit GAMLSS allows the selection of the most informative variables from a potentially high-dimensional ($p > n$) set of candidate variables, while simultaneously fitting the corresponding models for the different distribution parameters. Additionally, it can be adapted to carry out model choice, e.g. to choose between a simple linear or a nonlinear spline-based effect of a continuous covariate. Boosting as an inference scheme for GAMLSS could hence in general be favorable in data situations with large sets of explanatory variables, where no subject-matter variable selection is feasible or wanted. It might also be an alternative in situations where other methods do not work due to high numbers of effect estimates, or instability due to collinearity issues (Mayr and Hofner, 2018).

The application of boosting leads, however, not only to data-driven variable selection but also to shrinkage of effect estimates. For classical descriptive modeling approaches, where the aim is to estimate adequately the impact of a low to medium number of covariates on the conditional distribution (e.g. the estimation of treatment effects in clinical studies or the construction of reference growth charts) this might in fact not be desirable. Another issue is the absence of standard errors for effect estimates, which makes the classical approaches of confidence intervals for expressing statistical uncertainty and statistical tests to assess the significance of a covariate effect, infeasible. Potential work-arounds include confidence intervals based on re-sampling approaches or permutation tests (Hofner et al., 2016b; Mayr et al., 2015; Hepp et al., 2019).

7.3.1 The Algorithm for Boosting Distributional Regression

In order to fit additive predictors for all distribution parameters while simultaneously selecting the most informative covariates, the boosting concept was extended to optimize the corresponding loss function (e.g. the negative log-likelihood) with respect to multiple parameters $\theta_1, \ldots, \theta_K$ and hence over K dimensions (Mayr et al., 2012a). Since there is not only one (as in classical GAMs) but a set of K distribution parameters corresponding to K separate additive predictors η_k, we also have to specify K sets of base-learners for each of those additive predictors. These sets of base-learners typically contain one potential variable each and are often simply the same base-learners and variables for all parameters. The algorithm is left to select the best combination.

As in the classical statistical boosting approach, the base-learners are fitted to the negative gradient of the loss function, which can be seen as *pseudo-residuals*. In the case of statistical boosting for distributional regression, these pseudo-residuals now have multiple dimensions referring to the different parameters $\theta_1, \ldots, \theta_K$ of the corresponding likelihood. The algorithm hence cycles through those different parameters, and computes for every dimension $k = 1, \ldots, K$ the corresponding gradient vector

$$\boldsymbol{u}_k^{[m]} = \left(\frac{\partial}{\partial \eta_k} \ell(y_i, \eta_{i1}^{[m]}, \ldots, \eta_{iK}^{[m]}) \right)_{i=1,\ldots,n \quad k=1,\ldots,K}$$

The concept of component-wise boosting via gradient descent in function space is now extended to these multiple dimensions. Instead of fitting the base-learners one-by-one to one vector of pseudo-residuals and selecting the best-performing (as in classical statistical boosting), the procedure is carried out for all K distribution parameters. First, in each dimension the base-learners are fitted to the corresponding gradient vector $\boldsymbol{u}_k^{[m]}$. Afterwards, the potential best updates j_k^*, referring to the corresponding base-learners $h_{j_k^*}(\cdot)$ for $k = 1, \ldots, K$, are selected. Then the algorithm chooses the best parameter dimension to carry out only this overall best performing update, adding the selected base-learner solution to the additive predictor of the corresponding distribution parameter. The term *component-wise* in the case of distributional regression therefore does not refer only to the space of potential variables and base-learners, but also to the space of distribution parameters.

For illustration, we consider a model with $K = 5$ distribution parameters:

$$\frac{\partial}{\partial \eta_1} \ell\left(y, \hat{\theta}_1^{[m]}, \hat{\theta}_2^{[m]}, \hat{\theta}_3^{[m]}, \hat{\theta}_4^{[m]}, \hat{\theta}_5^{[m]} \right) \xrightarrow{\text{select}} j_1^*$$

$$\frac{\partial}{\partial \eta_2} \ell\left(y, \hat{\theta}_1^{[m]}, \hat{\theta}_2^{[m]}, \hat{\theta}_3^{[m]}, \hat{\theta}_4^{[m]}, \hat{\theta}_5^{[m]} \right) \xrightarrow{\text{select}} j_2^*$$

$$\frac{\partial}{\partial \eta_3} \ell\left(y, \hat{\theta}_1^{[m]}, \hat{\theta}_2^{[m]}, \hat{\theta}_3^{[m]}, \hat{\theta}_4^{[m]}, \hat{\theta}_5^{[m]} \right) \xrightarrow{\text{select}} j_3^* \ \textbf{best} \ \xrightarrow{\text{update}} \hat{\eta}_3^{[m+1]}$$

$$\frac{\partial}{\partial \eta_4} \ell\left(y, \hat{\theta}_1^{[m]}, \hat{\theta}_2^{[m]}, \hat{\theta}_3^{[m]}, \hat{\theta}_4^{[m]}, \hat{\theta}_5^{[m]} \right) \xrightarrow{\text{select}} j_4^*$$

$$\frac{\partial}{\partial \eta_5} \ell\left(y, \hat{\theta}_1^{[m]}, \hat{\theta}_2^{[m]}, \hat{\theta}_3^{[m]}, \hat{\theta}_4^{[m]}, \hat{\theta}_5^{[m]} \right) \xrightarrow{\text{select}} j_5^*.$$

The algorithm computes all potential updates for $\eta_1, \ldots, \eta_5$ by fitting the corresponding gradients to the base-learners. Afterwards, these five best solutions $j_1^*, \ldots, j_5^*$ are compared regarding their potential increase in the likelihood of the overall model. Then the overall best-performing update j_3^* over all potential base-learner solutions is selected, based on the induced increase in the likelihood, and $\hat{\eta}_3$ gets updated. This procedure is repeated, until the stopping iteration m_{stop} is reached.

The presented algorithm is a newer and extended variant compared to the originally proposed algorithm. The extension is called a *noncyclical* boosting algorithm (Thomas et al., 2018), because it does not carry out each potential best update cycling through the different dimensions. Instead it selects only the overall best update. This is particularly advantageous regarding the tuning of the algorithm, as it decreases the complexity by finding *one* best stopping iteration instead of K different versions as in the classical *cyclical* algorithm.

Potentially, the *noncyclical* algorithm as a special case could only carry out updates on one particular distribution parameter θ_k (e.g. the location, yielding a classical GAM) if that always leads to the highest increase in the likelihood and therefore the highest decrease in the empirical risk. In most situations, however, the algorithm will select the most influential explanatory variables for various distribution parameters.

The basic properties of the statistical boosting concept are hence carried over for distributional regression models. The algorithm selects the most informative covariates for the different distribution parameters simultaneously by choosing only the best performing base-learners at every iteration. The candidate base-learners can be the same for the different parameters $\theta_1, \ldots, \theta_K$ but could also differ. The algorithm imposes no hierarchy between the parameters (e.g. first fitting a good model for the location, then the scale) but basically treats all parameters equally. The selection of the best-performing update is solely based on the increase of the likelihood as corresponding objective functions.

A direct consequence of this strategy is that each of the prediction functions may depend on a different set of covariates at the final iteration, leading to variable selection in each predictor. Basically, any type of base-learner that can be used in classical gradient boosting (Section 7.2.1) can also be specified for the prediction functions for distributional regression.

7.3.2 Implementation in R

The framework for boosting distributional regression is implemented in the **R** package **gamboostLSS**, which builds on the fitting functions and base-learners implemented in **mboost**. For a tutorial article on **gamboostLSS** see Hofner et al. (2016a).

By relying on the **mboost** package, **gamboostLSS** incorporates a wide range of base-learners and hence offers great flexibility when it comes to the type of effects of explanatory variables on GAMLSS distribution parameters. Convenience functions to extract coefficients, plot the effects, make predictions and manipulate the model are available in **gamboostLSS**. Fitting is carried out with the functions `glmboostLSS()` (for linear models) and `gamboostLSS()` (for all kinds of structured additive models), referring to the corresponding fitting functions in **mboost**. A simple example is given next.

```
library("gamboostLSS")
m3 <- glmboostLSS(y ~ x1 + x2 + x3, data = data,
                  families = GaussianLSS())
```

The model m3 uses linear base-learners with candidate variables x1, x2, and x3 for both additive predictors η_μ and η_σ of a distributional Gaussian regression model.

If one wants to specify different candidate models for the distribution parameters, one can use a named list.

```
m3 <- glmboostLSS(list(mu = y ~ x1 + x2, sigma = y ~ x3),
                  data = data, families = GaussianLSS())
```

A limited number of GAMLSS distributions are implemented in the **gamboostLSS** package (see table 2 in Hofner et al., 2016a). Additionally, there also exists a wrapper function to incorporate distributions from the classical **gamlss** package into the boosting framework. For example, the same model could be also fitted by specifying families = as.families("NO"), where NO refers to the the normal distribution as implemented in **gamlss.dist**.

```
m4 <- glmboostLSS(list(mu = y ~ x1 + x2, sigma = y ~ x3),
                  data = data, families = as.families("NO"))
```

As for the single-parameter boosting case, there is also the possibility of defining formulas with different types of base-learners using gamboostLSS() instead of glmboostLSS(). The same base-learners as in **mboost** are available, for example bols() for ordinary linear models and bbs() for penalized B-splines.

```
m5 <- gamboostLSS(y ~ bols(x1) + bbs(x2) + bbs(x3),
                  data = data, families = GaussianLSS(),
                  method = "noncyclic")
```

There are different types of fitting schemes available. The previously described non-cyclical method, that automatically selects the most suitable distribution parameter to do an update, can be specified via method = "noncyclic". The default is a cyclical variant that cycles through all distribution parameters. The latter variant is sometimes more stable, but relies on a grid search to find the best combination of stopping iterations, which can be rather time-consuming.

7.3.3 Model Tuning

As for any statistical boosting algorithm, the main tuning parameter when applying boosting for distributional regression is the stopping iteration m_{stop}. It controls the variable selection properties as well as the resulting shrinkage of effect estimates.

Higher values of m_{stop} lead to larger models with many variables or base-learners selected for the different distribution parameters. From a statistical learning and prediction modeling point of view, larger values of m_{stop} increase the risk of overfitting (see Figure 7.3). While the model might fit well on the underlying sample, it could perform poorly on new observations. This is avoided by tuning the algorithm, that is by selecting smaller values of m_{stop} by resampling or cross-validation schemes. The model is fitted several times on slightly altered datasets and evaluated on observations that were left out of the fit (*out-of-bag* observations). One selects the m_{stop} that leads to the smallest risk, for example the highest likelihood, on these test observations.

This *early* stopping is one of the key concepts of boosting for statistical modeling, and is even more important in case of distributional regression, due to the complexity of the model class. Stopping the algorithm before convergence to the maximum likelihood estimate on the training data reduces the risk of overfitting and leads to a better generalization and hence better performance on test data (e.g. prediction accuracy). Stopping the algorithm too early (selecting a very small m_{stop}), however, increases the risk of underfitting; strong influential explanatory variables might not be selected for any update.

In the **gamboostLSS** package, the tuning of the stopping iteration m_{stop} is again implemented via the `cvrisk()` function. By default, it applies a bootstrap approach, using $B = 25$ bootstrap samples, to find the optimal values for m_{stop} by evaluating the predictive risk (the likelihood on the out-of-bag observations).

```
cvr <- cvrisk(m5)
mstop(m5) <- mstop(cvr)
```

In this case, the bootstrap (sampling with replacement) is carried out and model **m5** is fitted on each of these 25 bootstrapped samples. The call `mstop(cvrisk)` then evaluates the model on the 25 samples that had been left out and provides the m_{stop} that led to the highest average likelihood on these observations. The model **m5** is then set to the model corresponding to this particular value of m_{stop}.

Alternatives for the simple tuning of the algorithm could be the application of stability selection in combination with boosting (see Section 7.3.4) to determine a stable set of explanatory variables for each distribution parameter. This stable subset of variables could then also be fitted via classical maximum likelihood or Bayesian approaches. Also the probing procedure (Section 7.2.5) could be adapted for distributional regression, stopping the algorithm when the first probe is selected for any distribution parameter, or also for the $\theta_1, \dots, \theta_K$ separately.

7.3.4 Post Selection Inference and Stability Selection

One limitation of the boosting approach is the lack of estimators for standard errors of effect estimates. Owing to the iterative and adaptive design of the algorithm, in which the base-learners are fitted one-by-one and the best performing is selected at each iteration, there exists no closed-form solution to quantify the variability of final coefficients. As a result, there are no easy solutions for the construction of confidence intervals or significance tests corresponding to the effects of single covariates. This makes the use of boosting algorithms in the context of confirmatory data analyses (e.g. estimating treatment effects in clinical trials) problematic.

There exist some computational work-arounds based on resampling for this issue: Confidence intervals can be constructed making use of bootstrapping techniques (Hofner et al., 2016b) in order to quantify the statistical uncertainty. Statistical tests can be performed via permutation testing, that is by permuting the variables of interest and comparing the resulting effect estimates with the original one (Mayr et al., 2015; Hepp et al., 2019). While these approaches may be helpful in some situations, they clearly become computationally expensive in high-dimensional data settings, particularly when combined with tuning of the algorithm for each resampling step.

A naïve approach could be to apply boosting in order to select the best-performing variables, and then refit the final model using penalized maximum likelihood or Bayesian inference techniques, in order to report the resulting confidence intervals or p-values of statistical tests. The problem with this seemingly reasonable combination of different techniques to combine the best of these inference schemes is that the resulting estimates are severely biased: They do not take the actual selection process and the corresponding uncertainty into account (Rügamer and Greven, 2020). This procedure should hence be avoided in practice, at least for confirmatory analyses.

The stability selection approach (Meinshausen and Bühlmann, 2010; Shah and Samworth, 2013) follows a different mindset for this issue. Instead of trying to assess the statistical uncertainty or the significance of a single effect, the approach defines an upper bound for the per family error rate (PFER) which relates to the number of false positives. In the context of variable selection, *false positives* are variables which are not truly associated with the outcome variable, and are selected to be included in the statistical model. By controlling the expected number of false positives, one can hence provide statements about the statistical uncertainty of the selection process itself for the complete model. An upper bound for the PFER of one means that, on average, one of the selected variables might actually be unassociated with the outcome of interest.

This control of the PFER is achieved by including only particularly *stable* variables in the final model. Stability is measured via repeatedly splitting the dataset (subsampling) and carrying out the variable selection strategy on each of these subsamples. Only variables that were selected in many of these subsamples, passing a particular selection threshold, qualify as stable variables. In the case of boosting distributional regression, this means that we perform boosting with variable selection on each of the subsamples for all distribution parameters simultaneously and only include those variables which have proved to be stable, in additive predictors η_k (Thomas et al., 2018). For an illustrative example using stability selection, see the gene expression application in Chapter 13.

The relationship between invariance (stability), predictive robustness, and statements about causality is discussed in recent work of Bühlmann (2020).

Example: Birthweight Prediction Based on Ultrasound Data

To illustrate the abilities for predictive modeling and simultaneous variable selection by boosting, we revisit the ultrasound birthweight example of Section 7.2.6, also discussed in Chapter 8.

In contrast to Section 7.2.6 where we applied statistical boosting to select and estimate the predictors for different linear or additive models regarding the expected birthweight, we now fit the complete conditional distribution. We focus on the JSU distribution as in the in-depth analysis of this data in Chapter 8. The JSU distribution contains four parameters, location (μ), scale (σ), skewness (ν), and kurtosis (τ). (For details see Rigby et al., 2019, chapter 18.)

We apply statistical boosting to select the best-performing variables for the four different additive predictors. For each predictor, we consider the same four sets of base-learners (similar to Section 7.2.6) but now for all four parameters of the JSU distribution.

- **Set 1**: JSU model with linear base-learners for the seven potential explanatory variables on μ, σ, ν, and τ.

- **Set 2**: Linear JSU model containing base-learners for the seven potential explanatory variables from set 1, but additionally all pairwise interactions for each distribution parameter.

- **Set 3**: Linear JSU model containing base-learners for the seven potential explanatory variables from set 1, together with all interaction for each distribution parameter.

- **Set 4**: Additive JSU model containing smooth nonlinear base-learners for the seven explanatory variables of set 1.

Results of 10-Fold CV

To evaluate the prediction accuracy of the resulting models, we again follow a 10-fold cross-validation scheme to separate model training and evaluation. On the training folds, a model is fitted and tuned using $B = 25$ bootstrap samples. Therefore the number of selected variables (and the models) will vary over the 10 cross-validation folds. This way, the model validation also takes the variable selection into account. On the test-folds we always compute the median absolute percentage error (MAPE), the mean (signed) percentage error (MPE), and the random error (RE, standard deviation of the signed percentage error) of the point predictions for the expected value of the birthweight. To take the complete conditional distribution into account, we also computed the log-score (mean log-likelihood on the test samples, see Section 4.4.2). Additionally, we evaluate how many variables were included for the different additive predictors for the corresponding distribution parameters. The stopping iteration m_{stop} is selected via bootstrapping with $B = 25$ bootstrap samples, optimizing the predictive risk on the out-of-bag observations.

Analyzing the results of the 10-fold cross-validation (Table 7.3), we observe in general a very similar performance of the models compared to those obtained via L_2 boosting in Table 7.1. One has to keep in mind that for prediction accuracy, the expected value of the conditional distribution is the main influencing factor. However, it is noteworthy that for all four sets of potential sets of base-learners the algorithm tends to select varying numbers of predictor variables for the four distribution parameters. Regarding the performance of boosting with the base-learners sets 1–4, we notice that the best prediction accuracy regarding the median absolute percentage error (MAPE) was achieved with a linear model with main effects and interactions (set 3).

Table 7.3 *Results based on 10-fold cross-validation for fitting the JSU distribution via statistical boosting with different types of candidate models to the ultrasound data. MAPE = median absolute percentage error, MPE = mean percentage error, RE = random error (sd of percentage error), LS = log-score. For MAPE and RE smaller values indicate a better prediction, for LS larger values are better. The MPE should be close to 0. The number of selected base-learners or terms refer to the average (mean) number of the 10 folds.*

	MAPE	MPE	RE	LS	Potential base-learners	Number of terms selected			
						μ	σ	ν	τ
Set 1	6.09	−0.65	9.71	−7.12	7	6	5.9	4.4	4.9
Set 2	5.65	−0.96	9.30	−7.09	28	19.7	15.6	10.9	10.7
Set 3	5.47	−0.96	9.42	−7.09	126	20.9	16.5	14.4	7.5
Set 4	5.9	−1.14	9.95	−7.11	7	7	5.3	5.9	4.6

The inclusion of spline base-learners to account for potential nonlinear effects (set 4) does not increase the prediction accuracy. These results support the conclusions from the previous analyses, and are also borne out in the analyses performed in the penalized maximum likelihood paradigm, in Chapter 8. The overall effect of the predictor variables seems to be rather linear, and adding nonlinear effects to the prediction model does not increase its performance. Note that the performance measures MAPE, MPE, and RE all focus on the point prediction (and hence the location parameter), and not the overall fit of the model. Regarding the log-score (LS), which takes all parameters of the distribution into account, set 3 shows the best performance but is only slightly better than set 2. Also, for this measure including nonlinear base-learners does not increase the accuracy of the model.

7.3.5 Diagnostics and Selecting the Most Suitable Distribution

The application of boosting for model estimation can be helpful in the selection of the most informative variables. However, other modeling choices in the context of distributional regression become more challenging.

Tools exist for model diagnostics and the selection of the most suitable outcome distribution, for classical inference approaches based on penalized regression and Bayesian inference. In the case of boosting, however, the structure and properties of the algorithm have to be taken into account. Boosting algorithms are typically not run until convergence, and the resulting model incorporates shrunken effect estimates. This is a fundamental difference with classical model fitting, as we intentionally do not fit the best model for the underlying training sample. As a result, diagnostic tools that aim at quantifying the goodness-of-fit (e.g. by graphically analyzing the residuals) might not be directly meaningful in the context of boosting. A remaining structure in the residuals or a non optimal worm plot (see Section 4.7.2) might not in fact be a sign of a badly fitted model, but could be just the result of early stopping and the stronger focus on optimizing the prediction accuracy rather than finding the most suitable model for the corresponding data.

In the context of boosting, it is more common to investigate the stability of the fitting process and whether the model reflects a reasonable trade-off between underfitting and overfitting (see Figure 7.3). A valid way to achieve this graphically is to examine the trajectory of the empirical risk on the training data (which should be decreasing until convergence for larger number of iterations – as in the right part of Figure 7.5), and the corresponding results on test data (e.g. the results from bootstrapping and cross-validation). On test data, the minimum of the risk should not be at the end of the search grid of m_{stop}. In the case of the best model being that with the maximum m_{stop}, it is good practice to further increase the search grid until a true minimum is observed.

Another challenge is the selection of the most suitable outcome distribution. Common approaches focus on information criteria, such as the AIC. In the context of boosting, however, the estimation of the effective degrees of freedom are problematic due to the stepwise updates and the search for the best-performing base-learner. There are estimates for the degrees of freedom; however, results from the literature suggest that these might be heavily biased (Mayr et al., 2012b). As these degrees of freedom influence commonly used information criteria via the penalty term, the use of the information criteria is more difficult in the context of boosting.

An alternative approach is to compare other graphical model diagnostics for different distributions in order to select that for which the model fits best. As this might also be misleading for boosting with early stopping (as discussed earlier), it could make sense to firstly select a suitable distribution based on a very simple model via the functionalities of the classical penalized likelihood approach (e.g. via `chooseDist()`), and then fit a boosting model with the full set of potential candidate variables for the resulting distribution. This approach is used in the following example.

Example: High-Dimensional Cancer Cell Data

To illustrate variable selection in the context of distributional regression for high-dimensional data, we investigate the cancer cell panel of the National Cancer Institute (NCI). The so-called NCI-60 data contain gene expression profiles of 60 tumor cell lines and can be downloaded from `http://discover.nci.nih.gov/cellminer/`. These 60 human tumor cell lines represent different tumor entities: They were derived from patients with leukemia (blood cancer), melanoma (skin cancer), and lung, colon, central nervous system, ovarian, renal, breast, and prostate cancers. These data have been widely used to investigate cellular mechanisms in oncology and also in-vitro drug action (Gholami et al., 2013). In our case, the aim is to model the protein expression (outcome) measured via antibodies, with gene expression data as the predictors (Affymetrix HG-U133A-B chip). This dataset has previously been used to illustrate robust regression methods, because of particular characteristics of the underlying data with a highly skewed distribution (Speller et al., 2022). Although both protein and gene expression values were previously normalized and log-transformed, one can clearly observe that the distribution of this type of genomic or proteomic data can be heavily skewed or include heavy tails. As in Sun et al. (2020), Figure 7.7 displays the kurtosis for the protein expression levels (left) and gene expression levels

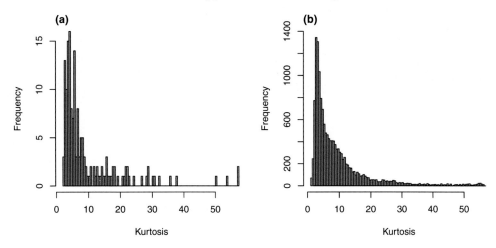

Figure 7.7 Histograms displaying (a) the kurtosis in protein expression values (the protein expression values of KRT19 will later serve as our outcome) based on antibodies and (b) the kurtosis in gene expression values (which will serve as predictor variable explaining the differences in the KRT19 protein of tumor cells) in the National Cancer Institute 60 tumor cell lines.

(right) of the NCI-60 data. While in general there are multiple protein expression values available in this dataset, we follow Sun et al. (2020) and Speller et al. (2022) and also focus on the keratin 19 antibody array (KRT19) due to the particularly skewed distribution (see Figure 7.8).

One tumor cell line had to be removed owing to missing observations. The remaining $n = 59$ cell lines from the different tumor entities serve as observations; the KRT19 protein expression is the response variable while $p = 14,951$ gene expression values are the potential explanatory variables. The aim of the analysis is to identify those genes whose expressions are most associated with KRT19 antibody. We will carry out a leave-one-out cross-validation (LOO-CV, see Section 4.4.3) to analyze the variable selection properties of the algorithm.

Note that for this small dataset it is rather uncommon to apply distributional regression. Generally, these complex models require larger sample sizes. However, the aim here is to illustrate that, via boosting, in this extremely high-dimensional situation with only $n = 59$ observations (cancer cell lines) and $p = 14,951$ potential predictors, a GAMLSS model can be selected and estimated.

To choose a suitable distribution for modeling the KRT19 antibody protein values (see Figure 7.8), we first applied the `chooseDist()` function of the **gamlss** package, using a univariate model with the KRT19 gene expression data as the single covariate. The AIC was used for selection. We searched through all distributions with a positive domain, which led to the generalized Gamma (`GG`) distribution with the smallest AIC, followed by the generalized Beta type 2 (`GB2`) and the generalized in-

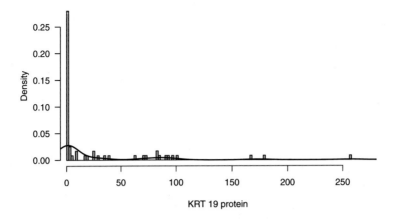

Figure 7.8 Histogram and density estimation displaying the heavily skewed outcome distribution of the KRT19 protein expression values based on antibodies in the National Cancer Institute 60 tumor cell lines.

verse Gaussian (`GIG`) distributions. We now use the generalized Gamma distribution in combination with statistical boosting to select the most informative gene expression values. The generalized Gamma is a three-parameter distribution, hence the boosting algorithm will be selecting predictors for the distribution parameters μ, σ, and ν. (Details of the distribution are given by Rigby et al. (2019, chapter 19).)

We apply the boosting algorithm with negative log-likelihood of the generalized Gamma distribution (`as.families(GG)`) in combination with linear base-learners for all available gene expression values via the `glmboostLSS()` function of the **R** package **gamboostLSS**. To determine the optimal stopping iteration m_{stop}, we use a subsampling approach to optimize the predictive risk on the observations not entering the corresponding subsamples.

To further illustrate variable selection, we carry out this procedure combined with a leave-one-out cross-validation scheme. Boosting algorithms have the tendency to yield rather sparse models in cases of $p \gg n$. This is observed in this illustration example: On average, for the μ component only 2–3 variables were selected (median 3, mean 2.45). More gene expression values were included for the σ component of the GAMLSS, in which on average 12 genes (median) were selected to be informative. For the ν component, most often only an intercept is included, and in only 2 of the 59 models was an additional gene selected for ν. This data example hence nicely illustrates how boosting can help to apply distributional regression with small samples and high numbers of potential predictors, selecting sparse models containing only the most informative variables in a purely data-driven way. In Chapter 13 we present another, more detailed, analysis of a high-dimensional gene expression dataset using statistical boosting.

Part III

Applications and Case Studies

8

Fetal Ultrasound

8.1 Data and Research Question

The use of ultrasound during pregnancy for the purpose of identification of fetal abnormalities and prediction of birthweight is a feature of standard obstetric care. In the last stage of pregnancy, focus is on predicting the birthweight, which has a high clinical relevance as both very low as well as very high birthweight are associated with a greater risk of complications during labor and the first days after birth. We introduce the `ultrasound` dataset, which comprises data from 1,038 births at the Royal Hospital for Women, Sydney, Australia, between 2008 and 2013 (Leader. L.R. personal communication; Yang (2014)). Each fetus was scanned twice, the first a median 60 days before delivery, and the second a median 24 days before delivery. As the purpose of this analysis is the prediction of birthweight, we base our analysis on the second scans.[1] A histogram of birthweight, with nonparametric density estimate, is shown in Figure 8.1. The marginal distribution looks somewhat left-skewed, although not too far from the normal. We investigate the use of GAMLSS for the prediction of birthweight, and also use this relatively straightforward dataset to demonstrate how GAMLSS models are fitted and interpreted.

8.2 Model Building

A variety of formulae for the computation of predicted birthweight (BW) are in use in the clinical literature, most notably the Spinnato (Spinnato et al., 1988) formulae and variants of them; these mostly comprise the fetal biometric measurements: abdominal circumference (AC), biparietal diameter (BPD), head circumference (HC) and femur length (FL), as well as maternal factors such as age, BMI and parity (number of previous births). These formulae were developed on the basis of classical linear modeling. They use log of birthweight as the response variable and fetal biometric measurements, with some interactions, as predictors. Notably predictions need to adjust for the time delay till birth. Naturally this is not known at the time of prediction; however, delivery at, for example, term (40 weeks) could be assumed when predicting, and the actual time delay used when evaluating predictions. For

---

[1] This dataset is also analyzed using statistical boosting (Chapter 7).

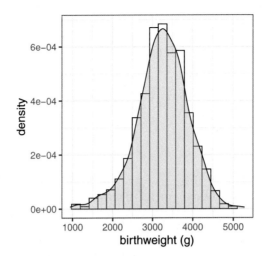

Figure 8.1 Histogram of birthweights, with nonparametric density estimate.

example the Spinnato equation is (Spinnato et al., 1988):

$$\log_{10} \text{BW} = 1.0009(1.326 + 0.0107\,\text{HC} + 0.0438\,\text{AC} + 0.158\,\text{FL}$$
$$- 0.00326\,\text{AC} \times \text{FL}) + 0.0043\,\text{DBD}, \qquad (8.1)$$

where DBD (days before delivery) is the time delay in days till birth.

We model the response variable birthweight using four fetal ultrasound measurements: abdominal circumference (AC), biparietal diameter (BPD), head circumference (HC) and femur length (FL), as well as three maternal variables: parity, age of the mother, and days before delivery DBD. In this chapter in order to demonstrate the effect of a factor on distribution parameters of a GAMLSS model, we treat parity as a factor with four levels: 0, 1, 2 and 3 or greater. (The original parity variable has seven distinct values (0, 1, ..., 6) and could also be treated as a numeric term, as is done in Section 7.2.6.)

Scatterplots of the explanatory terms against the response together with smoothing fitted curves, to help identify patterns in the data, are shown in Figure 8.2. All the ultrasound measurements show a positive relationship with birthweight, apart possibly from age, where the relationship is rather flat and possibly negative. The factor parity shows little evidence of contribution to birthweight. The shapes of the smoothing curves superimposed on the scatterplots indicate that some of the relationships between explanatory terms and the response may not be linear. This evidence is mostly at the extremes of the ranges of the explanatory terms, where the observations are sparse. At this stage of the analysis we will assume linearity in the relationships between the parameters of the response distribution and the explanatory terms, but we will reconsider this assumption later. Since birthweight is a

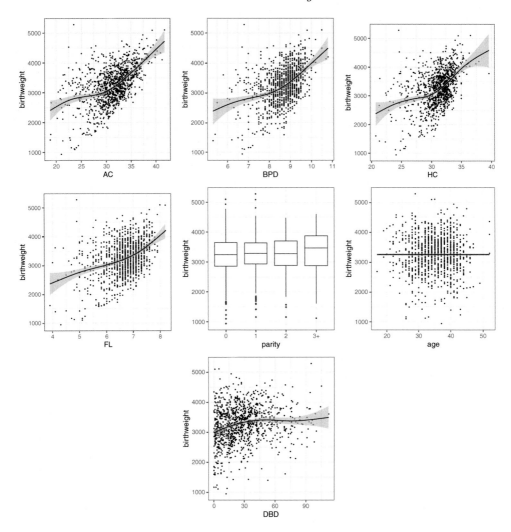

Figure 8.2 Ultrasound data: scatterplots with added smoothing curves.
The response variable `birthweight` is plotted against abdominal
circumference (`AC`), biparietal diameter (`BPD`), head circumference (`HC`),
femur length (`FL`), `parity`, maternal `age` and days before delivery (`DBD`).

continuous variable we start by assuming that birthweights are normally distributed;
we will challenge this assumption later. Our task here is to highlight the difference
between GAMLSS modeling, where all the parameters of the distribution could potentially be affected by the explanatory terms, and the standard linear regression
model, familiar to most of our readers, where only shifts in the mean are modeled.

Note that, at a preliminary regression analysis stage, it is always worth checking
the pairwise relationships between the continuous explanatory terms, since highly
correlated continuous terms in a model could possibly create troubles in either fitting

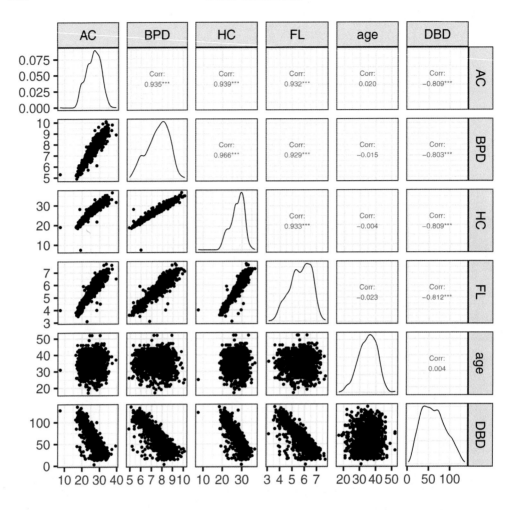

Figure 8.3 Pairwise inspection of the continuous explanatory terms used in modeling birthweights. The lower triangular cells show scatterplots of the terms while the upper triangular cells show their correlation coefficients. The diagonal plots are density estimates.

or interpreting the model. The reason for this is that they make the model unstable through multicollinearity or concurvity (for smoothers). Figure 8.3 shows the pairwise relationships between all of the continuous terms available for modeling. It is obvious that only `age` seems to be uncorrelated with the rest of the terms. The terms `AC`, `BPD`, `HC` and `FL` are highly positively correlated, while `DBD` is negatively correlated with the rest. This is worth bearing in mind when we are analyzing and interpreting the fitted models. To start off we use three candidate models. The first is a linear main effects model for both μ and σ, which we refer to as **linear model 1** or LM1.

LM1 : $\texttt{birthweight} \sim \mathcal{N}(\mu, \sigma^2)$

$$\mu = \beta_0^\mu + \beta_1^\mu \mathtt{AC} + \beta_2^\mu \mathtt{BPD} + \beta_3^\mu \mathtt{HC} + \beta_4^\mu \mathtt{FL} + \beta_5^\mu z_1 + \beta_6^\mu z_2 + \beta_7^\mu z_3$$
$$+ \beta_8^\mu \mathtt{age} + \beta_9^\mu \mathtt{DBD}$$
$$\log \sigma = \beta_0^\sigma + \beta_1^\sigma \mathtt{AC} + \beta_2^\sigma \mathtt{BPD} + \beta_3^\sigma \mathtt{HC} + \beta_4^\sigma \mathtt{FL} + \beta_5^\sigma z_1 + \beta_6^\sigma z_2 + \beta_7^\sigma z_3$$
$$+ \beta_8^\sigma \mathtt{age} + \beta_9^\sigma \mathtt{DBD}$$

z_1, z_2, and z_3 are dummy variables for $\texttt{parity}$ levels 1, 2 and ≥ 3, respectively. Note that this model uses 20 degrees of freedom, 10 for μ and 10 for σ.

The second candidate model, referred to as **linear model 2** (LM2), contains all linear main effects and all first-order interactions. Note that with r main effects, there are in general $r(r-1)/2$ first-order interaction terms.[2] LM2 has nine main effect terms (plus a constant), as we count each of the $\texttt{parity}$ dummy variables as a main effect; however, we do not consider interactions between the $\texttt{parity}$ dummy variables. This means each parameter in model LM2 has $36 - 3 = 33$ first-order interaction terms plus 9 main effects, plus a constant, that is, 43 in total. It is easier to define LM2 using a software-based formula rather than its mathematical form, as follows.

LM2: $\texttt{birthweight} \sim \mathcal{N}(\mu, \sigma^2)$

$$\mu \sim (\mathtt{AC} + \mathtt{BPD} + \mathtt{HC} + \mathtt{FL} + \texttt{parity} + \texttt{age} + \mathtt{DBD})^2$$
$$\log \sigma \sim (\mathtt{AC} + \mathtt{BPD} + \mathtt{HC} + \mathtt{FL} + \texttt{parity} + \texttt{age} + \mathtt{DBD})^2 .$$

The degrees of freedom for LM2 are 86.

The third model of interest at this stage is the *null* model, which does not involve any explanatory variables. In the classic regression analysis the null model fits a constant to the mean, namely $\mu \sim 1$. Equivalently in GAMLSS, the null (normal) model fits constants to all distribution parameters, namely μ and $\log \sigma$, with two degrees of freedom. We refer to the null model in GAMLSS as the *marginal* distribution model because we are not conditioning on any of the explanatory variables.

The global deviance and the AIC and BIC criteria from the three candidate models are shown in Table 8.1. Note that from the AIC we can conclude that LM2 is the "best" model, having the minimum AIC value, while according to the BIC LM1 seems "best". Contradictions between different information-based criteria such as this are common in statistical modeling. It has been argued that the AIC is best for prediction while the BIC has a higher probability of leading to the "correct" model, if the "correct" model is included in the candidate models. Given that "all models are wrong", the latter assessment is not something we should take very seriously. Nevertheless, BIC is best for simplicity of interpretation, since it generally leads to simpler models than the AIC because of its higher penalty for model complexity. Note that at this stage of the analysis we are dealing with only three candidate models; in practice we have to deal with far more.

---

[2] The general formula is that there are $\binom{r}{i+1}$ ith order interactions.

Table 8.1 *The deviance, AIC, and BIC from the three candidate models: the null, linear main effects (LM1), linear with first-order interactions (LM2).*

	df	Deviance	AIC	BIC
null	2	16,339	16,343	16,353
LM1	20	14,696	15,736	14,835
LM2	86	14,545	14,717	15,143

Table 8.2 *Coefficient estimates and standard errors from the linear model 1 (LM1) for birthweight. P-value < 0.0001 is denoted as "***", < 0.001, as "**", < 0.01 as "*" and < 0.05 as ".".*

Parameter		μ			σ	
Link		identity			log	
	$\hat{\beta}^{\mu}$	$SE(\hat{\beta}^{\mu})$		$\hat{\beta}^{\sigma}$	$SE(\hat{\beta}^{\sigma})$	
(intercept)	−4624.3	(170.5)	***	5.71	(0.47)	***
AC	152.8	(6.2)	***	0.037	(0.017)	*
BPD	97.2	(33.6)	**	0.27	(0.098)	**
HC	24.5	(10.4)	*	−0.092	(0.035)	**
FL	120.7	(34.2)	***	−0.105	(0.090)	
parity1	59.5	(20.59)	**	0.021	(0.051)	
parity2	61.3	(30.6)	*	0.40	(0.076)	
parity3	141.5	(39.5)	***	−0.085	(0.107)	
age	−3.7	(1.6)	*	−0.004	(0.004)	
DBD	28.5	(0.59)	***	−0.006	(0.001)	***

In Section 8.3 we use model LM1 to illustrate the application of hypothesis testing within a parametric GAMLSS model.

8.3 Hypothesis Testing

Testing a hypothesis about regression coefficients is usually appropriate for the parametric part of a GAMLSS model. LM1 is fully parametric, so all of its parameters can be tested using the asymptotic normality property of the MLE (equation (5.19)). Table 8.2 shows the fitted coefficients and their standard errors for LM1. For the μ model, all of the coefficients for the linear continuous terms AC, BPD, HC, FL, age and DBD and the dummy variables for parity, are either significant or highly significant (given the rest of the variables are in the model). For the σ model, the coefficients of DBD BPD and HC are highly significant, while the AC coefficient is significant.

A reliable way of checking the significance of any linear (or smooth) term is to include and exclude the term from the model, and check whether the difference in deviance is significant by performing a χ^2 test. The degrees of freedom for the test is the difference in degrees of freedom between the two models. For factors and smooth terms this is the recommended method. The function drop1() in the **gamlss** package produces the ANOVA-type Table 8.3, which shows the χ^2 test for dropping each term in turn in the σ model. Notice however, that for larger datasets and for a large number of terms in the model, the computation time to produce this table could be

Table 8.3 *The ANOVA-type table for the σ model from the linear model 1 (LM1) for birthweight. Df = degrees of freedom, AIC = Akaike information criterion, LRT = likelihood ratio test statistic, and Pr(Chi)* = $\mathbb{P}(\chi^2_{Df} > LRT)$.

	Df	AIC	LRT	Pr(Chi)
<none>.		14,736		
AC	1	14,739	4.76	0.02
BPD	1	14,742	7.53	0.00
HC	1	14,741	6.86	0.00
FL	1	14,736	1.36	0.24
parity	3	14,732	1.18	0.75
age	1	14,736	1.40	0.23
DBD	1	14,752	17.92	0.00

considerable. Note also that the selection of terms for models using multiple testing is in general discouraged because of the danger of using inappropriate p-values for the tests. (This is discussed in the context of stepwise selection in Section 4.5.2.)

We now investigate alternative methods for hypothesis testing and confidence interval generation, at least for continuous linear terms, and refer to Section 5.3.1. Without loss of generality, we concentrate on the AC term for the σ model. We have computed 95% intervals using alternative methods; the results are summarized and displayed in Figure 8.4. A 95% confidence interval for β_1^σ using the standard error provided by Table 8.2 is $0.037 \pm 1.96 \times 0.017 = (0.0037, 0.0703)$. This interval does not include zero and therefore we would reject the null hypothesis $H_0 : \beta_1^\sigma = 0$ and conclude that the linear term for AC contributes to the explanation of the standard deviation (σ) model of birthweight. A robust interval obtained using the asymptotic normality of MLEs used by equation (5.16) is given by $(-0.0072, 0.0813)$; the 95% confidence interval obtained from nonparametric bootstrap sampling, as described in Section 5.3.3 (using $B = 1000$ bootstrap samples), is $(-0.0076, 0.0875)$. The latter two intervals both include zero, and therefore do not reject $\beta_1^\sigma = 0$. Figure 8.5(a) shows the density estimate of the $\hat{\beta}_1^\sigma$'s obtained from nonparametric bootstrap samples. The Bayesian bootstrap procedure, as described also in Section 5.3.3, produces the 95% confidence interval $(-0.0052, 0.0786)$, which again includes zero. Section 5.3.2 discusses profile likelihood confidence intervals: Figure 8.5(b) shows the profile global deviance (equation (5.20)) of β_1^σ, which results in a 95% profile likelihood CI of $(0.0037, 0.0701)$. A 95% credible interval from a fully Bayesian analysis, obtained using the function `bamlss()`, is $(0.0002, 0.0665)$.

This discussion has highlighted the problems associated with a rigid approach to hypothesis testing: Some of the 95% confidence intervals include zero, while others do not. Different approaches could lead to different conclusions; Heller et al. (2019) demonstrate that, with a Poisson–inverse Gaussian response distribution, the various likelihood-based methods yield confidence intervals varying in some cases fairly far from nominal coverage, and with profile likelihood yielding the most accurate confidence intervals. There are mainly two cautionary tales we can draw from this. The first is that because AC is highly correlated with HC, BPD, FL, and DBD, its co-

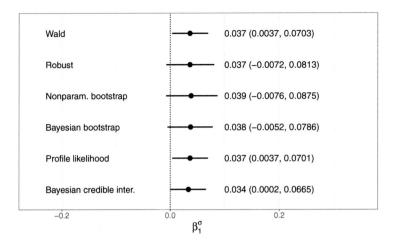

Figure 8.4 Estimates and 95% confidence intervals (credible interval in the Bayesian case) for the coefficient β_1^σ in the linear model 1 (LM1) for birthweight, using different methods.

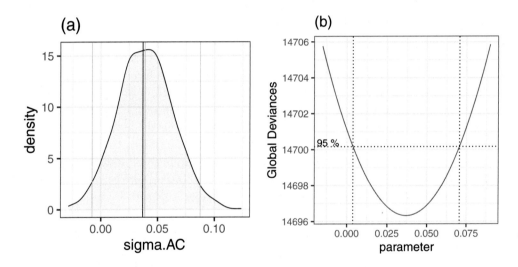

Figure 8.5 Plots for the coefficient of AC from the σ model (β_1^σ): (a) density estimate of the $\hat{\beta}_1^\sigma$'s obtained from $B = 1000$ nonparametric bootstrap samples, and (b) the profile deviance.

efficients are very sensitive to the values of the rest of the terms. They are rather unstable. The second is that all of the above approaches assume that the model used for obtaining the confidence intervals is correct or at least approximately correct. We use the parametric GAMLSS model LM1 here merely as a demonstration of how confidence intervals may be obtained for any coefficient of the model. We still have some distance to go to achieve an "adequate" model for the data, from which reliable confidence intervals may be constructed.

Table 8.4 *Log-likelihoods and degrees of freedom, models LM1 and LM1a for ultrasound data.*

Model	log-likelihood	df
LM1	−7348.16	20
LM1a	−7357.13	17

Checking the contribution of a factor to a distribution parameter differs slightly from checking that of a continuous covariate. For a factor having L levels, there are $L - 1$ dummy variable terms in the model. (One of levels of the factor, the reference category, is dropped to ensure identifiability.) We wish to test the $L - 1$ coefficients simultaneously. In model LM1 for both the μ and σ models, parity $=$ 0 is the reference category and the three dummy variables are associated with the remaining levels of parity. For example, in the μ model in Table 8.2, parity1, parity2, and parity3, which correspond to levels 2, 3, and 4 of the parity factor respectively, are significantly different from the reference category. Because all levels of parity are significant we would expect that overall any test performed on the factor parity would show a contribution to the explanation of the μ model. A likelihood ratio test can be used to check whether the three parity dummy variables are jointly significant for the μ model. This is performed using two fitted models: one with parity included in the μ model (LM1), the other with parity excluded from the μ model (LM1a). (Note that the σ model contains parity at this stage.) The elements needed for the computation are shown in Table 8.4. We have

$$\text{LRT} = -2[\ell(\text{LM1a}) - \ell(\text{LM1})] = -2(-7357.13 + 7348.16) = 17.94$$

on 3 degrees of freedom and p-value < 0.001, so parity is highly significant for the μ model. Notice however, that parity does not contribute significantly to the explanation of the parameter σ.

As a final note on hypothesis testing, we emphasize that linear coefficients affect the distribution parameters differently, depending on the link function used. While $\beta = 0$ results in no contribution of the corresponding term to the regression of the parameter distribution, the actual effect differs. For example, for the identity link, the effect is additive ($x\beta = 0$), while for the log link the effect is multiplicative ($\exp(x\beta) = 1$). In both cases the result is no effect. Link functions are discussed in more detail in Section 1.5.3.

8.4 Variable Selection

As we have discussed in Section 4.5, GAMLSS provides a very flexible and general framework, and therefore it is almost impossible to find a unified procedure for selecting an appropriate model for the data. In GAMLSS the model selection is an iterative process, requiring the selection of both the response distribution and the predictors for the distribution parameters. In the ultrasound dataset we start the model selection by assuming initially that the response distribution is normal. We then use a stepwise procedure and shrinking procedures for the selection of appropriate terms. The selection of an appropriate response distribution is considered in

Section 8.5, conditional on the variables selected under the normality assumption; the variables selected are then revisited and revised.

Stepwise Procedure

Any stepwise procedure needs a *criterion* to minimize (typically AIC is used), and a *scope* on how to look for the different available terms. Common practice is for the scope to consist of three different (GAMLSS) models: (a) the *lower* or minimal permitted model, (b) the *current* model, and (c) the *upper* or maximal permitted model. One starts with the current model, which in terms of complexity lies between the lower and the upper models and goes in either direction. For example, the `direction` for the search can be *forward, backward,* or *stepwise*. In the last case, at each step, the algorithm checks whether to drop or add terms. Stasinopoulos et al. (2017, chapter 11) gives a comprehensive review of the stepwise procedures available in the **gamlss** package in **R**. For the ultrasound data, the function `stepGAICAll.A()` was used to select terms for all the distribution parameters, namely μ and σ of the normal distribution. The lower model in the search was the null model, the upper model the first-order linear interaction model `LM2`, and the current model the main effect linear model `LM1`. The search starts with μ, then goes to σ and returns to μ. The direction of the search for both μ and σ is set to `"both"`. When the selection returns to μ the direction changes to `"backward"`. The procedure was performed twice: first with $\kappa = 2$ (AIC) as the criterion and then with $\kappa = \log(n)$ (BIC). The resulting models are as follows.

$$
\begin{aligned}
\texttt{LM3.aic:} \quad & \texttt{birthweight} \sim \mathcal{N}(\mu, \sigma^2) \\
& \mu \sim \texttt{AC} + \texttt{BPD} + \texttt{HC} + \texttt{FL} + \texttt{parity} + \texttt{age} + \texttt{DBD} \\
& \quad + \texttt{AC:FL} + \texttt{AC:age} + \texttt{FL:parity} \\
& \log \sigma \sim \texttt{AC} + \texttt{BPD} + \texttt{HC} + \texttt{FL} + \texttt{DBD} \\
& \quad + \texttt{FL:DBD} + \texttt{AC:BPD} + \texttt{BPD:FL}
\end{aligned}
$$

$$
\begin{aligned}
\texttt{LM3.bic:} \quad & \texttt{birthweight} \sim \mathcal{N}(\mu, \sigma^2) \\
& \mu \sim \texttt{AC} + \texttt{BPD} + \texttt{FL} + \texttt{DBD} + \texttt{AC:FL} \\
& \log \sigma \sim \texttt{BPD} + \texttt{DBD} + \texttt{BPD:DBD}
\end{aligned}
$$

Notice that the use of BIC inevitably leads to simpler models which are easier to interpret; however, important terms could be left out.

Shrinking procedures: the Lasso and PCR

The use of shrinking procedures for covariate selection is discussed in Section 5.4. Here for the ultrasound data we firstly use lasso and then PCR. For lasso we use the function `gnet()` of the **R** package **gamlss.lasso** (Ziel et al., 2021), using a scaled design matrix X containing all linear main effects and first-order interactions. The package provides several methods for the estimation of the lasso smoothing parameter λ. This includes using the BIC in a simple or adaptive lasso; bagging and cross-validation. Here we report the model which is fitted using adaptive lasso with BIC.

Table 8.5 *AIC, χ^2 (GAIC($\kappa = 3.841$)) and BIC for models LM1, LM2, LM3.aic, LM3.bic, LM4.ad.lasso.bic and LM5.pcr. All models assume a normally distributed response.*

	df	AIC	χ^2	BIC
Null	2	16,343	16,347	16,353
LM1	20	14,736	14,773	14,835
LM2	86	14,718	14,876	15,144
LM3.aic	24	14,676	14,720	14,794
LM3.bic	10	14,708	14,727	14,758
LM4.ad.lasso.bic	27	14,740	15,180	15,214
LM5.pcr	20	14,715	14,752	14,814

The resulting model `LM4.ad.lasso.bic`, shown below, has a rather complicated σ model, while the μ model has eight non-zero coefficients. Note that some variables interact with different levels of the factor `parity`. The algorithm we have used does not treat the dummy variables for `parity` as a group.[3]

$$\text{LM4.ad.lasso.bic:} \quad \text{birthweight} \sim \mathcal{N}(\mu, \sigma^2)$$
$$\mu \sim \text{AC} + \text{BPD} + \text{FL} + \text{parity1} + \text{parity2} + \text{age} + \text{DBD}$$
$$+ \text{BPD:parity2}$$
$$\log \sigma \sim \text{AC} + \text{BPD} + \text{HC} + \text{FL} + \text{parity} + \text{age} + \text{DBD}$$
$$+ \text{BPD:parity3} + \text{HC:parity1} + \text{HC:parity2}$$
$$+ \text{FL:parity3} + \text{FL:DBD} + \text{parity1.DBD}$$
$$+ \text{parity2.DBD} + \text{parity3.DBD}$$

The lasso procedure does not preserve hierarchy in the model selection of interaction terms. Fortunately our fitted model does not suffer from this problem. By "hierarchy" we mean that if a first-order interaction is included in the model, the main effects are also included even if they are not significant. One could force the algorithm to comply with the hierarchy principle, see for example Bien et al. (2013), but we will not pursue this further.

To fit a PCR model within GAMLSS we use the package **gamlss.foreach**, which uses the methodology described in Stasinopoulos et al. (2022). Two methods for fitting a PCR within GAMLSS are available: (i) the singular value decomposition (SVD) of the design matrix is performed once, at the beginning of the GAMLSS algorithm; and (ii) the SVD is performed at each IWLS iteration. Here we report the results from the second case. This model is as follows.

$$\text{LM5.pcr:} \quad \text{birthweight} \sim \mathcal{N}(\mu, \sigma^2)$$
$$\mu \sim \boldsymbol{Z}[17]$$
$$\log \sigma \sim \boldsymbol{Z}[1].$$

---

[3] In a "group effect lasso" the group of dummy variables can enter the model as a group, or be eliminated as a group.

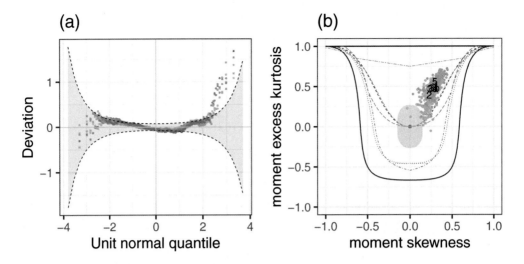

Figure 8.6 (a) Worm plot and (b) bucket plot of normalized quantile residuals from models: LM1, LM2, LM3.aic, LM3.bic, LM4.ad.lasso.bic and LM5.pcr, represented by different colours. In the bucket plot the models are presented as points 1, 2, 3a, 3b, 4 and 5, respectively; unfortunately their points are too close to each other to be distinguished. None of the models provides an adequate fit.

Here, the notation $\boldsymbol{Z}[p]$ indicates that p principal components were used in the fit (plus a constant). For example, model LM5.pcr used a total of 20 degrees of freedom: one for the constant and 17 principal components for μ; and one for the constant and one principal component for σ.

Table 8.5 shows the AIC and BIC for all models fitted thus far. The fourth column (χ^2) uses a penalty $\kappa = 3.841$, as discussed in Section 4.4.1. From Table 8.5 the model LM3.aic performs best according to AIC (and χ^2), and LM3.bic perform best according to BIC. This is not surprising since the algorithm uses AIC and BIC as its model selection criteria, respectively. A possibly fairer comparison could be one using prediction-based criteria as in Section 8.6.6.

8.5 Selection of Distribution

We now check the normality assumption of the models using residual diagnostics.

Residual Diagnostics

The use of normalized quantile residuals, worm plots and bucket plots as diagnostic tools for GAMLSS is described in Sections 4.7.1, 4.7.2, and 4.7.3, respectively. Figure 8.6(a) shows the worm plots of the residuals from all models in Table 8.5 (apart from the null model). Note that all of the models assume a normally distributed response. Almost all of the points on the left side of the worm plot fall within the point-wise

95% confidence region. Unfortunately this is not true for the points on the right side, and we conclude that the assumed distribution fails to capture the right tail of the response distribution. Figure 8.6(b) shows a bucket plot of the same models. Note that models LM1, LM2, LM3.aic, LM3.bic, LM4.ad.lasso.bic and LM5.pcr are represented with the symbols 1, 2, 3a, 3b, 4 and 5 respectively, and that they are surrounded by the corresponding clouds of their bootstrap samples. Unfortunately the model points are not very clear in the plot since they fall close to each other. All models fall in the upper right region of the plot, indicating that they have not accounted properly for the kurtosis and skewness in the response variable.

Fitting Distributions

Next we search for a better distribution for the response, conditional on covariates. The model LM3.bic seems to be a reasonable covariate model to start the process. The **gamlss** function chooseDist() fits all possible distributions from a predetermined set of distributions, conditional on the covariate model specified, and finds the "best" according to a GAIC(κ) criterion. It does this by fitting the starting covariate model using a set of relevant distributions. The set of distributions depends on the type of response variable, as shown in Figure 4.3.

The relevant set of distributions in the search for the birthweight response is all continuous distributions defined on the real line, and the positive real line. A plot of a standardized version of the AIC values obtained is shown in Figure 8.7. The standardization is done by setting the worst fitted distribution model's scaled AIC to zero, the best to one, and the remainder scaled accordingly. The best five distributions are JSU, ST3, SST, ST5, and ST4. Those distributions all have four parameters, indicating that there is reasonable skewness and kurtosis in the data. Figure 8.8 shows the worm plot for the fitted JSU distribution and clearly the distribution appears adequate. The JSU is a reparametrization of the original Johnson's S_u distribution having $\mathbb{E}(y) = \mu$ and $\mathbb{V}(y) = \sigma^2$. The parameter ν is a "true" skewness parameter, with $\nu > 0$ for positive skewness and $\nu < 0$ for negative skewness, respectively; τ affects the kurtosis of the distribution. (See Rigby et al. (2019, p. 391) for more details.) The current fitted JSU model, which we call LM6.jsu, has covariates specified as in model LM3.bic.

LM6.jsu: birthweight $\sim$ JSU(μ, σ, ν, τ)

$$\mu \sim \text{AC} + \text{BPD} + \text{HT} + \text{FL} + \text{parity} + \text{DBD} + \text{AC:BPD}$$
$$\log \sigma \sim \text{AC} + \text{DBD}$$
$$\nu \sim 1$$
$$\log \tau \sim 1.$$

Now that we have found a suitable distribution for the data, we select terms for the JSU model, using the function stepGAICAll.A() and starting with the covariate

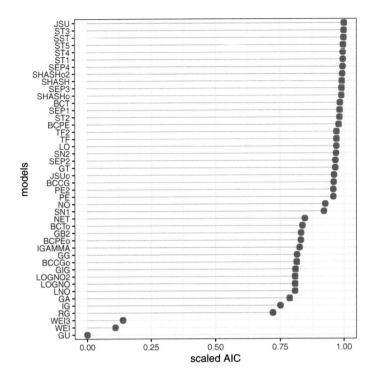

Figure 8.7 The standardized AIC produced from the function `chooseDist()`, comparing fitted distributions on $\mathbb{R}$ and $\mathbb{R}_+$.

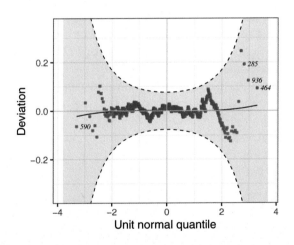

Figure 8.8 The worm plot from the best fitted distribution (`JSU`).

model of `LM6.jsu`. The models selected using criteria AIC and BIC are, respectively:

$$\texttt{LM7.jsu.aic:} \quad \texttt{birthweight} \sim \texttt{JSU}(\mu, \sigma, \nu, \tau)$$

$$\mu \sim \texttt{AC} + \texttt{DBD} + \texttt{HC} + \texttt{FL} + \texttt{parity} + \texttt{age} + \texttt{BPD}$$

$$+ \ \texttt{FL:parity} + \texttt{AC:age} + \texttt{AC:BPD}$$

$$\log \sigma \sim \texttt{DBD} + \texttt{AC} + \texttt{FL} + \texttt{DBD:FL}$$

$$\nu \sim \texttt{AC} + \texttt{DBD}$$

$$\log \tau \sim 1$$

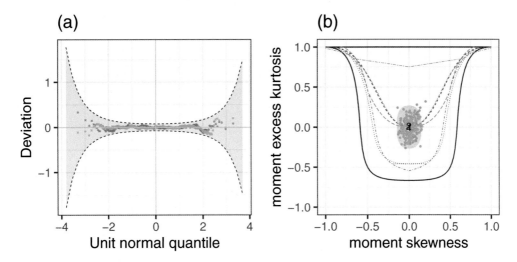

Figure 8.9 Worm plots and bucket plots for models `LM7.jsu.aic` and `LM7.jsu.bic`. The worm plots indicate that the JSU distribution fits adequately, in both models. Note that the models in the bucket are indicated as 1 and 2, respectively.

$$\texttt{LM7.jsu.bic:} \quad \texttt{birthweight} \sim \text{JSU}(\mu, \sigma, \nu, \tau)$$

$$\mu \sim \text{AC} + \text{DBD} + \text{HC} + \text{FL} + \text{AC:HC}$$

$$\log \sigma \sim \text{DBD}$$

$$\nu \sim 1$$

$$\log \tau \sim 1 \, .$$

The worm plots and bucket plots for models `LM7.jsu.aic` and `LM7.jsu.bic`, shown in Figure 8.9, indicate that both models provide adequate fits to the data.

8.6 The Partial Effects

In Section 4.8 we stated that the interpretation of a GAMLSS model is different from that of GAM and GLM. Particularly the question "how $\boldsymbol{x}$ affects the response" has a completely different meaning. Here it is not only a question of how the explanatory variable x affects the mean (or location), but also how it affects the shape of the response distribution. Generally, in a GAMLSS other parts of the distribution besides the center may also be of interest. Partial effects are designed to help with the interpretation of a GAMLSS model and to show how characteristics of the distribution are affected by the independent variables. Partial effects were defined in Section 4.8 as

$$\text{PE}_{\omega(\mathcal{D})}\left(\boldsymbol{x}_j | g(\boldsymbol{x}_{-j}), \mathcal{S}\right),$$

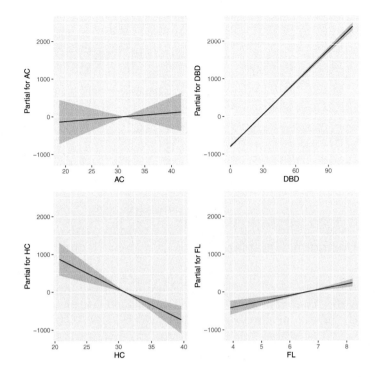

Figure 8.10 Term plots using the `fitted_terms()` for μ for model
`LM7.jsu.bic`. The plot shows all main effects but no interactions.

where $\omega(\mathcal{D})$ is a characteristic of the response distribution, $\boldsymbol{x}_j$ is the variable of interest, $\boldsymbol{x}_{-j}$ denotes all other terms in the GAMLSS model not including $\boldsymbol{x}_j$, and $g(\cdot)$ is a function of interest applied at a given scenario $\mathcal{S}$. Next we investigate the partial effects for different ω characteristics of the response distribution.

8.6.1 Partial Effects on the Predictors η_θ

The partial effects on the predictor $\boldsymbol{\eta}_\theta$ is what is called an additive *term plot* in **R**. It is a way of checking how a term in the model affects the predictor $\boldsymbol{\eta}_\theta$. It is defined as:

$$\mathrm{PE}_{\boldsymbol{\eta}_\theta}\left(\boldsymbol{x}_j | \mathrm{mean}(\boldsymbol{x}_{-j}), \boldsymbol{X}_\theta\right), \qquad (8.2)$$

where $\boldsymbol{\eta}_\theta$ is the predictor of a specified parameter θ, and where all other terms $\boldsymbol{x}_{-j}$ are fixed at their mean value taken from the design matrix $\boldsymbol{X}_\theta$ of the parameter $\boldsymbol{\theta}$.[4]

Figure 8.10 shows the partial effects of the linear main effect terms fitted for the μ model in the model `LM7.jsu.bic`. The plots are useful for identifying trends in the

---

[4] *Technical note*: The **gamlss** partial effects functions use the `model.matrix` (that is, the design matrix $\boldsymbol{X}_\theta$) rather than the training `data.frame` (or `model.frame`), $\boldsymbol{D}_{\mathrm{train}}$. An implication of this is that terms declared as factors are represented in the `model.matrix`, $\boldsymbol{X}_\theta$, as dummy variables and therefore the mean of those binary variables is taken for fixing the scenarios.

relationships between the explanatory terms and $\boldsymbol{\eta}_\theta$. Their interpretation is "what will happen to the predictor $\boldsymbol{\eta}_\theta$ if I change the term $\boldsymbol{x}_j$" given that all other terms, $\boldsymbol{x}_{-j}$, are fixed at their means. If additive smoothers are fitted in the model, the plots are also useful for identifying nonlinearities in the relationships.

However, there are occasions in which the plots can be misleading. The more obvious case is when *interactions* are present in the model. Note that term plots, at least in the way they are implemented in **R**, do not show interactions. The problem with this is that a trend of a main effect can be reversed if a strong interaction exists in the model, namely a negative main effect can become positive. We recommend avoiding the interpretation of main effects in the presence of interactions. For example, for the `LM7.jsu.bic` model do not over-interpret the terms `AC` and `HC`.

The second case in which term plots can be misleading is when there is high correlation between the explanatory terms. High correlation between terms is related to the problems of *multicollinearity* for linear terms and *concurvity* for smoothing terms. High linear or nonlinear correlation means that observations on two highly correlated variables are concentrated in only a small region of the two-dimensional space. That makes the estimation of terms unstable, whether a linear or a nonlinear (smoothing) term is used. In this example, we know from Figure 8.3 that `AC` and `HC` have a very high correlation coefficient of 0.939. That means that a small change in any direction in the space of those variables could have a huge effect on the estimated parameters. That effect can be dramatic not only for linear terms, but even possibly for additive smoothers of the type $s(\texttt{AC})$ and $s(\texttt{HC})$.

The functions `term.plot()` and `fitted_terms()` (in the **gamlss** and **gamlss.ggplots** packages, respectively) can be used for plotting the partial effects (8.2). In the presence of interactions one should consider the function `pe_param()` described next.

8.6.2 Partial Effects on the Distribution Parameters θ_k

Terms affecting the predictor $\boldsymbol{\eta}_\theta$ obviously also affect the distribution parameter $\boldsymbol{\theta}$. If one uses the identity link the effect is identical, but not so for non-identity links. The function `pe_param()` allows partial effect plotting for both the predictor $\boldsymbol{\eta}_\theta$ and the parameter $\boldsymbol{\theta}$, and it does this slightly differently from the term plots described in Section 8.6.1. This could be beneficial in certain circumstances. The scenarios in the partial effects in `pe_param()` use the medians of the training data $\boldsymbol{D}_{\text{train}}$ to fix $\boldsymbol{x}_{-j}$. Therefore their definition of the partial effect for the parameters $\boldsymbol{\theta}_k$ is given by:

$$\text{PE}_{\boldsymbol{\theta}_k}\left(\boldsymbol{x}_j | \text{median}(\boldsymbol{x}_{-j}), \boldsymbol{D}_{\text{train}}\right),$$

where the training data frame $\boldsymbol{D}_{\text{train}}$, is used rather than the design matrix $\boldsymbol{X}_\theta$. The $\boldsymbol{x}_{-j}$ variables are fixed at their medians for continuous terms and their modes for factors. Equivalently the effect for the predictor η_k is given by

$$\text{PE}_{\eta_k}\left(\boldsymbol{x}_j | \text{median}(\boldsymbol{x}_{-j}), \boldsymbol{D}_{\text{train}}\right).$$

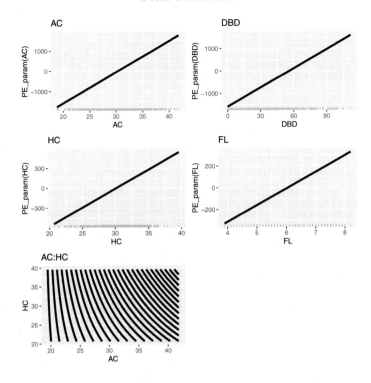

Figure 8.11 Partial effects for the parameter μ of the model LM7.jsu.bic. The main effect terms for AC, DBD, HC, and FL and the first-order interaction AC:HC, are shown.

The main advantage of using the training data frame $\boldsymbol{D}_{\text{train}}$ rather than the design matrix $\boldsymbol{X}_\theta$ is that it is easier to obtain partial interactions and also to extend to the test dataset $\boldsymbol{D}_{\text{test}}$ when interest lies in prediction.

Note that the distribution JSU has as default the identity link for μ so in this case μ and η_μ are identical. Figure 8.11 shows the partial effect of terms on μ for the fitted model LM7.jsu.bic.

$$\hat{\mu} = -678.2 + 11.29 \text{ AC} + 28.4 \text{ DBD} - 84.9 \text{ HC} + 152.7 \text{ FL} + 4.50 \text{ AC:HC}. \qquad (8.3)$$

Note that the main effects of AC and HC, in the presence of their interaction AC:HC, are correctly represented here.

Nonlinearities in the relationship between the explanatory terms and any of the distribution parameters $\boldsymbol{\theta}_k$ can be explored by fitting smoothing terms. In the current analysis thus far, for simplicity we have avoided using smoothers and assumed that only linear relationships exist for all parameters $\boldsymbol{\theta}$. At any stage of the analysis, nonlinearities can be checked by fitting smoothers instead of linear terms. As a demonstration, we refit model LM7.jsu.bic using smoothing nonparametric terms

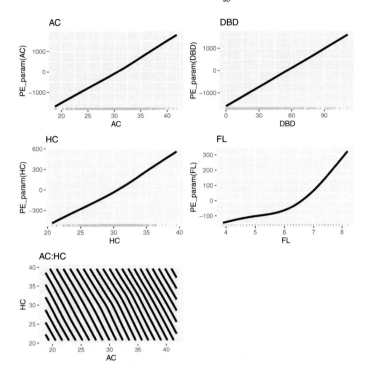

Figure 8.12 Partial effects for the parameter μ of model LM7.jsu.smooth, in which all main effects are fitted using one-dimensional smoothers, and the first-order interaction between AC and HC is fitted by a two-dimensional smoother.

for the μ parameter.

LM7.jsu.smooth: $\texttt{birthweight} \sim \text{JSU}(\mu, \sigma, \nu, \tau)$

$$\mu \sim s(\texttt{AC}) + s(\texttt{DBD}) + s(\texttt{HC}) + s(\texttt{FL}) + s(\texttt{AC}, \texttt{HC})$$

$$\log \sigma \sim \texttt{DBD}$$

$$\nu \sim 1$$

$$\log \tau \sim 1$$

The resulting partial effects for μ are shown in Figure 8.12. In the main effects, only FL shows a possible nonlinearity, while the rest of the terms appear linear. For the first-order interaction between AC and HC, the partial effect of the two-dimensional smoother is not very different to the linear interaction shown in Figure 8.11. The GAIC for the model with smoothing terms is worse than that of LM7.jsu.bic; therefore we do not pursue the nonlinearities further.

Because the terms in the μ model of LM7.jsu.bic are fitted linearly, the coefficients in equation (8.3) represent *elasticities*. They show how the parameter μ (expected birthweight) changes with the explanatory variables. For example, if we compare observations that differ in DBD by one unit (a day) then we expect an associated

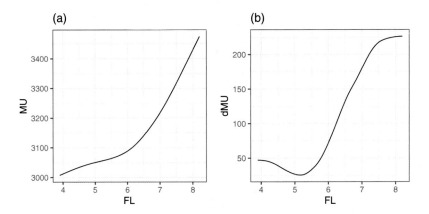

Figure 8.13 Plot for the shape of the (partial) smooth function $s(\text{FL})$ from the fitted μ of model `LM7.jsu.smooth`: (a) the fitted smooth function $s(\text{FL})$ (estimated using P-splines) is shown as the partial effect that FL has on μ, given all other terms in the model are fixed at their median of the training dataset $\boldsymbol{D}_{\text{train}}$ (or mode for factors); (b) the first derivative $ds/d(\text{FL})$ of (partial) estimated smooth function $s(\text{FL})$.

increase in birthweight of 28.4. Note, however, that when a link function other than the identity link is used, the linear coefficients represent elasticities for η_k rather than for θ_k. In order to get the elasticities for θ_k in the presence of an a non-identity link or when smoothers are used in the model, the **gamlss** function `getPEF()` can be used. It uses a cubic spline approximation at a specified grid of the covariate to create the required partial effect. This approximate function can be used for plotting or evaluating the partial effect (or its derivatives) at different covariate values. The elasticity at a specific x value is the first derivative of the partial effect function. As an example, consider the smooth function of the partial effect for FL produced in the middle right panel of Figure 8.12. Figure 8.13(a) plots the partial effect of FL for the μ model, given that all other variables are held at their median values. The first derivative of the function is shown in Figure 8.13(b). The elasticity of FL at FL = 5 can be calculated from the first derivative function as 26.75.

8.6.3 Partial Effects on the Moments

While μ and σ (or θ_1 and θ_2) are location and scale parameters for almost all of the **gamlss.dist** distributions defined on $\mathbb{R}$, they are *not* necessarily the mean and standard deviation of the distribution. In order to see the effect that any explanatory term has on the first two moments of the response distribution, the function `pe_moment()` can be used. For the function to work the first two moments of the distribution should be defined. Not all distributions have defined moments; however, all distributions have centile-based summary statistics, namely median and semi-interquartile range (SIR). The partial effect of the mean is defined as

$$\text{PE}_{\mathbb{E}(y)}\left(\boldsymbol{x}_j|\text{median}(\boldsymbol{x}_{-j}), \boldsymbol{D}_{\text{train}}\right)$$

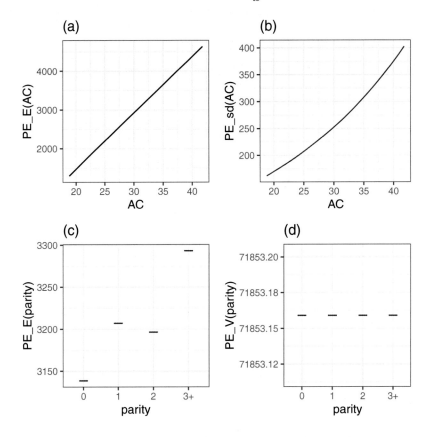

Figure 8.14 Partial effects of the continuous variable `AC` and the factor `parity` on the mean and standard deviation of `birthweight` in model `LM7.jsu.aic`. (a) Partial effect of `AC` on the mean; (b) partial effect of `AC` on the standard deviation; (c) partial effect of `parity` on the mean; and (d) partial effect of `parity` on the standard deviation. Note that the contribution of `parity` to the standard deviation is flat as it is not in the predictor for σ.

and of the variance as

$$\text{PE}_{\mathbb{V}(y)}\left(\boldsymbol{x}_j|\text{median}(\boldsymbol{x}_{-j}), \boldsymbol{D}_{\text{train}}\right).$$

Plotting the standard deviation rather than the variance is a better visual option. Figure 8.14 shows the partial effect on the mean and standard deviation of the continuous term `AC` and the factor `parity`, for the model `LM7.jsu.aic`. (Here we use the results from model `LM7.jsu.aic` since we wish to demonstrate how factors are plotted.) Note that `parity` does not affect the standard deviation because it is not in the model for σ. The plots can be used to gain understanding about how explanatory terms affect the trend of mean and standard deviation, respectively. Unfortunately since the plots do not incorporate standard errors of the estimated mean and variance, it is difficult to draw any conclusions on their statistical significance. For distributions

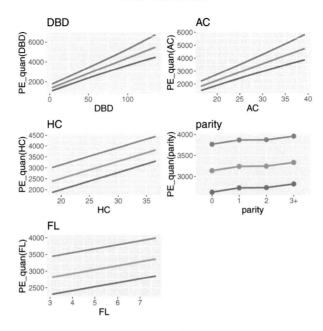

Figure 8.15 Plots of the partial effects on quantiles for DBD, AC, HC, parity and FL at 0.05 (red), 0.50 (green), and 0.95 (blue) levels for the parameter μ of model LM7.jsu.aic.

in which moments do not exist, the function pe_moment() will fail. Rigby et al. (2019, Part III) should be consulted for the moments of the distributions implemented in the GAMLSS packages.

8.6.4 Partial Effects on the Quantiles y_p

We may be interested in assessing the effect that explanatory terms have on the quantiles of the response distribution. This is closely related to quantile regression (QR), where interest lies in the effect of covariates on specific quantiles of the distribution. GAMLSS estimates all quantiles simultaneously, rather than estimating each quantile separately as QR does, but the influence of any explanatory term on the quantiles is usually indirect. This because a specific term may affect one or more different distribution parameters, which consequently affects the corresponding quantile of the response distribution. Potentially any term considered in a GAMLSS model selection could affect a specific quantile of the fitted distribution, either directly or indirectly. The functions getQuantile() and pe_quantile() (of the **gamlss** and **gamlss.ggplots** packages, respectively) can be used to check the effect a term has on specified quantiles. Both functions create the partial effect:

$$\text{PE}_{y_p}\left(\boldsymbol{x}_j|\text{median}(\boldsymbol{x}_{-j}),\boldsymbol{D}_{\text{train}}\right)$$

where y_p is the p quantile and $0 < p < 1$. Other related functions exploring the *exceedance* (defined as $1 - F(y|x)$) are also present in the package **gamlss.ggplots**.

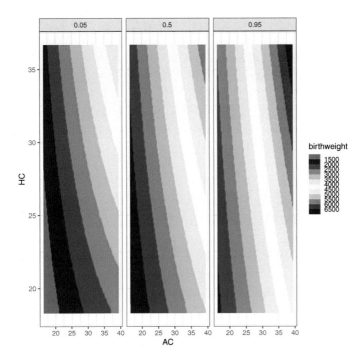

Figure 8.16 Plots of the 5%, 50% and 95% level partial quantiles for terms AC and HC for the μ parameter of model LM7.jsu.bic.

The function `pe_quantile()` can take one or two explanatory variables at a time. Note that a partial quantile plot behaves differently, depending on whether the explanatory terms are continuous variables or factors. Figure 8.15 uses the function `pe_quantile_grid()`, which allows multiple quantile plots for the partial quantiles $y_{0.05}$, $y_{0.50}$ and $y_{0.95}$, for all main effect terms fitted in model LM7.jsu.aic for the parameter μ. (As in Section 8.6.3, we use model LM7.jsu.aic to demonstrate the plotting of factors.)

When first-order interactions are present, in order to show more than one partial quantile we need either a three-dimensional plot or several two-dimensional contour plots. Figure 8.16 shows the first-order interaction AC:HC for the parameter μ of model LM7.jsu.bic (chosen here for display because of its single interaction term). It shows three contour plots for the 5%, 50% and 95% quantiles, respectively.

8.6.5 Partial Effects on the Distribution $\mathcal{D}()$

It is of interest to check how the response distribution itself varies according to the different explanatory terms in the model. In some respects this is the essence of GAMLSS. The functions `pe_pdf()` and `pe_pdf_grid()` are designed to display the

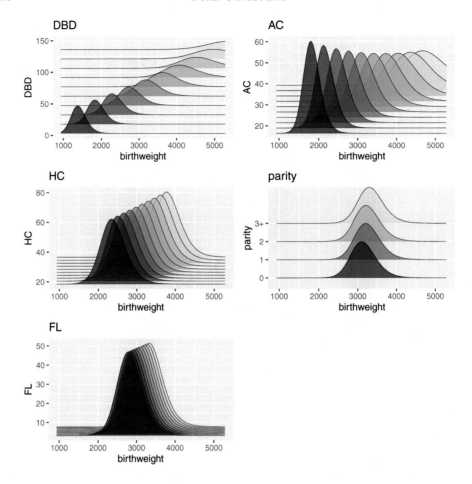

Figure 8.17 The partial effects of the terms DBD, AC, HC , parity and FL on the response distribution, in model LM7.jsu.bic.

partial effect:

$$\mathrm{PE}_{\mathcal{D}(y)}\left(\boldsymbol{x}_j|\mathrm{median}(\boldsymbol{x}_{-j}), \boldsymbol{D}_{\mathrm{train}}\right),$$

where $\mathcal{D}(y)$ is the fitted probability function of the response variable. The behavior of pe_pdf() varies depending on whether the term argument is continuous or a factor. For continuous variables, the function takes equally spaced values in the x variable, and predicts the estimated probability function. For a factor, it does this at each level of the factor.

Figure 8.17 shows the resulting plots from using the function pe_pdf_grid() on the model LM7.jsu.aic. Note particularly the effect of DBD on the response distribution: increasing DBD (that is, observations made further from birth) is associated with a wider fitted distribution of birthweight. In other words, the further out from birth the ultrasound measurements are taken, the more variability and less certainty we can attach to the distribution of birthweight. The term DBD appears to be the most

Table 8.6 *Comparison of the predictive ability of models LM1, LM2, LM3.aic, LM3.bic, LM7.jsu.aic and LM7.jsu.bic, based on 10-fold cross-validation. MAPE = median absolute percentage error, LS = logarithmic score, CRPS = continuous ranked probability score. The "best" model according to each criterion is shaded: for MAPE, a low score indicates a better model, whereas for LS and CRPS higher is better.*

Model	MAPE	LS	CRPS
LM1	5.84	−7.10	−1.87
LM2	5.89	−7.97	−0.62
LM3.aic.	5.63	−7.07	−0.60
LM3.bic	5.60	−7.08	−1.36
LM7.jsu.aic	5.42	−7.05	−0.20
LM7.jsu.bic	5.58	−7.07	−1.30

important term affecting the shape of the distribution. This is indicated by the use of the function `term_importance()` of the package **gamlss.ggplots**, which shows that DBD on its own accounts for more than 50% of contribution of terms in the model.

Partial effects on the distribution can also be plotted using the **R** package **distreg.vis** (Stadlmann and Kneib, 2021). This is demonstrated in Chapter 10.

8.6.6 Prediction

As birthweight prediction is the primary focus of this analysis, we now evaluate the predictive ability of the fitted models LM1, LM2, LM3.aic, LM3.bic, LM7.jsu.aic and LM7.jsu.bic. We use three measures: the median absolute percentage error (MAPE), defined in equation (7.2); and the logarithmic score (LS) and continuous ranked probability score (CRPS), both described in Section 4.4.2. Ten-fold cross-validation was used. The basic idea here is to answer whether (i) the models provide useful predictions of birthweight; and (ii) whether by moving from the normal to the Johnson's S_u response distribution, the predictive capability of the model has improved. Note that the performance measure MAPE evaluates point predictions, while LS and CRPS evaluate the predictive distributions as a whole.

The results are summarized in Table 8.6. The LS, CRPS and MAPE all indicate that the best model for prediction is LM7.jsu.aic, thus justifying the added complexity of the JSU model in comparison with the normal model. The MAPEs are at an acceptable level for useful prediction; note that the boosting approach using the JSU response distribution (Section 7.3.4) achieved comparable scores.

The reader would have noticed that we have not taken a clear-cut position on what would be the "best" model for the modeling the birthweight response. Instead we have used the ultrasound data to demonstrate how GAMLSS can be used, and how it can enhance our knowledge.

9

Speech Intelligibility Testing

In this chapter we develop a statistical model for a bounded continuous outcome, with inflated endpoint probabilities. The model incorporates random effects and is estimated in the penalized maximum likelihood paradigm.

9.1 Data and Research Question

Speech intelligibility tests are conducted on hearing-impaired people, for the purpose of evaluating the performance of hearing devices under controlled listening conditions in a sound laboratory. The simplest version is that the subject listens to a pre-recorded word, mixed with noise, and repeats it back. Here the stimulus is the signal-to-noise ratio (SNR) and the subject's response is either correct or incorrect. More common usage is for the subject to listen to a short sentence (of length N words) and repeat it back. The number of correctly identified words w is recorded and the proportion correct ($y = w/N$) is the response of interest. Typically data are collected in tracks of 20 sentences each, in which experimental conditions are the same but the SNR is varied over its range. Using the sentence as the unit of observation, an obvious model for the proportion correct is the binomial distribution; however, the assumption of independent recognition of words in a sentence is unlikely to be satisfied.

Hu et al. (2015) compared the speech recognition of seven recipients of cochlear implants in the presence of an interfering talker. The primary factor of interest was the sound-processing algorithm in the device (A, B, C), and the research question was whether the three algorithms yielded differences in the subjects' speech recognition. The other factors were the direction of the interfering talker ("noise direction"), which was either from the front or from both sides, and the gender of the talker ("noise gender"). Variables recorded are given in Table 9.1. The histogram of the proportion correct for all observations (sentences) is shown in Figure 9.1, in which the striking features are the high frequencies at zero and one. This is explained by the use of context in word recognition, which results in a large proportion of sentences which are either completely recognized, or not recognized at all.

Table 9.1 *Variables: speech recognition study.*

Variable name	Description
N	Number of words in sentence
w	Number of words correctly identified
y	Proportion correct (w/N)
snr	Signal-to-noise ratio
algorithm	Sound-processing algorithm (A, B, C)
noise_dir	Noise direction (F = front, S = side)
noise_gender	Noise gender (M, F)
subject	Subject identifier

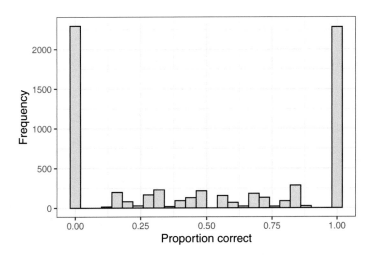

Figure 9.1 Histogram of proportion of words correctly recognized.

9.2 Model Building

9.2.1 Response Distribution

Hu et al. (2015) proposed the use of the zero-and-one-inflated beta distribution (Ospina and Ferrari, 2010), as in model (2.4) with $f_c(y|\boldsymbol{\theta})$ the pdf of the beta distribution. For the continuous part of the model, we considered all distributions on $\mathbb{R}_{(0,1)}$ available in **gamlss**, discussed in Section 2.2.3, namely the beta, logit-normal, simplex, and generalized beta type I. We used the function `gamlss::chooseDist()` to choose the response distribution, conditional on all covariates present in the models for all distribution parameters. (This approach to model selection is discussed in Section 4.5.1.) The simplex distribution was preferred based on both the AIC and BIC. Combining the simplex with zero-and-one-inflation results in the following mixed discrete–continuous probability model, for which the **gamlss** notation is

`SIMPLEXInf0to1:`

$$f(y; \mu, \sigma, \xi_0, \xi_1) = \begin{cases} \frac{\xi_0}{1+\xi_0+\xi_1} & y = 0 \\ \frac{1}{1+\xi_0+\xi_1} f_c(y; \mu, \sigma) & y \in (0,1) \\ \frac{\xi_1}{1+\xi_0+\xi_1} & y = 1, \end{cases} \qquad (9.1)$$

where

$$f_c(y; \mu, \sigma) = \frac{1}{[2\pi\sigma^2 y^3(1-y)^3]^{1/2}} \exp\left[-\frac{(y-\mu)^2}{2\sigma^2 y(1-y)\mu^2(1-\mu)^2}\right], \quad y \in (0,1)$$

is the pdf of the simplex distribution, parametrized in terms of its mean μ and shape parameter σ; $\mu \in (0,1)$ and $\sigma, \xi_0, \xi_1 > 0$ (Rigby et al., 2019). Note that the probability of zero correct is

$$\mathbb{P}(y = 0) = \frac{\xi_0}{1+\xi_0+\xi_1}, \qquad (9.2)$$

and the probability of all correct is

$$\mathbb{P}(y = 1) = \frac{\xi_1}{1+\xi_0+\xi_1}. \qquad (9.3)$$

While the mean of the "middle part" of the distribution is

$$\mathbb{E}(y|0 < y < 1) = \mu,$$

the mean of the entire distribution (9.1) is

$$\mathbb{E}(y) = \frac{\mu + \xi_1}{1+\xi_0+\xi_1}. \qquad (9.4)$$

In the distributional regression model, the parameters μ, σ, ξ_0, and ξ_1 are modeled with covariates, as well as random intercepts to account for within-subject correlation as subjects vary substantially in their degree of hearing loss. The logit link is used for μ and log links for σ, ξ_0, and ξ_1. Assuming m subjects having n_i observations each, the model is

$$\begin{aligned} y_{ij} &\sim \texttt{SIMPLEXInf0to1}(\mu_{ij}, \sigma_{ij}, \xi_{0,ij}, \xi_{1,ij}) \\ \log \tfrac{\mu_{ij}}{1-\mu_{ij}} &= \eta_{ij}^{\mu} \\ \log \sigma_{ij} &= \eta_{ij}^{\sigma} \\ \log \xi_{0,ij} &= \eta_{ij}^{\xi_0} \\ \log \xi_{1,ij} &= \eta_{ij}^{\xi_1} \end{aligned} \qquad (9.5)$$

for $i = 1, \ldots, m$; $j = 1, \ldots, n_i$, and where the linear predictors η_{ij}^{θ} comprise appropriate linear combinations of the predictors and their interactions, and random intercepts.

9.2.2 Variable Selection

Selection of covariates for the models for the distribution parameters μ, σ, ξ_0, and ξ_1 can be approached in a rigorous way, as there are only four potential covariates: algorithm, noise direction, noise gender and SNR. In low-dimensional variable selection settings in which variables are being selected for a single parameter (i.e. small p and $K = 1$), best-subset selection (Hastie et al., 2009) is an attractive option. In the case of p potential covariates and models having only main effects, for $h = 0, \ldots, p$ all $\binom{p}{h}$ combinations (subsets) of h covariates are compared, according to a chosen criterion. The total number of models compared is then $\sum_{h=0}^{p} \binom{p}{h} = 2^p$. While enumeration of the 2^p models is computationally unattractive for even moderate p, for the current problem it is a feasible approach. (For large p, generally the maximum number of covariates to be considered in the model is set to $\ell < p$, which means that $\sum_{h=0}^{\ell} \binom{p}{h}$ models are compared. Also, there are algorithms for finding solutions that do not involve complete enumeration, for example Zhu et al. (2020) and Hastie et al. (2020).)

In the current application, for variable selection for a single distribution parameter in which only main effects are considered, the evaluation of $2^4 = 16$ models is required. As the device algorithm is the focus of this study, interaction terms of algorithm with the other three covariates should also be considered, making the number of models to be compared 35. However, we are selecting models for four distribution parameters, making the number of models $35^4 \approx 1.5$ million, a substantial computational effort. In order to reduce the computational burden, a GAMLSS modification of the best subset selection method was used, in which the distribution parameters were cycled through and best subset selection performed for each, conditional on the models previously chosen for the other parameters. The process was performed twice, making the number of models examined far more feasible ($35 \times 4 \times 2 = 280$).

We use the generalized AIC ($\text{GAIC}(\kappa)$) for selection. The selection was repeated for $\kappa = 2, 3, \ldots, 8, 8.81$; $\kappa = 2$ gives the AIC, and $\kappa = \log n = 8.81$ gives the BIC. Random effects (intercepts) for subject were specified for all distribution parameters; however, examination of the models confirmed that, for all values of κ attempted, the model for σ never contained covariates; accordingly it was thought that the random effect in the model for σ was also unnecessary and the model selection process was repeated omitting this term. The total number of terms in the models thus obtained, over all distribution parameters and including random effects, are plotted against κ in Figure 9.2. Models selected using the AIC were judged to be overly complex, and those using the BIC too simple; $\kappa = 4$ was considered a reasonable compromise. Results for this model are given in Table 9.2. The corresponding worm plot, shown in Figure 9.3, indicates a well-fitting model.

In Section 4.4.1 we discussed the use of model selection criteria in the presence of random effects; specifically that the selection criteria do not take into account the uncertainty involved in estimation of the variance parameters; and the effective sample size used in the BIC is less than the sample size n. In the absence of implemented solutions for these issues for the general GAMLSS model, we have used the "naïve

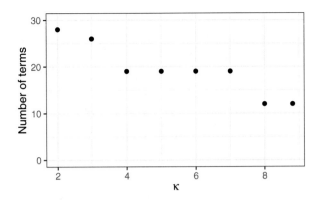

Figure 9.2 Total number of terms in models chosen according to GAIC(κ), plotted against κ.

Table 9.2 *Parameter estimates, zero-and-one-inflated simplex model for proportion correct. Subject random intercept standard deviations are given in the last row.*

Parameter	μ		σ		ξ_0		ξ_1	
Link	logit		log		log		log	
	$\hat{\beta}^\mu$	SE($\hat{\beta}^\mu$)	$\hat{\beta}^\sigma$	SE($\hat{\beta}^\sigma$)	$\hat{\beta}^{\xi_0}$	SE($\hat{\beta}^{\xi_0}$)	$\hat{\beta}^{\xi_1}$	SE($\hat{\beta}^{\xi_1}$)
(Intercept)	−0.211	0.038	0.826	0.089	1.139	0.023	−0.878	0.064
algorithmB					−0.338	0.108	−0.100	0.038
algorithmC					−0.389	0.028	−0.371	0.007
noise_genderM	0.109	0.036			−0.319	0.106	0.308	0.077
snr	0.058	0.007			−0.293	0.065	0.224	0.078
snr:algorithmB					0.121	0.086		
snr:algorithmC					0.034	0.012		
random(subject)	0.015				0.029		0.036	

approach", that is, our model selection based on GAIC(κ) does not correct for the complexities introduced by the random effects.

9.3 Model Interpretation

Recall that the point of this study was to compare the performance of the three sound-processing algorithms A, B, and C. Examination of the coefficients in Table 9.2 does not yield an obvious answer to this question. Algorithm appears in the equations for ξ_0 and ξ_1, and in an interaction term with SNR for ξ_0, but the pattern of its influence on proportion correct is not clear. Instead, in order to compare the algorithms we examine:

- the fitted probability of zero correct,

- the fitted probability of all correct, and

- the fitted overall mean proportion correct

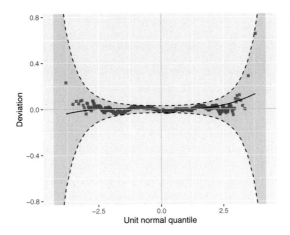

Figure 9.3 Worm plot for zero-and-one-inflated simplex model for proportion correct.

using fitted values $\hat{\mu}$, $\hat{\xi}_0$, and $\hat{\xi}_1$ derived from the model (9.5), estimates in Table 9.2, and equations (9.2), (9.3), and (9.4), for given values of noise gender, and over the range of SNR. In computing the fitted values all random effects were taken to be zero. (Very similar results were obtained for random effects taken as other values in the range $(-2, 2)$.) Note that noise direction does not appear in any of the model equations so does not need to be considered here. The results are shown as the fitted curves, over SNR, with 95% parametric bootstrap confidence regions, in Figure 9.4. The overall pattern is for proportion correct to increase with increasing SNR (as is well known). In addition, for noise gender both female and male, we have the following.

- For fitted proportion zero correct (*low* is good):
  - at the high end of the SNR range, algorithm B performs worst; A and C are practically indistingushable from each other;
  - in the middle of the SNR range, the three algorithms are indistinguishable;
  - at the low end of the SNR range, algorithm B performs best and A worst.

- For fitted proportion all correct (*high* is good):
  - at the middle and high end of the SNR range, algorithm A performs best; B and C are practically indistingushable;
  - at the low end of the SNR range, algorithm B performs best; A and C are practically indistingushable.

- For fitted overall mean proportion correct (*high* is good):
  - at the high end of the SNR range, algorithm A performs best; B and C are practically indistingushable;

- in the middle of the SNR range, the three algorithms are indistinguishable;

- at the low end of the SNR range, algorithm B performs best and A worst.

Choice of algorithm is therefore dependent on the environment in which the device is to be used. Algorithm A would be a good choice for use of the device in middle to high SNR settings, but would be a poorer choice in low SNR settings.

We consider that this detailed assessment of the algorithms over the full range of the subjects' scores would not have been possible without the flexibility of distributional regression modeling. While the number of predictors available to us was small, the model could easily have accommodated multiple subject-specific or other factors.

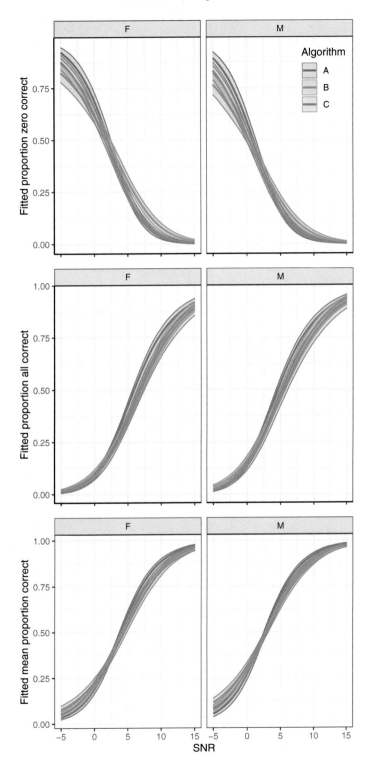

Figure 9.4 Fitted proportions, with 95% parametric bootstrap confidence regions, for noise gender female (left-hand column) and male (right-hand column); algorithms A, B, and C, over the range of signal-to-noise ratio (SNR). Top row: proportion zero correct; middle row: proportion all correct; bottom row: overall mean proportion correct. Random effects are set to zero.

10

Social Media Post Performance

In this chapter we present an example of an unbounded count response variable, the number of shares of social media posts. The model incorporates smooth terms (cyclic P-splines) and is estimated using penalized maximim likelihood.

10.1 Data and Research Question

Moro et al. (2016) present a study on the performance of Facebook posts, with the aim of prediction of a post's performance, or impact, on the basis of its characteristics. The metrics used for post performance (the output) were the number of likes, shares or comments on the post, as well as a number of metrics used by Facebook. Predictors (the inputs) were characteristics of the post, which are given in Table 10.1. All posts on the page of a particular cosmetic brand in 2014 were analyzed ($n = 495$). Moro et al. (2016) used a data-mining approach, in particular support vector machines. In the following we analyze the *number of shares*, an unbounded count variable, as outcome in a distributional regression model. Figure 10.1 shows the histogram of the number of shares, truncated for the purpose of display at 100 shares and overlaid with a kernel density estimate. (The maximum number of shares is 790.) The purpose of the analysis is to predict the number of shares on the basis of the post characteristics.

10.2 Model Building

The first step in model building is selection of the response distribution for the number of shares. It is clear that the shape of the observed marginal response distribution, as shown in Figure 10.1, does not resemble the unbounded count distributions discussed in Chapter 2, such as those shown in Figure 2.12. We will follow the pragmatic "Recipe for Model Selection" given in Section 4.5.3.

(1) The set of appropriate distributions is the unbounded count distributions (Section 2.3.1).

(2) Assuming that the distribution could have up to four distribution parameters (denoted in what follows as μ, σ, ν, and τ), based on preliminary experimentation with the data we specify the following covariates.

Table 10.1 *Predictor variables: characteristics of Facebook posts.*

Variable name	Description
Type	Type of content (Link, Photo, Share, Video)
Category	Content categorization: 1 = action, 2 = product, 3 = inspiration
Post.Month	Month posted: 1 = Jan, ..., 12 = Dec
Post.Weekday	Day posted: 1 = Sunday, ..., 7 = Saturday
Post.Hour	Hour posted $(0, ..., 23)$
Paid	Paid advertisement (0 = no, 1 = yes)

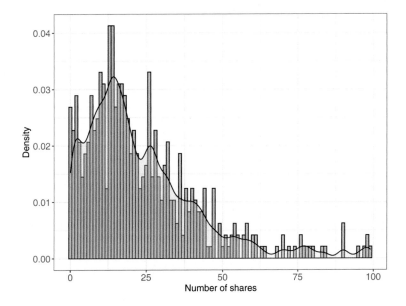

Figure 10.1 Facebook data: histogram of the number of shares, truncated for display at 100 and overlaid with density estimate.

μ	Category, Post.Month, Post.Weekday
σ	Category, Post.Month, Post.Weekday
ν	Category
τ	Category

Note that the distribution parameters may have different meanings in the distributions that will be compared; however, usually μ will be a location parameter and σ will be a scale parameter. Because of the cyclic nature of the time covariates Post.Month, Post.Weekday, and Post.Hour, we specify cyclic P-spline terms for these.[1]

(3) Using the function gamlss::chooseDist(), we fit all discrete response distribu-

---

[1] Cyclic P-splines are generally used for temporal covariates for which it is assumed that there is continuity between the end of a cycle (e.g. Sunday) and the beginning of the next cycle (e.g. Monday). More details are given in Section 3.1.

tions in Table B.1 with nonnegative integer support ($\mathcal{S} = \mathbb{N}$), with covariates specified as above. The distributions are then compared using an information criterion. According to the AIC, the first three distributions in decreasing order of preference are: zero-adjusted Sichel (`ZASICHEL`) , zero-inflated Sichel (`ZISICHEL`) and zero-adjusted beta-negative binomial (`ZABNB`). The BIC yields the Sichel distribution variants `SICHEL`, `SI`, and `ZASICHEL` as its first three choices. Based on this we commence model building using the `ZASICHEL` distribution, but bear the alternatives in mind should this fail.

As background, the Sichel distribution, also known as the generalized inverse Gaussian-Poisson (GIGP), is a three-parameter mixed Poisson distribution having the generalized inverse Gaussian as mixing distribution. (See Section 2.3.1, and Rigby et al. (2019) for more details.) In **gamlss**, two parametrizations are available: `SICHEL` and `SI`; the `SICHEL` parametrization has the advantage that $\mathbb{E}(y) = \mu$. However the interpretation of distribution parameters σ and ν is not as simple. Plots of the `SICHEL` probability function over a range of values of μ, σ, and ν are shown in Figure 10.2.

(4) The next step in model building is variable selection for each of the model parameters. A strategy[2] of stepwise selection of covariates for each distribution parameter sequentially, based on an information criterion, was used with the zero-adjusted Sichel as response distribution. As computational problems of non-convergence resulted, the next two preferred response distributions according to the AIC (zero-inflated Sichel, zero-adjusted beta-negative binomial) were attempted and they also resulted in convergence problems. It was then decided to implement the simpler alternative preferred by the BIC: the `SICHEL` distribution. The resulting model is as follows.

$$y_i \sim \texttt{SICHEL}(\mu_i, \sigma_i, \nu_i) \tag{10.1}$$
$$\log \mu_i = \beta_0^\mu + \beta_1^\mu x_{i1} + \beta_2^\mu x_{i2} + \beta_3^\mu x_{i3} + \beta_4^\mu x_{i4}$$
$$\quad + s_1^\mu(\texttt{Post.Month}_i) + s_2^\mu(\texttt{Post.Weekday}_i) + s_3^\mu(\texttt{Post.Hour}_i)$$
$$\log \sigma_i = \beta_0^\sigma + \beta_1^\sigma x_{i1} + \beta_2^\sigma x_{i2}$$
$$\quad + s_1^\sigma(\texttt{Post.Month}_i) + s_2^\sigma(\texttt{Post.Weekday}_i) + s_3^\sigma(\texttt{Post.Hour}_i)$$
$$\nu_i = \beta_0^\nu + \beta_1^\nu x_{i1} + \beta_2^\nu x_{i2} + \beta_3^\nu x_{i3} + \beta_4^\nu x_{i4}$$
$$\quad + s_1^\nu(\texttt{Post.Month}_i) + s_2^\nu(\texttt{Post.Weekday}_i) + s_3^\nu(\texttt{Post.Hour}_i).$$

Here, x_1 and x_2 are dummy variables for the factor `Category`; x_3 and x_4 are dummy variables for `Type` (with the levels Photo and Video combined); and $s_j^\mu(\cdot)$, $s_j^\sigma(\cdot)$ and $s_j^\nu(\cdot)$ are cyclic P-spline terms.

(5) The worm plot of the randomized normalized quantile residuals of model (10.1), shown in Figure 10.3, indicates a well-fitting model. (See Sections 4.7.1 and 4.7.2 for details on quantile residuals and the worm plot, respectively.) The Sichel seems a reasonable choice of response distribution, given that it is suitable for

---

[2] Implemented in the function `gamlss::stepGAICAll.A()`.

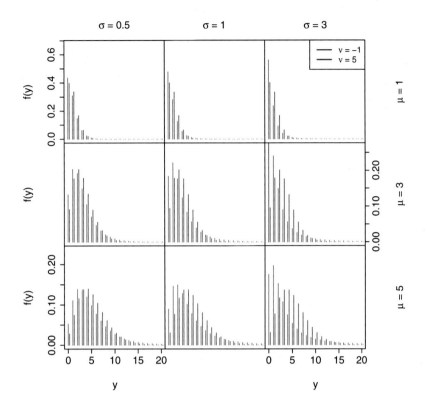

Figure 10.2 Probability function of the `SICHEL` distribution, for $\mu = 1, 3, 5$, $\sigma = 0.5, 1, 3$ and $\nu = -1, 5$.

modeling heavy-tailed or overdispersed responses, which we observe in these data. We prefer the `SICHEL` parametrization over the `SI`, because of better parameter interpretability. (μ is the mean.) We therefore take (10.1) as our final model.

10.3 Prediction

As the aim of the original analysis was prediction of the performance of the Facebook posts, we examine the predictive ability of model (10.1) and compare this with the predictive ability of two other models, based on 10-fold cross-validation. The first is a GAMLSS model with a Sichel response distribution, using neural network predictors for all distribution parameters; the second is a boosting model, also using a Sichel response distribution. (Details are given on the accompanying website.) The results are summarized in Table 10.2. The median absolute percentage error (MAPE) shows clearly that, while the neural network model performs the best, in fact all models perform poorly, and we would conclude that the predictors are not useful for accurate

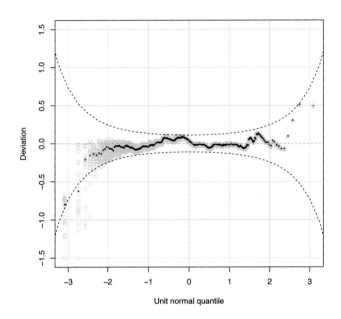

Figure 10.3 Facebook data: worm plot, Sichel response distribution model (10.1). Twenty realizations of randomized quantile residuals are shown.

prediction of post performance. In particular, the boosting algorithm stopped on many folds before updating much more than the intercept: This indicates that the available explanatory variables (and the corresponding base-learners in the boosting context) do not add much to the prediction accuracy. The logarithmic score (LS) and continuous ranked probability score (CRPS) show model (10.1) to be superior to the neural network model; however, we should bear in mind that these measures are comparative only, and are not indicative of the quality of the model predictions. The mean squared error of prediction (MSEP) is best for the boosting and neural network models.

Despite its poor predictive performance we can, however, glean useful information from interpretation of the fitted model, which we address in Section 10.4.

10.4 Model Interpretation

In general parameter interpretation is complex for models in which covariates appear in more than one parameter. In Section 4.8, partial effects of terms on various aspects of a GAMLSS model are introduced, with the notation

$$\mathrm{PE}_{\omega(\mathcal{D})} \left(\boldsymbol{x}_j | g(\boldsymbol{x}_{-j}), \mathcal{S} \right) \ ,$$

Table 10.2 *Comparison of the predictive ability of Sichel model* (10.1), *a neural network model, and a boosting model, based on 10-fold cross-validation. MSEP = mean squared error of prediction, LS = logarithmic score, CRPS = continuous ranked probability score, MAPE = median absolute percentage error. The "best" model according to each criterion is shaded; note that for MSEP and MAPE, a low score indicates a better model, whereas for LS and CRPS higher is better.*

Model	MSEP	LS	CRPS	MAPE
Sichel (10.1)	1810	−2064	−6259	59.7
Neural network	1785	−2155	−6467	59.6
Boosting	1785	−	−	67.9

and are illustrated in detail in the fetal ultrasound application in Section 8.6.5. Here we examine partial effects on the Sichel distribution parameters μ, σ, and ν, and on the distribution.

10.4.1 Partial Effects on the Distribution Parameters

Term plots give the partial effects of specified terms on either the distribution parameters or predictors. For the purpose of illustration, we focus here on the effects of the terms `Category` and `Post.Month` on the distribution parameters μ, σ and ν. Figure 10.4 shows the term plots in which, for example, the top left panel is the partial effect of the term `Category` on μ, with all other covariates held fixed at their modes (since they are all factors):

$$\text{PE}_\mu \left(\texttt{Category} | \text{mode}(\boldsymbol{x}_{-\texttt{Category}}), \boldsymbol{D}_{\text{train}}\right) \ .$$

We see in Section 10.4.2 that these plots are of limited use in this situation of covariates appearing in multiple distribution parameters.

10.4.2 Partial Effects on the Distribution $\mathcal{D}()$

The effect of each covariate on the overall response distribution is not easily assessable from the term plots. In order to gain better insight we visualize the partial effects of specified terms over the entire distribution, holding other covariate values constant, using, for example, the partial effect:

$$\text{PE}_{\mathcal{D}(y)} \left(\texttt{Category} | \text{mode}(\boldsymbol{x}_{-\texttt{Category}}), \boldsymbol{D}_{\text{train}}\right) \ .$$

We illustrate this in Figure 10.5, in which the fitted Sichel response distribution probability functions are shown. In Figure 10.5(a), the three levels of `Category` are compared, with other covariates fixed at their modes. `Category = 1` (action) has by far the lowest distribution of the number of shares, being sharply right-skewed with a long right tail. `Category = 2, 3` (product, inspiration) have very similar distributions with higher means. In Figure 10.5(b), `Post.Month` January, May and October are compared, with other covariates fixed at their modes. It is apparent that the number of shares tends to be lowest in October, and highest in January. While for `Category` the insight on the location of these distributions is consistent with the

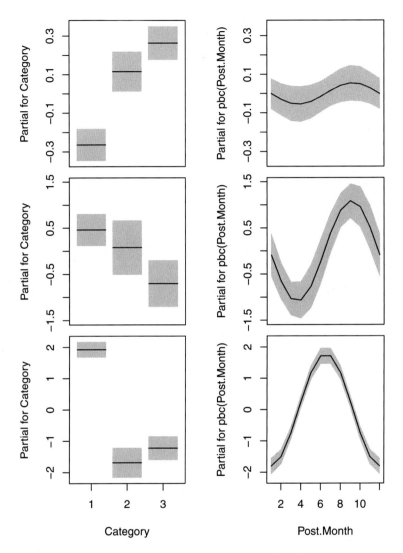

Figure 10.4 Sichel response distribution model (10.1): Term plots for
`Category` (left) and `Post.Month` (right), for parameters μ (top row), σ
(middle row), and ν (bottom row).

term plot for μ (top left panel, Figure 10.4), for `Post.Month` it is quite different. The
term plot for μ (top right panel, Figure 10.4) shows October to have a high fitted
mean; however October also has a high fitted value for σ and from Figure 10.2 we see
that, for fixed μ, increasing σ has the effect of pushing the distribution towards zero.
This is the effect that we see in Figure 10.5(b), and it illustrates that (i) term plots
have the potential to be misleading when a covariate is in the predictor for more than

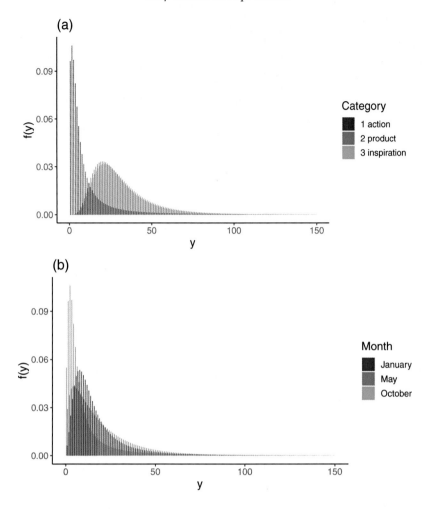

Figure 10.5 Plots of fitted response distribution for Sichel model (10.1): (a) categories action, product and inspiration; (b) months January, May, and October. Other covariates are fixed at their modes, that is, category = action; type = Photo/Video; month = October; weekday = Friday; hour = 3. These plots were created using the **R** package **distreg.vis** (Stadlmann and Kneib, 2021).

one distribution parameter, and (ii) partial effects on the entire distribution are a very effective tool for GAMLSS model interpretation. Note also that any conclusions regarding the distribution are conditional on the values for all covariates in the model, and different features may be present for other combinations.

In conclusion, while we found that distributional regression and machine learning approaches were not successful in achieving accurate prediction of the outcome number of shares, the distributional regression model provides potentially useful interpretation of the effects of the predictor variables on the outcome. Such insights are not immediately possible with machine learning approaches.

Childhood Undernutrition in India

In this case study, we exemplify the Bayesian approach to GAMLSS. In addition to points relevant to any GAMLSS-type analysis such as the choice of the response distribution, this includes various aspects specific to a Bayesian inferential approach, such as

- assessment of convergence and mixing of the Markov chain in MCMC-based inference,

- utilizing the advantages of sampling-based inference, and

- investigating hyperprior sensitivity.

Further, we highlight how GAMLSS can be extended from univariate to multivariate responses based on copula specifications.

11.1 Data and Research Question

Childhood undernutrition is among the most urgent public health challenges in developing and transition countries, as reflected in Goal 2 ("Zero hunger") of the Sustainable Development Goals. Studying the determinants of childhood undernutrition therefore receives considerable attention in development economics. Associating extreme forms of undernutrition with potential determinants then provides more information rather than studying the average nutritional status of children alone. This hints at the fact that a distributional analysis of undernutrition has the potential to offer additional, policy-relevant insights.

Typically, childhood undernutrition is measured by Z-scores comparing the nutritional status of children in the population of interest with the nutritional status in a reference population. Specifically, the Z-scores are defined as

$$Z_i = \frac{\mathrm{AI}_i - \mu_{\mathrm{AC}}}{\sigma_{\mathrm{AC}}},$$

where AI_i is some anthropometric characteristic of child $i = 1, \ldots, n$ reflecting the child's nutritional status (such as height or weight), while μ_{AC} and σ_{AC} represent measures of location and variability from a reference population (typically the median and some robust measure of the standard deviation), stratified with respect to characteristics such as age and sex. Different anthropometric indicators then reflect different aspects of undernutrition. In particular, `wasting` (measuring weight for height) reflects acute undernutrition, `stunting` (measuring height for age) reflects chronic undernutrition, while `underweight` (measuring weight for age) reflects a mixture of both acute and chronic undernutrition.

In this chapter, we will study determinants of both the marginal and the joint distribution of `wasting` and `underweight` as two different undernutrition Z-scores. With our analysis, we illustrate and apply several concepts for Bayesian inference introduced in Chapter 6 as well as the possibility of fitting joint models for more than one response variable using copulas (introduced in Section 2.4.3).

In Section 11.2, we start our investigations with the analysis of regional, age-, and gender-specific differences in the (univariate) distribution of acute undernutrition (i.e. using `wasting` as the response variable). Our analysis is based on a nationally representative survey conducted in 2015/2016 in India (see `www.measuredhs.com` for details). The dataset contains information on a total of $n =$26,055 children (after deleting implausible and incomplete observations). For illustration purposes, we will use age, gender and region in India as the only available covariate information, even though the full dataset contains far more child-, mother-, and family-related factors. See, for instance, Klein et al. (2015b) for a more complex analysis in terms of the predictor structure.

In Section 11.3, we extend our model to a bivariate regression model, where we consider `wasting` and `underweight` as measures of undernutrition jointly. Here we focus on the strength of the association between the two undernutrition indicators and how this dependence changes with the age of the child. Such an association is not only plausible to assume since both responses reflect undernutrition, but in particular since both are related to acute forms of undernutrition.

11.2 Univariate Analysis

11.2.1 Model Specification

To study regional, age-, and gender-specific differences in the (univariate) distribution of acute undernutrition, we consider the wasting score $Z_{\mathrm{wa},i}$, $i = 1, \ldots, n$ as the response variable while the covariate vector $\boldsymbol{x}_i$ comprises the age of the child in months ($\mathtt{age}_i$), the dummy-coded gender of the child ($\mathtt{sex}_i$, 0 = girl, 1 = boy), and the state a child lives in ($\mathtt{state}_i$). For the effect of age, we will assume a potentially nonlinear effect represented as a cubic penalized spline with 10 inner knots and second-order random walk prior (see Section 3.1 for details), while for the spatial effect we employ an intrinsic Gaussian Markov random field specification where two states are treated as neighbors if they share a common boundary (see Section 3.5). The possibility of gender-specific differences in the effect of age can be investigated using a varying coefficient term with age as effect modifier and gender as interaction variable (see Section 3.7), where again we employ a penalized spline with the same settings as for the nonlinear effect of age. In total, a typical predictor in our analysis will look as follows:

$$\eta_i = \beta_0 + \textbf{sex}_i \beta_1 + s_1(\textbf{age}_i) + \textbf{sex}_i s_2(\textbf{age}_i) + s_{\text{spat}}(\textbf{state}_i). \tag{11.1}$$

As explained in Section 6.7, we use the inverse gamma distribution with hyperparameters $a = 0.001$ and $b = 0.001$ as a prior for all variance components of nonlinear and spatial effects as our baseline choice. To evaluate the sensitivity of the results to the chosen prior, we compare different hyperpriors in Section 11.2.4.

The first step in any distributional regression analysis is the determination of an appropriate response distribution. In our case, we could choose from a large number of candidates for continuous responses with values on the real line; see Section 2.2.1 for an overview on the most prominent examples. Instead of conducting a thorough search among the candidate distributions, we rely on previous analyses and consider the normal distribution as a natural candidate for the distribution of Z-scores. This is particularly attractive for the bivariate models discussed in Section 11.3.

Assuming a constant variance, that is a homoscedastic distribution, leads to a geoadditive model with Gaussian responses, that is

$$M_1: \; Z_{\text{wa},i} \sim \mathcal{N}(\mu(\boldsymbol{x}_i), \sigma^2)$$

with

$$\mu(\boldsymbol{x}_i) = \eta_i^\mu = \beta_0^\mu + \textbf{sex}_i \beta_1^\mu + s_1^\mu(\textbf{age}_i) + \textbf{sex}_i s_2^\mu(\textbf{age}_i) + s_{\text{spat}}^\mu(\textbf{state}_i). \tag{11.2}$$

GAMLSS now allows us to go one step further and investigate whether the assumption of homoscedastic variances is indeed reasonable. We therefore consider a second model specification in which the variance is also covariate-dependent, leading to

$$M_2: \; Z_{\text{wa},i} \sim \mathcal{N}(\mu(\boldsymbol{x}_i), \sigma^2(\boldsymbol{x}_i)).$$

We assume the same predictor as before for the mean and also apply this predictor to the log-variance, that is

$$\log(\sigma^2(\boldsymbol{x}_i)) = \eta_i^\sigma = \beta_0^\sigma + \textbf{sex}_i \beta_1^\sigma + s_1^\sigma(\textbf{age}_i) + \textbf{sex}_i s_2^\sigma(\textbf{age}_i) + s_{\text{spat}}^\sigma(\textbf{state}_i), \tag{11.3}$$

where we use the logarithmic link function to ensure positivity of the variances.

11.2.2 MCMC Diagnostics

We estimate both M_1 and M_2 with the **R** package **bamlss** (Umlauf et al., 2018), utilizing 12,000 MCMC iterations, a burn-in of 2000 iterations and a thinning of 10, which leaves us with a total of 1000 samples. Before we analyze the results achieved from these simulations, we have to validate the convergence of the posterior distribution to the stationary distribution, and the achievement of satisfactory mixing. (See Section 6.4.3 for details.)

We start with a visual check similar to that in Figure 6.4, visualizing the traces of the MCMC samples for all model parameters. However, the number of parameters involved in complex GAMLSS specifications such as ours is usually prohibitively

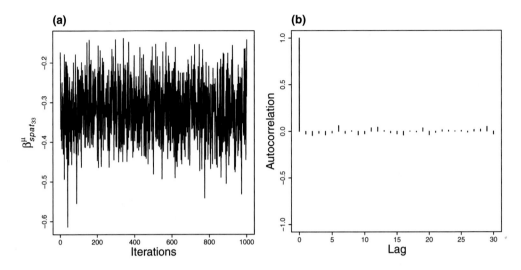

Figure 11.1 MCMC sampling path after thinning and removal of burn-in (panel (a)) and autocorrelation (ACF) plot (panel (b)) for $\beta^{\mu}_{\text{spat}_{33}}$ (the spatial effect of the state Uttar Pradesh, which is the state with the most observations in our dataset) in model M_1.

large, too large to show all of them in this chapter. Since we did not find any indication against convergence, we show one exemplary sampling path for the regression coefficient for the **state** Uttar Pradesh in Figure 11.1(a). Figure 11.9(b) shows the geographical location of this Indian region. The sampling path is non-persistently moving around the same mean.

In addition to the visual check, one can do further output analyses and diagnostics for MCMC, facilitated, for example, by the implementations in the **R** package **coda** (Plummer et al., 2006). Here, we focus on autocorrelation plots and effective sample size for the coefficients in the model. High autocorrelations are an indication of poor mixing and slow convergence of the corresponding chain, which also implies low effective sample sizes. In agreement with the visual appearance of the sampling path in Figure 11.1(a), panel (b) shows fast declining autocorrelations between the final thinned sample of the coefficient. The corresponding effective sample size is 1,000, indicating the best possible mixing of this particular trajectory.

11.2.3 Model Comparison

To compare the two rival model specifications M_1 (homoscedastic variances) and M_2 (heteroscedastic variances), we employ normalized quantile residuals (see Section 4.7.1) and Bayesian information criteria (see Section 6.8.3). The normalized quantile residuals shown in Figure 11.2 should be close to the diagonal line if the considered model provides a good fit to the data. We find that for the majority of observations the heteroscedastic model M_2 seems to provide a somewhat better fit.

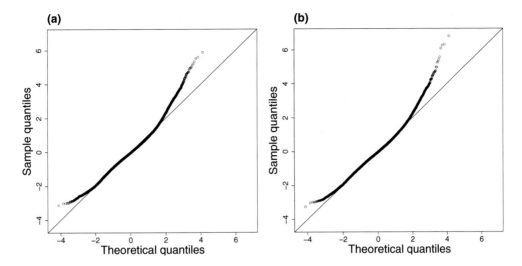

Figure 11.2 Normalized quantile residuals for M_1 (panel (a)) and M_2 (panel (b)), where the y-axis shows the empirical normalized quantile residuals, while the x-axis shows the theoretical quantiles of a standard normal.

Both models deviate from the ideal diagonal line in the tails, indicating that a better fit could be achieved with more complex models, enabling regression effects beyond mean and variance or allowing for different distributional shapes by construction. For the illustrations in this chapter, we ignore these deviations and accept the normal distribution as a reasonable working assumption.

To supplement the graphical assessment with some quantitative evidence, we use the DIC and WAIC. The results (DIC_{M_1} =82,156, DIC_{M_2} =81,128 and WAIC_{M_1} =82,159, WAIC_{M_2} =81,151) both favor M_2, that is, the more flexible distribution model, which we therefore use for the following interpretation of results.

If predictive ability is of interest, it is also possible to evaluate proper scoring rules on some new data. Since in our example no new data are present, we illustrate the usage of the log-score based on 10-fold cross-validation. The values for models M_1 and M_2 are −41082 and −40589 respectively, which are in line with the DIC/WAIC and graphical checks based on quantile residuals.[1]

11.2.4 Regression Effects and Hyperprior Sensitivity

As mentioned in Section 6.7, the inverse gamma is a popular choice for the variance components in semiparametric terms with a multivariate Gaussian prior for the basis coefficients, owing to its conjugacy to the multivariate normal distribution. However,

---

[1] Note that we defined proper scores with positive orientation in Section 4.4.2 such that larger scores are better. This is in contrast to the negative orientation of DIC and WAIC, where smaller values indicate a better fit.

this is a convenience argument rather than an objective reason to follow the standard approach. In particular, for random effects models there has been considerable debate about the sensitivity with respect to hyperpriors for the variance parameters; see for example, Gelman (2006) for details. In the light of this it is good practice to study the results of Bayesian approaches to GAMLSS under different hyperpriors. In the following, we illustrate this with the results for the mean of the normal distribution in the heteroscedastic model specification M_2, and consider the following hyperpriors:

- inverse gamma with $a = 0.001$ and $b = 0.001$,

- scale-dependent with $b = 0.009$,

- half-Cauchy with $b = 0.01$, and

- uniform with $b = 0.27$.

The scale parameters b for the scale-dependent, half-Cauchy and uniform priors correspond to the default values in `bamlss` and are based on $c = 3$ and $\alpha = 0.1$ in the scaling criterion in (6.5). To visualize the age-specific variation in the mean of the wasting score, we do not show the raw estimates $\hat{s}_1^\mu(\text{age}_i)$ (age effect for girls) and $\beta_1^\mu + \hat{s}_1^\mu(\text{age}_i) + \hat{s}_2^\mu(\text{age}_i)$ (age effect for boys) but rather apply an effect display where we fix all but the covariate age at pre-specified values and then determine the expected Z-score for a variety of ages. In our simple setting, we fix the spatial effect to the one associated with Uttar Pradesh (state 33), the state with the most observations. For gender, we consider separate representations for boys and girls to facilitate comparison of their age-specific variation in the wasting score. Based on the MCMC samples, we can also determine samples for the predicted means such that exact inference in general and uncertainty assessments in particular can be conducted. More specifically, the posterior mean estimates and quantile-based pointwise credible intervals are easily computed, see Figure 11.3.

From Figure 11.3, it is apparent that boys and girls equally are slightly undernourished at a very early age, with a strong deterioration until the age of one year. From then their wasting levels stay approximately constant, at a very low level. We also find that the choice of the hyperprior has no significant impact on the results, which is the result of the large sample size in our example. This of course cannot be generalized to other settings. Based on these results, we stick with our baseline choice $(\text{IG}(a = 0.001, b = 0.001))$ in the following.

Figure 11.4 presents the predicted standard deviation, applying a similar procedure as for the mean. In this case, however, we additionally have to apply the exponential response function to map all samples to the scale of the standard deviation. Like the effect on the mean, the predicted standard deviations are similar for boys and girls. For both genders, variation decreases with increasing age. This insight could only be gained in the heteroscedastic model M_2 in which the parameter $\sigma^2(\boldsymbol{x}_i)$ also depends on the covariates. As mentioned in Section 6.8.1, the plotted pointwise 95% credible intervals in Figures 11.3 and 11.4 do not address the problem of quantifying uncertainty of the complete estimated curve. Rather, pointwise credible intervals are

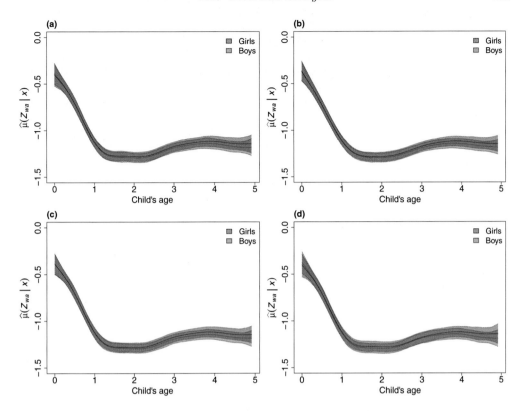

Figure 11.3 Model M_2: Predicted mean of `wasting` separately for boys and girls as a function of the children's ages with different hyperpriors for the variance component τ^2. (a) Inverse gamma $\mathrm{IG}(a = 0.001, b = 0.001)$, (b) inverse gamma $\mathrm{IG}(a = 0.01, b = 0.01)$, (c) half-Cauchy $\mathcal{C}^+(b = 0.01)$, and (d) uniform $\mathcal{U}(b = 0.27)$. The evaluated state is Uttar Pradesh (state 33), which is the state with the most observations. The areas around the lines indicate the 95% pointwise credible intervals. Red corresponds to girls, blue corresponds to boys, and the darker color indicates the largely overlapping parts of both.

applied separately for each covariate value (e.g. `age = 3`), while simultaneous credible bands should contain 95% of the complete curves, see Krivobokova et al. (2010) and Section 6.8.1 for details. In Figure 11.5, simultaneous credible bands are added to the pointwise intervals. The figure shows that the simultaneous credible regions are, as expected, slightly wider than their pointwise counterparts, but do not change the discussed interpretation of the results.

11.3 Bivariate Analysis

We now extend the univariate model to a bivariate model with a particular interest in the dependence structure between wasting and underweight as two important dimensions of childhood malnutrition.

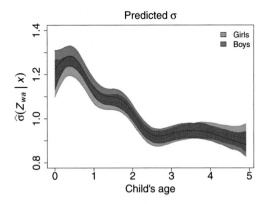

Figure 11.4 Model M_2: Predicted standard deviation together with 95% pointwise credible intervals separately for boys and girls as a function of the children's ages. The evaluated state is Uttar Pradesh (state 33), which is the state with the most observations.

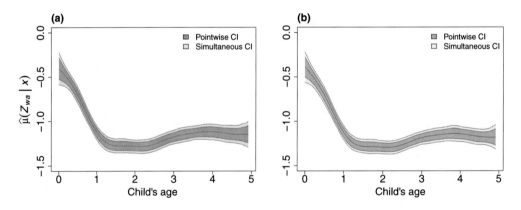

Figure 11.5 Model M_2: Predicted mean of `wasting` separately for girls (panel (a)) and boys (panel (b)) as a function of the children's ages. The evaluated state is Uttar Pradesh (state 33), which is the state with the most observations. The areas around the line indicate the 95% pointwise credible intervals and the 95% simultaneous credible bands.

11.3.1 Model Specification

In an earlier analysis of a survey from 1998 to 1999 in India, Klein et al. (2015b) showed that the strength of the dependence between different measures for childhood malnutrition varies depending on covariates such as the age of the child. The authors suggest a bivariate Gaussian GAMLSS with all distribution parameters (the two means, the two variances, and the correlation parameter) related to these covariates. Furthermore, the results of Klein and Kneib (2016b) indicate that the dependence between wasting and underweight is likely to be nonlinear, which cannot be captured by a bivariate Gaussian distribution but by copulas enabling the consideration of more general dependence types. (See Section 2.4.3 for a short introduction to

Table 11.1 *Response functions $h(\eta)$ transforming the predictors η to the distribution parameters for M_3 and M_4. The last row shows the formula for Kendall's τ.*

Model Parameter	M_3	M_4
μ_1	η	η
μ_2	η	η
σ_1	$\exp(\eta)$	$\exp(\eta)$
σ_2	$\exp(\eta)$	$\exp(\eta)$
θ	$\eta/(1+\eta^2)^{1/2}$	$\exp(\eta)$
Kendall's τ	$\frac{2}{\pi}\arcsin(\xi)$	$\frac{\xi}{\xi+2}$

copulas.) Specifically, the Clayton copula allows for lower tail dependence, that is, higher dependence for both scores being low, which is not possible with a Gaussian model where dependence in the lower and the upper tail is of the same strength by construction.

It is therefore our aim to compare a bivariate Gaussian GAMLSS (M_3) with a bivariate GAMLSS based on the Clayton copula and univariate normal distributions as marginals (M_4) for the joint bivariate responses $\boldsymbol{Z}_i = (Z_{\mathrm{uw},i}, Z_{\mathrm{wa},i})^\top$. In more detail, the models are specified as follows:

$$M_3: \; \boldsymbol{Z}_i \sim \mathcal{N}_2\left[\begin{pmatrix}\mu_{\mathrm{uw}}(\boldsymbol{x}_i)\\ \mu_{\mathrm{wa}}(\boldsymbol{x}_i)\end{pmatrix}, \begin{pmatrix}\sigma_{\mathrm{uw}}^2(\boldsymbol{x}_i) & \rho(\boldsymbol{x}_i)\sigma_{\mathrm{uw}}(\boldsymbol{x}_i)\sigma_{\mathrm{wa}}(\boldsymbol{x}_i)\\ \rho(\boldsymbol{x}_i)\sigma_{\mathrm{uw}}(\boldsymbol{x}_i)\sigma_{\mathrm{wa}}(\boldsymbol{x}_i) & \sigma_{\mathrm{wa}}^2(\boldsymbol{x}_i)\end{pmatrix}\right]$$

and

$$M_4: \; \boldsymbol{Z}_i \sim C_{\mathrm{Clayton}}(F(Z_{\mathrm{uw},i}), F(Z_{\mathrm{wa},i}); \xi(\boldsymbol{x}_i))$$
$$Z_{\mathrm{wa},i} \sim \mathcal{N}(\mu_{\mathrm{wa}}(\boldsymbol{x}_i), \sigma_{\mathrm{wa}}^2(\boldsymbol{x}_i))$$
$$Z_{\mathrm{uw},i} \sim \mathcal{N}(\mu_{\mathrm{uw}}(\boldsymbol{x}_i), \sigma_{\mathrm{uw}}^2(\boldsymbol{x}_i)).$$

As a result, M_3 has parameters $\mu_{\mathrm{uw}}(\boldsymbol{x}_i)$, $\mu_{\mathrm{wa}}(\boldsymbol{x}_i)$, $\sigma_{\mathrm{uw}}(\boldsymbol{x}_i)$, $\sigma_{\mathrm{wa}}(\boldsymbol{x}_i)$, and $\rho(\boldsymbol{x}_i)$ representing the marginal means, standard deviations and the linear correlation coefficient. In contrast, the Clayton copula has a more general dependence parameter $\xi(\boldsymbol{x}_i)$. All five distribution parameters (for both M_3 and M_4) are modeled through the general predictor specification given in equation (11.1). Each of the parameters is again related to the predictor by a response function $h(\cdot)$, see Table 11.1 for the specifications. For estimating both models, we use the software **BayesX** (Belitz et al., 2015) with 12,000 MCMC iterations, a burn-in of 2000 iterations and a thinning of 10 as before.

11.3.2 Estimated Contour Lines and Model Comparison

Figures 11.6 and 11.7 show contour lines representing the estimated densities of the bivariate response for M_3 and M_4, for four selected ages and for both genders. We observe a positive dependence between wasting and underweight and decreasing marginal means up to an age of two years, in both figures. Additionally, the

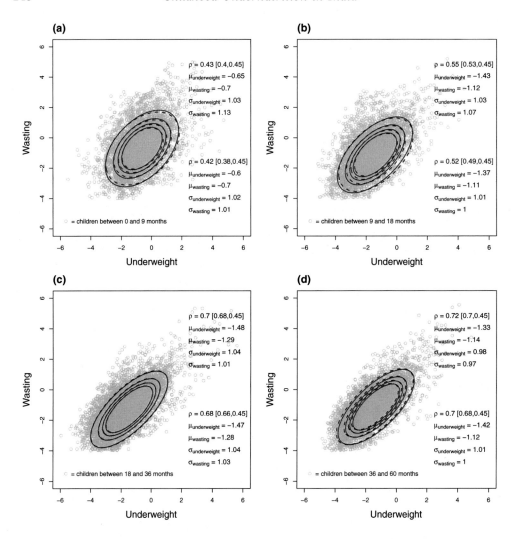

Figure 11.6 Contour lines of the estimated densities of M_3 (bivariate normal model) for boys (dashed blue) and girls (solid red) at four selected ages and in Uttar Pradesh (state 33): children's age is (a) six months, (b) one year, (c) 2 years, and (d) 4 years.. The upper and lower legend in each subfigure show the corresponding posterior mean estimates for the distribution parameters for girls and boys, respectively.

dependence increases with the child's age. However, Figure 11.7 reveals a stronger dependence in the lower tail, that is for both scores being low – an observation that cannot be captured by M_3. To decide between M_3 and M_4, we first use bivariate quantile residuals (Hohberg et al., 2021) in Figure 11.8. Normalized bivariate quantile residuals reveal a good model fit if they are approximately bivariate standard normally distributed (see Section 4.7.1).

Panel (a) in Figure 11.8 shows a decent fit for M_3, while the contour lines of M_4 in

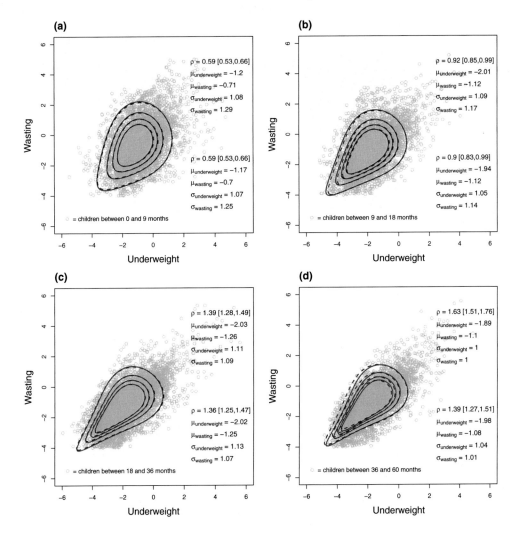

Figure 11.7 Contour lines of the densities of M_4 (Clayton copula with normal marginals) for boys (dashed blue) and girls (solid red) at four selected ages and in Uttar Pradesh (state 33): children's age is (a) six months, (b) one year, (c) 2 years, and (d) 4 years. The upper and lower legend in each subfigure show the corresponding posterior mean estimates for the distribution parameters for girls and boys, respectively.

panel (b) deviate more strongly from the reference of a bivariate standard normal. To further compare the two models, we use the DIC and WAIC as before. Especially in complex bivariate models, one should include the WAIC, since it does not rely on point estimates of the posterior distribution and is applicable even if the posterior distribution is far from a Gaussian distribution. In fact, both the DIC and WAIC clearly favor M_4 (DIC_{M_4} =133,041, WAIC_{M_4} =133,200) over M_3 (DIC_{M_3} =146,572, WAIC_{M_3} =146,641), such that we focus on M_4 in the following with a closer look at the dependence structure between the two scores. The disagreement between in-

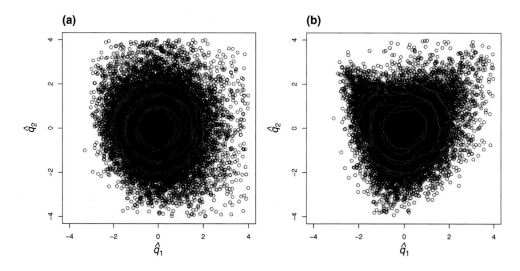

Figure 11.8 Contour lines of the bivariate quantile residuals of M_3 (panel (a)) and M_4 (panel (b)). Here, a kernel density estimate of the bivariate quantile residuals (blue) is compared to the theoretical residuals of a bivariate standard normal distribution (red). This can be used to evaluate the fit of the model. Close contour lines indicate a good fit of the model,

formation criteria and quantile residuals that we find here is sometimes observed especially for large datasets (see the corresponding comment in Section 4.7). Basically, in large datasets the information criteria are dominated by the majority of points in the center of the distribution while the residual plots give more visual attention to the tails.

11.3.3 Effects on the Dependence

We now use model M_4 to predict the dependence structure of the two malnutrition scores. As a measure for dependency we choose Kendall's τ. In contrast to for example, Pearson's correlation coefficient, Kendall's τ can be used to measure dependency beyond the linear case, since it measures the similarity based on an ordered association of the variables and not their actual values. It ranges between -1 and 1, where 1 can be interpreted as perfect association of the measures. As mentioned in Section 2.4.3, there exists a closed-form transformation of the Clayton copula parameter ξ to Kendall's τ, namely $\tau = \xi/(\xi + 2)$.

In Figure 11.9 one can see the estimated posterior means for both the spatial and nonlinear effects of age on τ. It can be concluded from that figure that there is indeed substantial spatial variation in the dependence structure of wasting and underweight, which may provide important information to decision makers when designing interventions combatting malnutrition. Moreover, the increasing dependence structure for older children suggests the urgency of tackling undernutrition in early childhood.

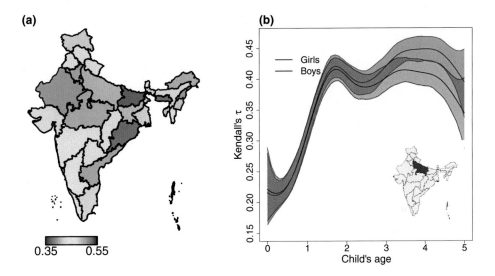

Figure 11.9 Predicted Kendall's τ for M_4 as a function of `state` (panel (a)) and as a function of child's age (panel (b)), separately for girls and boys. In (a), `age` was set to two years and the prediction was made for girls, although the spatial predictions for boys do not differ notably. In (b), the predictions were made for the `state` having the most observations (`state 33` = Uttar Pradesh, shown in red on the map in the corner.). The shaded areas mark the pointwise 95% credible intervals.

While M_3 assumes symmetry of the dependence for lower and upper tail, M_4 was able to model the stronger dependence of wasting and underweight at the lower tail.

We finally look at predicted joint probabilities of falling below the usual undernutrition cutoffs of -2 for malnutrition and -3 for severe malnutrition. This should also illustrate that we can easily derive quantities of particular interest from a fitted GAMLSS specfication. In particular, the Bayesian paradigm, with sampling-based inference, immediately allows the determination of uncertainty estimates for such derived quantities as well.

More precisely, we calculate the probabilities $\mathbb{P}(Z_{\mathrm{wa}} < -2$ and $Z_{\mathrm{uw}} < -2)$ (simultaneous malnourishment in both indicators) and $\mathbb{P}(Z_{\mathrm{wa}} < -3$ and $Z_{\mathrm{uw}} < -3)$ (simultaneous severe malnourishment in both indicators). Figure 11.10 shows that in those states where a strong dependence between underweight and wasting has already been identified in our previous results, the risk of moderate and severe underweight is also particularly large. However, Figure 11.10 also shows that moderate and severe undernutrition is a problem across India, with only states in the Northeast and far North not facing this problem to the same extent as the other states.

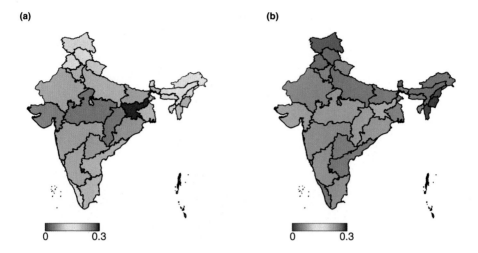

Figure 11.10 Posterior mean probabilities of a two-year-old girl being
(a) "undernourished" ($\mathbb{P}(Z_{\mathrm{wa}} < -2$ and $Z_{\mathrm{uw}} < -2)$) and
(b) "severely undernourished" ($\mathbb{P}(Z_{\mathrm{wa}} < -3$ and $Z_{\mathrm{uw}} < -3)$).
Results are similar for two-year-old boys.

11.4 Summary

To sum up, the GAMLSS regression framework allowed us to gain insights into how undernutrition in India is related to age and gender while spatial variation across states could also be identified. While standard models mostly focus on the mean of a univariate distribution, we have seen that it is also worth linking other distribution parameters to a structured predictor. The bivariate analysis showed that dependence between responses can be flexibly modeled within the GAMLSS framework. Especially in cases when researchers do not want to restrict their model to the Gaussian case, that is to better capture tail-dependence, a model such as M_4 can be a useful alternative.

12

Socioeconomic Determinants of Federal Election Outcomes in Germany

In this chapter, we provide a second example on Bayesian inference, complementary to Chapter 11. We consider a multivariate response model for compositional data, that is, multiple continuous fractions summing to one, based on the Dirichlet distribution.

12.1 Data and Research Question

We illustrate the analysis of compositional responses with a study on the outcome of Germany's federal election in 2017. We focus on an ecological regression setup where we associate aggregate characteristics of the 413 districts on NUTS-3 (Nomenclature des Unités Territoriales Statistiques) level with the proportions of votes achieved for the six main political parties in Germany. This provides us with a multivariate extension of beta regression utilizing the Dirichlet distribution (see Section 2.4.2 for a formal definition) as the response distributions. Our analysis is based on a similar study conducted in Klein et al. (2015b) and we will also follow a Bayesian approach for inference as suggested there.

The data utilized in this chapter are available from the German Federal Statistics Office ("Statistische Ämter des Bundes und der Länder," www.destatis.de). For each of the 413 districts on NUTS-3 level ("Landkreise") in Germany, the data contain proportions of the electorate voting (the response variable) for the six main parties (Christian Democratic Union and Christian Social Union "CDU/CSU," the Social Democratic Party "SPD," the Liberals "FDP," the Left "Die Linke," the Greens "Die Grünen," and the Alternative for Germany "AfD") plus "others," which is the combined voting share for all other parties. Consequently, the proportion of votes for the seven response components sums to one in each district. As covariates we consider district-specific quantities:

- the percent of electorates (PoE) compared with the population entitled to vote (the turnout),

- the rate of unemployment in 2017 in percent,

- the gross domestic product per capita (GDPpC) in 2016 (measured in thousand Euros), and

- the 38 administrative `region`s on NUTS-2 level within which the NUTS-3 level districts are nested.

We do not use the NUTS-3 level information as a spatial covariate since this would yield quite unstable estimates due to the large number of regression effects (six spatial effects parameters per NUTS-3 level unit).

The $D = 7$ dimensional response vector $\boldsymbol{y} = (y_1, \ldots, y_D)$ is assumed to follow a Dirichlet distribution $\text{Dir}(\boldsymbol{\alpha})$ with $\boldsymbol{\alpha} = (\alpha_1, \ldots, \alpha_D)$ and $\alpha_d > 0$, $d = 1, \ldots, D$. For the interpretation of Dirichlet regression results, we will make use of the relation

$$\mathbb{E}(y_d) = \frac{\alpha_d}{\sum_{r=1}^{D} \alpha_r},$$

that is, the expectation of response category d is proportional to α_d. As usual in distributional regression, each α_d is linked to an additive predictor with a logarithmic link function, that is, $\eta_d = \log(\alpha_d)$. The predictor for the dth response category (i.e. the dth party in our study) is given by η_d, where

$$\eta_{d,i} = \beta_0 + s_{d,1}(\texttt{PoE}_i) + s_{d,2}(\texttt{GDPpC}_i) + s_{d,3}(\texttt{unemployment}_i) + s_{d,\text{spat}}(\texttt{region}_i).$$

The nonlinear effects $s_{d,1}, s_{d,2}, s_{d,3}$ are represented as cubic penalized splines (see Section 3.1 for details) with 20 inner knots and second-order random walk prior. For the spatial effects, we assumed Gaussian Markov random field priors (see Section 3.5). For the smoothing variances, we consider inverse gamma priors with hyperparameters $a = b = 0.001$. We estimate the model with the **R** package **bamlss** (Umlauf et al., 2018) based on 12,000 MCMC iterations, a burn-in of 2000 iterations and afterwards apply thinning of 10.

12.2 Estimation Results

Since we are relying on simulation-based Bayesian inference, the first step after fitting the model is to evaluate mixing and convergence of the MCMC chains. Here we can rely on the same tools as in Chapter 11. As an illustration, Figure 12.1 shows (a) the MCMC sampling path of the spatial effect for the liberals in Berlin and (b) the corresponding autocorrelation function. Both do not indicate any convergence problems. In fact, the effective sample size of this trajectory is not 1,000, which would indicate perfect mixing, but only 218. To achieve a better effective sample size, one could now increase the iterations and the thinning parameter and rerun the model. For the sake of illustration, we stick to the results from a relatively short chain.

We now have a closer look at results from the model. Owing to the exponential response function and the complex relation between the regression effects and the expected proportions predicted by the model, we resort to effect displays where we vary only one covariate while keeping all other covariates fixed at pre-specified values. Specifically, Figure 12.2 shows the spatial variation in election outcomes induced by the spatial effects $s_{d,\text{spat}}(\texttt{region}_i)$ when fixing all other covariates at their

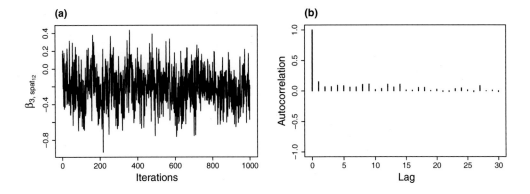

Figure 12.1 (a) MCMC sampling path of the spatial effect for the liberals in Berlin and (b) the autocorrelation function of the same coefficient after thinning and burn-in.

global means and then determining the expected election outcome under this common covariate scenario for all districts. Note that this does not represent the spatial variation of actually observed election outcomes but rather the spatial variation represented by the spatial effect. The actually observed proportions also depend on the spatial variation in the covariates, which is neglected in Figure 12.2 for the sake of illustration.

The spatial effect for the CDU/CSU is highest in a large number of regions in Germany, reflecting a strong preference for the conservative party under our "means" scenario. The spatial impact for the other parties is less equally distributed. In the former German Democratic Republic, the eastern part of Germany, the Left is traditionally strong, where in other parts they play a more minor role. Conversely, the SPD, the Liberals, and the Greens are traditionally weaker in East Germany than in West Germany. In the Southwest of Germany, there is a strong tendency towards the Greens, which may be related to the popular Green prime minister of the time.

In Figure 12.3, we proceed similarly for the three nonlinear effects of the percent of electorates, unemployment rate, and gross domestic product. More precisely, for each party we plot the predicted proportion of votes as a function of the respective covariate while setting all other covariates to their averages (GDPpC = 35913, PoE = 75.9, unemployment = 5.3) and restricting the spatial effect to be equal to zero. In addition to point estimates, we provide 95% pointwise credible intervals which can be easily obtained from the MCMC samples. The unemployment rate tends to have a positive effect for the votes of SPD and a negative effect for the votes of CDU/CSU, which matches with usual expectations. The impact of the GDP per capita is also not surprising and is in line with the prejudice of the wealthy Green voters and the "working class" SPD voters. The vote participation suggests that the AfD especially is recruiting voters where the general vote participation is low. On

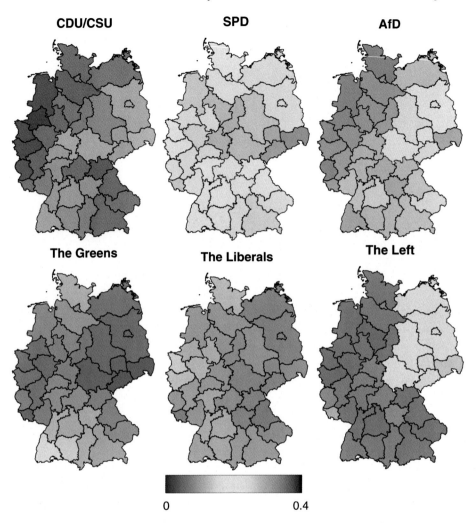

Figure 12.2 Predicted proportions of votes in each region based on posterior mean spatial effect (all other effects are evaluated at global average covariate values).

the contrary, the Greens and Liberals have relatively low predicted proportions of votes when participation is low, but both profit from higher participation rates.

12.3 Scenario Analysis

While the discussion of results in Section 12.2 allowed us to derive some general associations of socio-economic quantities with expected election outcomes, we can go one step further and employ the model for scenario analyses. We illustrate this using the two large German cities Berlin and Munich. Not only are the two cities far apart geographically, but their socio-economic indicators (see Table 12.1) and voting

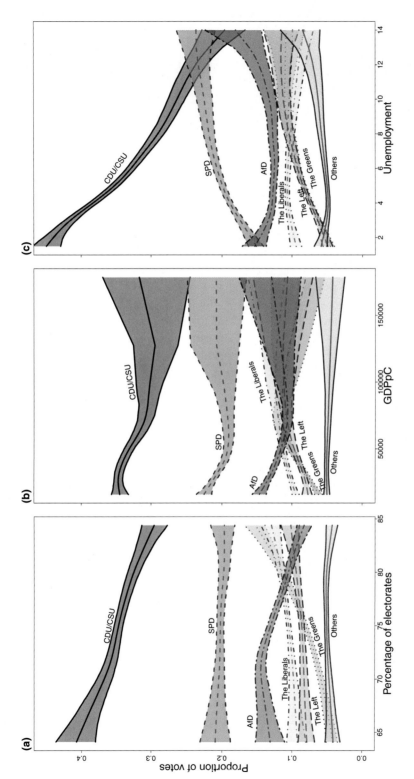

Figure 12.3 Predicted federal election proportions of votes as functions of the covariates. In each individual plot the spatial effect is excluded and all other covariates are set to their global average value ((a) percent of Electorates = 75.9, (b) GDPpC =35,913, and (c) unemployment = 5.3). The plots show the predicted posterior mean together with their 95% pointwise credible interval.

Table 12.1 *True covariates and assumed covariates in the scenario analysis for the cities Berlin and Munich.*

	Berlin		Munich	
	True	Scenario	True	Scenario
unemployment	9	5.5	2.6	5.5
GDPpC	36,798		100,475	
PoE	75.6		83.9	

patterns are also very different (see Table 12.2). In the following, we study the change in expected voting outcomes if, rather than having quite distinct unemployment rates (Berlin: 9%, Munich: 2.6%) both cities had a common unemployment rate of 5.5% (see also Table 12.1).

Table 12.2 summarizes the predicted election outcomes under the observed covariates as well as under the hypothetical covariates. While for the former the predicted values are quite similar to the true voting shares, assuming the same unemployment rate in both cities changes the predicted voting shares considerably. The predicted voting shares are indeed much closer in this scenario, highlighting the strong association of unemployment rate and voting behavior. While, of course, our model is not able to identify causal effects from observational data on an aggregate level, it nonetheless allows us to conduct interesting thought experiments concerning hypothetical voting outcomes under changes in the covariate setup.

Table 12.2 *Observed voting shares, predicted voting shares for the true covariates and predicted voting shares for the described scenario in Berlin and Munich, together with posterior 95% credible intervals.*

	Berlin			Munich		
	True	Predicted	Pred. scenario	True	Predicted	Pred. scenario
CDU/CSU	0.23	0.23 (0.19;0.27)	0.25 (0.23;0.28)	0.37	0.34 (0.32;0.37)	0.29 (0.24; 0.34)
SPD	0.18	0.18 (0.14;0.22)	0.14 (0.12;0.16)	0.14	0.12 (0.11;0.14)	0.17 (0.14;0.21)
AfD	0.12	0.13 (0.10;0.16)	0.07 (0.06;0.09)	0.09	0.09 (0.08;0.10)	0.13 (0.10;0.16)
The Greens	0.13	0.12 (0.09;0.16)	0.24 (0.21;0.27)	0.13	0.17 (0.14;0.19)	0.10 (0.07;0.13)
The Liberals	0.09	0.09 (0.06;0.12)	0.14 (0.12;0.16)	0.15	0.14 (0.13;0.16)	0.10 (0.07;0.13)
The Left	0.19	0.19 (0.15;0.23)	0.10 (0.08;0.12)	0.05	0.07 (0.06;0.08)	0.16 (0.12;0.20)
Others	0.07	0.07 (0.05;0.10)	0.06 (0.05;0.07)	0.06	0.07 (0.06;0.08)	0.06 (0.04;0.09)

13

Variable Selection for Gene Expression Data

13.1 Data and Research Question

The following application illustrates the use of boosting to estimate GAMLSS on datasets with more explanatory variables than observations. The drawback of such high-dimensional genetic examples is that the underlying research question is not easily assessable. Also the interpretation of the combined effect of different genes is far less interpretable than more classical applications with well-known variables such as age or gender.

In our case, the research question involves the analysis of riboflavin (vitamin B2) production with *Bacillus subtilis* in relation to gene regulation and expression. *Bacillus subtilis* is a bacterium that is often used in research as model organism to study cell differentiation and chromosome replication, but in industry is also used for the synthetic production of biological products such as riboflavin, vitamin B6, and amino acids. The expression levels, which serve as explanatory variables, were measured with an antisense oligonucleotide microarray using Affymetrix technology. The dataset is available as `riboflavin` in the **R** package **hdi** for high-dimensional inference (Bühlmann et al., 2014). The dataset contains $p = 4{,}088$ explanatory variables (gene expression data) for only $n = 71$ samples.

The logarithm of the riboflavin production rate serves as the response variable, while the logarithm of the normalized expression levels of the 4,088 genes are the potential explanatory variables. The aim is to determine the most informative genes in a distributional regression setting, identifying genes that are responsible for the mean riboflavin (vitamin B_2) rate as well as its variability (location scale model). For this high-dimensional example with $p \gg n$, the only feasible solution to estimate GAMLSS based on the inference procedures presented in this book, is to apply statistical boosting for model fitting and variable selection. Statistical boosting is discussed in detail in Chapter 7.

13.2 Model Comparison and Variable Selection via Boosting

In a first step, we consider a simple Gaussian response distribution and let the boosting algorithm select the predictors for both the location parameter μ and the variance parameter σ. Figure 13.1 displays the coefficient paths for the first 100 iterations from

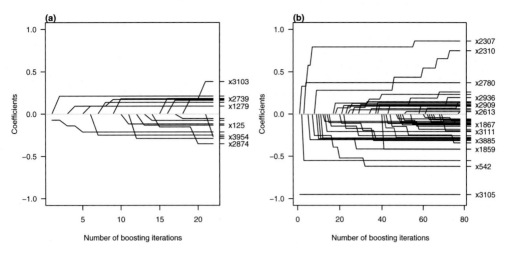

Figure 13.1 Coefficient paths from boosting 100 iterations for a Gaussian location scale model with linear base-learners for the high-dimensional gene expression example on riboflavin production. The algorithm chose to update the μ component 22 times (a), and the σ part 78 times (b).

a Gaussian location and scale model. One can clearly observe how the complexity of the model increases with the number of boosting iterations as more and more coefficient paths rise from zero, as the corresponding base-learners are updated. It's also noteworthy that many updates are actually carried out for the scale parameter (78 iterations for σ and only 22 for μ).

To put these results into perspective, we also apply the classical L_2 boosting for the mean. Further, we consider the use of non-Gaussian response distributions and use the function `chooseDist()` to select the best distribution based on information criteria. As classical inference via the algorithms implemented in the **gamlss** package is infeasible for this $p > n$ scenario, we use here an intercept-only model, leading to the choice of the Gumbel distribution.

To evaluate the prediction performance as well as variable selection properties of boosting with these three modeling options, we apply leave-one-out cross-validation (LOO-CV): The model is fitted n times after leaving out one observation i, $i = 1, \ldots, n$, and evaluating the prediction model on this particular observation afterwards (see also Section 4.4.3). Prediction accuracy is assessed via the mean squared error of prediction (MSEP) and the median absolute error (MAE), as these are common statistical evaluation criteria for the prediction of continuous outcomes. We use simple linear models as base-learners for all potential covariates (both for μ and for σ) and determine the stopping iteration via bootstrapping, using $B = 25$ bootstrap samples on each of the folds of LOO-CV. For the Gumbel distribution we had to set the step-length to 0.01 (not the default 0.1) to ensure stability.

Analyzing the results with respect to the prediction performance of the final models (Table 13.1), we observe the best predictions for the models based on a Gaussian

Table 13.1 *Results of leave-one-out cross-validation (LOO-CV) on the riboflavin data. Numbers represent the mean squared error of prediction (MSEP) the median absolute error (MAE), and the log-score (LS) on the corresponding test observations. For the MSEP and MAE lower values are better; for the LS a higher value indicates a better model. The numbers of selected variables are averaged (mean) over all $n = 71$ folds.*

Model	MSEP	MAE	LS	Potential variables	Selected variables	
					μ	σ
L_2 mean regression	0.204	0.241	−0.962	4,088	33.2	-
Gaussian (μ, σ) regression	0.214	0.234	−0.764	4,088	19.4	3.0
Gumbel (μ, σ) regression	0.375	0.358	−1.043	4,088	24.5	9.0

distribution. The lowest mean squared error of prediction (MSEP) is observed for the classical L_2-boosting (mean regression), while the lowest median absolute error (MAE) is achieved by Gaussian distributional regression. Also the log-score (LS, log-likelihood on test data) is best for Gaussian distributional regression. However, the MSEP and MAE values are quite stable across the two Gaussian approaches. The Gumbel distribution performs worse for all three considered criteria. The number of selected variables for the mean parameter is the highest for L_2-boosting (33 selected variables) and decreases for both location scale models (19.4 for the Gaussian distribution and 24.5 for the Gumbel). Fewer variables were selected in this case for the scale parameter (3.0 and 9.0).

Overall, one can conclude that this small example with $n = 71$ and $p = 4,088$ nicely illustrates how boosting can be applied for the selection of sparse models for distributional regression.

13.2.1 *Comparison with Cyclical Boosting Approach*

The description of the fitting scheme of **gamboostLSS** in Chapter 7 referred to the so-called *noncyclical* fitting scheme that was proposed by Thomas et al. (2018). The original version of statistical boosting for GAMLSS incorporated a *cyclical* update scheme (Mayr et al., 2012a).

While the cyclical version updates the additive predictors for the distribution parameters one-by-one, the noncyclical method selects the best-performing update with respect to the distribution parameters, at each iteration. The advantage of the newer noncyclical update scheme is faster tuning, as only one overall stopping iteration m_{stop} has to be selected. In the case of the classical cyclical update, in order to allow for different complexities of the various parameter dimensions, a grid search for the best combination of $\left(m_{\text{stop}}^{\theta_1}, \ldots, m_{\text{stop}}^{\theta_K}\right)$ is performed. On the other hand, the cyclical version tends to lead in some situations to more stable results and is less dependent on the offset (starting) values of the parameters.

In order to illustrate the different approaches, we performed the same LOO-CV analysis as above, with the classical cyclical update scheme for the Gaussian distributional regression model, and a separate grid search for each left-out observation. In order to ensure stable results, we had to decrease the step-length to 0.01 and perform more boosting iterations.

While, in general, both types of algorithms lead to similar performance, in this illustrative example there were some minor differences. The prediction accuracy of the point decreased slightly (MSEP = 0.244) while the log-score remained the same (LS = –0.764). The number of selected variables increased, likely also because of the smaller step-length. In this case, the algorithm selected on average 28.2 variables for μ and 6.7 variables for σ.

13.2.2 Additional Approaches towards Enhanced Variable Selection

There have been various attempts to further improve the variable selection properties of statistical boosting, particularly for settings where the algorithm tends to select too many variables (see also Section 7.2.5). This is much more likely to occur in low-dimensional settings ($p < n$), particularly for large datasets (Strömer et al., 2022). The reason for this behavior is that the algorithm aims to optimize a prediction problem. The fact that while doing this, variable selection is performed, is basically just a side-effect of the component-wise updates in combination with early stopping. In the case of larger datasets with a relatively low number of potential predictors, the prediction will not become worse when some variables are falsely selected and updated a few times. As the resulting effect sizes might be small, and the size of the model is not directly regularized, the selection of m_{stop} will lead to a rather late stopping.

In these settings it can make sense not to use a stopping iteration m_{stop} that was selected to optimize the prediction accuracy (e.g. based on a validation set or resampling procedures) but actually a smaller one (e.g. via the one standard error rule). An approach to completely disentangle the stopping of the algorithm and prediction accuracy is *probing*, where m_{stop} is determined based on falsely selected noninformative variables. While both approaches typically lead to earlier stopping and therefore also stronger shrinkage of effect estimates (smaller coefficients), a different concept avoids this side-effect by actively deselecting base-learners that had been originally selected by the algorithm if they do not contribute enough to the final model fit.

In the following we illustrate these additional approaches with the Gaussian location and scale model via the LOO-CV analysis. The results are summarized in Table 13.2.

One Standard Error Rule (1se-rule)

The idea of the 1se-rule is not the use the optimal m_{stop}, but the smallest one that still leads to a test risk which is only one standard error larger that the minimal one (for a more detailed description see Section 7.2.5). In our case, applying the 1se-rule led to a decrease of m_{stop} from 49 iterations (median) to 36 iterations. As a result of the earlier stopping, the algorithm selected on average only 15 variables for μ and 1.4 for σ. However, this also affects the prediction accuracy: The MSEP for the point

Table 13.2 *Results of leave-one-out cross-validation (LOO-CV) on the riboflavin data comparing boosting Gaussian (μ, σ) distributional regression with additional approaches for enhanced variable selection. The noncyclical boosting is identical to the Gaussian (μ, σ) regression from Table 13.1. Numbers represent the mean squared error of prediction (MSEP), the median absolute error (MAE), and the log-score (LS) on the corresponding test observations. For MSEP and MAE lower values indicate higher accuracy; while for the LS higher values are better. The numbers of selected variables are averaged (mean) over all $n = 71$ folds.*

Variant	MSEP	MAE	LS	Selected variables	
				μ	σ
Noncyclical boosting	0.214	0.234	−0.764	19.4	3.0
Cyclical boosting	0.244	0.295	−0.764	28.2	6.7
1se-rule	0.284	0.288	−0.792	15	1.4
Probing	0.545	0.473	−1.136	5.3	0.4
Stability selection	0.438	0.358	−1.065	1.9	0
Deselection	0.217	0.247	−0.798	18.4	2

prediction increases from 0.214 to 0.284 while the log-score also got slightly worse, from −0.764 to −0.792.

Probing

The idea of probing was introduced for boosting by Thomas et al. (2017) and completely disentangles the tuning of the algorithm from prediction accuracy, focusing directly on the variable selection properties. Not only are the original p explanatory variables are included via their base-learners for potential updates, but the algorithm can also choose from p additional shadow variables without an effect on the outcome. These noninformative artificial shadow variables (*probes*) are constructed via permutation (shuffling) of the original variables. The algorithm is then simply stopped when it first chooses one of the noninformative probes as an update. The final model is then that of the previous iteration, before the first probe was selected.

In our case, probing for the riboflavin data led to a median m_{stop} of only 16 iterations (compared to the median optimal m_{stop} of 49). As a result, the algorithm selected on average only 5.3 genes for μ and 0.4 genes for σ. However, this led to a notably lower prediction accuracy (MSEP = 0.545, LS = −1.136).

Stability Selection

Stability selection is a general approach to enhance variable selection techniques and was introduced by Meinshausen and Bühlmann (2010) and Shah and Samworth (2013). The concept is based on performing variable selection in combination with subsampling. Via this resampling approach, one can identify variables that are *stable* in the sense that they are selected on most of the subsamples. As a result, only variables that pass a threshold with respect to their selection rates on the subsamples are included in the final model. This general concept was extended to boosting by Hofner et al. (2015); for further discussion see also Section 7.3.4. Stability selection is controlled via the number of variables q that are selected on each subsample and the per-family-error-rate which then leads to a threshold value.

In case of the riboflavin data we chose $q = 20$ and a PFER of 1, leading to a threshold of $\pi_{\text{stabsel}} = 0.524$. With these settings, stability selection leads to very sparse models, identifying on average only 1.9 variables for the location parameter as *stable* and none for the scale parameter σ. Refitting a boosting model with only these stable predictors resulted in slightly better prediction accuracy than the earlier stopping via probing.

Deselection

The concept of deselection is considerably newer and was particularly developed for statistical boosting (Strömer et al., 2022). Instead of stopping the algorithm earlier, which inevitably leads to more shrinkage of effect sizes, the m_{stop} is chosen as usual. Afterwards, selected variables are deselected again, if their contribution to the model fit is only negligible. To identify variables that were selected but are of minor importance for the prediction, the idea is to evaluate the overall risk reduction of the model. If only a small proportion of the risk reduction can be attributed to updates from the corresponding base-learner of this variable (the authors propose a threshold of 0.01, i.e. 1%) the variable is *deselected*. Afterwards, the model is refitted with the same m_{stop} but only with the initially selected variables which passed the threshold.

This pragmatic procedure works fairly well in practice, and leads typically to smaller models with a very similar prediction accuracy. In our case with the riboflavin data the differences are not huge, because the classical boosting approach already led to very sparse models. Deselection led to slightly smaller models (on average 18.4 variables selected for μ and 2.0 for σ), while showing nearly the same prediction accuracy the classical approach (MSEP = 0.215, LS = –0.798).

Appendix A

Continuous Distributions

Table A.1: Continuous distributions implemented within the **gamlss** package, with default link functions. (id = `identity`, log-2 = `logshiftto2`)

Distribution	**gamlss** name	Parameter link functions			
		μ	σ	ν	τ
$\mathcal{S} = \mathbb{R}$					
Exponential Gaussian	`exGAUS`	id	log	log	-
Exponential gen beta 2	`EGB2`	id	log	log	log
Generalized t	`GT`	id	log	log	log
Gumbel	`GU`	id	log	-	-
Johnson's SU	`JSU, JSUo`	id	log	id	log
Logistic	`LO`	id	log	-	-
NET	`NET`	id	log	fixed	fixed
Normal	`NO, NO2`	id	log	-	-
Normal family	`NOF`	id	log	-	-
Power exponential	`PE`	id	log	log	-
Reverse Gumbel	`RG`	id	log	-	-
Sinh-arcsinh	`SHASH`	id	log	log	log
Sinh-arcsinh original	`SHASHo, SHASHo2`	id	log	id	log
Skew t type 1	`ST1`	id	log	id	log
Skew t type 2	`ST2`	id	log	id	log
Skew t type 3	`ST3`	id	log	log	log
Skew t type 3 repar	`SST`	id	log	log	log-2
Skew t type 4	`ST4`	id	log	log	log
Skew t type 5	`ST5`	id	log	id	log
Skew normal type 1	`SN1`	id	log	id	-
Skew normal type 2	`SN2`	id	log	log	-
Skew power exp type 1	`SEP1`	id	log	id	log
Skew power exp type 2	`SEP2`	id	log	id	log
Skew power exp type 3	`SEP3`	id	log	log	log
Skew power exp type 4	`SEP4`	id	log	log	log
t family	`TF`	id	log	log	-
t family repar	`TF2`	id	log	log-2	-
$\mathcal{S} = \mathbb{R}_+$					
Box–Cox t	`BCT`	id	log	id	log
Box–Cox t orig	`BCTo`	log	log	id	log

Table A.1 – continued from previous page

Distribution	gamlss	Parameter link functions			
	name	μ	σ	ν	τ
Box–Cox Cole Green	BCCG	id	log	id	-
Box–Cox Cole Green orig	BCCGo	log	log	id	-
Box–Cox power exp	BCPE	id	log	id	log
Box–Cox power exp orig	BCPEo	log	log	id	log
Exponential	EXP	log	-	-	-
Gamma	GA	log	log	-	-
Gamma family	GAF	log	log	id	-
Generalized beta type 2	GB2	log	log	log	log
Generalized gamma	GG	log	log	id	-
Generalized inv Gaussian	GIG	log	log	id	-
Inverse Gamma	IGAMMA	log	log	-	-
Inverse Gaussian	IG	log	log	-	-
Log-normal	LOGNO	id	log	-	-
Log-normal 2	LOGNO2	log	log	-	-
Log-normal (Box–Cox)	LNO	id	log	fixed	-
Pareto 2	PARETO2, PARETO2o, GP	log	log	-	-
Weibull	WEI, WEI2, WEI3	log	log	-	-
$\mathcal{S} = \mathbb{R}_{(0,1)}$					
Beta	BE	logit	logit	-	-
Generalized beta type 1	GB1	logit	logit	log	log
Logit-normal	LOGITNO	id	log	-	-
Simplex	SIMPLEX	logit	log	-	-
$\mathcal{S} = \{y > \mu - (\sigma/\nu)\}$					
Reverse gen extreme	RGE	id	log	log	-

Appendix B

Discrete Distributions

Table B.1: Discrete distributions implemented within the **gamlss** package, with default link functions. (id = `identity`)

Distribution	**gamlss** name	μ	σ	ν	τ
		\multicolumn{4}{c}{}			
$\mathcal{S} = \mathbb{N}$					
Beta neg binomial	BNB	log	log	log	-
Geometric	GEOM	log	-	-	-
Geometric (original)	GEOMo	logit	-	-	-
Discrete Burr XII	DBURR12	log	log	log	-
Delaporte	DEL	log	log	logit	-
Double Poisson	DPO	log	log	-	-
Negative binomial type I	NBI	log	log	-	-
Negative binomial type II	NBII	log	log	-	-
Neg binomial family	NBF	log	log	id	-
Poisson	PO	log	-	-	-
Poisson–inv Gaussian	PIG	log	log	-	-
Poisson–inv Gaussian 2	PIG2	log	log	-	-
Poisson shifted GIG	PSGIG	log	log	logit	logit
Sichel	SI	log	log	id	-
Sichel (μ the mean)	SICHEL	log	log	id	-
Waring (μ the mean)	WARING	log	log	-	-
Yule (μ the mean)	YULE	log	-	-	-
Zero-adj beta neg binom	ZABNB	log	log	id	logit
Zero-adj logarithmic	ZALG	logit	logit	-	-
Zero-adj neg binomial	ZANBI	log	log	logit	-
Zero-adj neg binom fam	ZANBF	log	log	log	logit
Zero-adj PIG	ZAPIG	log	log	logit	-
Zero-adj Sichel	ZASICHEL	log	log	id	logit
Zero-adj Poisson	ZAP	log	logit	-	-
Zero-adj Zipf	ZAZIPF	log	logit	-	-
Zero-inf beta neg binom	ZIBNB	log	log	log	logit
Zero-inf neg binomial	ZINBI	log	log	logit	-
Zero-inf neg binom fam	ZINBF	log	log	log	logit
Zero-inf Poisson	ZIP	log	logit	-	-
Zero-inf Poisson (μ the mean)	ZIP2	log	logit	-	-
Zero-inf PIG	ZIPIG	log	log	logit	-

Table B.1 – continued from previous page

Distribution	gamlss	Parameter link functions			
	name	μ	σ	ν	τ
Zero-inf Sichel	ZISICHEL	log	log	id	logit
$\mathcal{S} = \mathbb{N}_+$					
Logarithmic	LG	logit	-	-	-
Zipf	ZIPF	log	-	-	-
$\mathcal{S} = \{0, 1, \ldots, n\}$					
Binomial	BI	logit	-	-	-
Beta binomial	BB	logit	log	-	-
Double binomial	DBI	logit	log	-	-
Zero-adj beta binomial	ZABB	logit	log	logit	-
Zero-adj binomial	ZABI	logit	logit	-	-
Zero-inf beta binomial	ZIBB	logit	log	logit	-
Zero-inf binomial	ZIBI	logit	logit	-	-

References

Aeberhard, W. H., Cantoni, E., Marra, G., and Radice, R. 2021. Robust fitting for generalized additive models for location, scale and shape. Statistics and Computing, **31**(1), 1–16.

Aitkin, M. 1987. Modelling variance heterogeneity in normal regression using GLIM. Applied Statistics, **36**, 332–339.

Aitkin, M. 2010. Statistical Inference: an Integrated Bayesian/Likelihood Approach. CRC Press.

Aitkin, M. 2018. A History of the GLIM Statistical Package. International Statistical Review, **86**(2), 275–299.

Aitkin, M. 2019. The Universal model and prior: multinomial GLMs. arXiv:1901.02614.

Akaike, H. 1973. Maximum likelihood identification of Gaussian autoregressive moving average models. Biometrika, **60**, 255–265.

Akaike, H. 1983. Information measures and model selection. Bulletin of the International Statistical Institute, **50**(1), 277–290.

Anderson, N. G., Jolley, I. J., and Wells, J. E. 2007. Sonographic estimation of fetal weight: comparison of bias, precision and consistency using 12 different formulae. Ultrasound in Obstetrics and Gynecology, **30**(2), 173–179.

Barndorff-Nielsen, O. 2014. Information and Exponential Families: in Statistical Theory. John Wiley & Sons.

Belitz, C., Brezger, A., Klein, N., Kneib, T., Lang, S., and Umlauf, N. 2015. BayesX – Software for Bayesian inference in structured additive regression models. www.bayesx.org. Version 3.0.2.

Belkin, M., Hsu, D., Ma, S., and Mandal, S. 2019. Reconciling modern machine-learning practice and the classical bias–variance trade-off. Proceedings of the National Academy of Sciences, **116**(32), 15849–15854.

Bien, J., Taylor, J., and Tibshirani, R. 2013. A lasso for hierarchical interactions. The Annals of Statistics, **41**(3), 1111.

Bondell, H., Reich, B., and Wang, H. 2010. Noncrossing quantile regression curve estimation. Biometrika, **97**, 825–838.

Box, G. E. P. 1979. Robustness in the strategy of scientific model building. Robustness in Statistics, **1**, 201–236.

Breiman, L. 2001. Statistical modeling: The two cultures (with comments and a rejoinder by the author). Statistical Science, **16**(3), 199–231.

Breiman, L., Friedman, J. H., Olshen, R. A., and Stone, C. J. 2017. Classification and Regression Trees. Milton Park, Oxfordshire: Routledge.

Bühlmann, P. 2020. Invariance, causality and robustness. Statistical Science, **35**(3), 404–426.

Bühlmann, P., and Hothorn, T. 2007. Boosting algorithms: Regularization, prediction and model fitting (with discussion). Statistical Science, **22**, 477–522.

Bühlmann, P., Kalisch, M., and Meier, L. 2014. High-dimensional statistics with a view toward applications in biology. Annual Review of Statistics and Its Application, **1**(1), 255–278.

Carlan, M., Kneib, T., and Klein, N. 2023. Bayesian conditional transformation models. Journal of the American Statistical Association. 10.1080/01621459.2023.2191820.

Chambers, J. M., and Hastie, T. J. 1992. Statistical Models in S. London: Chapman & Hall.

Chen, T., He, T., Benesty, M., Khotilovich, V., Tang, Y., Cho, H., Chen, K., Mitchell, R., Cano, I., Zhou, T., Li, M., Xie, J., Lin, M., Geng, Y., Li, Y., and Yuan, J. 2023. xgboost: Extreme Gradient Boosting. R package version 1.7.3.1.

Chernozhukov, V., Fernández-Val, I., and Galichon, A. 2009. Improving point and interval estimators of monotone functions by rearrangement. Biometrika, **96**(3), 559–575.

Cho, S.-J., Wu, H., and Naveiras, M. 2022. The effective sample size in Bayesian information criterion for level-specific fixed and random effects selection in a two-level nested model. arXiv. 10.48550/ARXIV.2206.11880.

Claeskens, G., and Hjort, N. L. 2008. Model Selection and Model Averaging. Cambridge University Press.

Cleveland, W. S., Grosse, E., and Shyu, W. M. 2017. Local regression models. Pages 309–376 of: Chambers, J. M., and Hastie, T. J. (eds), Statistical Models in S. Routledge.

Cole, T. J., and Green, P. J. 1992. Smoothing reference centile curves: The LMS method and penalized likelihood. Statistics in Medicine., **11**, 1305–1319.

Cover, T. M., and Thomas, J. A. 2007. Elements of Information Theory Second Edition: Solutions to Problems. `https://cpb-us-w2.wpmucdn.com/sites.gatech.edu/dist/c/565/files/2017/01/solutions2.pdf`.

Cox, D. R., and Reid, N. 1987. Parameter orthogonality and approximate conditional inference. Journal of the Royal Statistical Society, Series B (Statistical Methodology), **49**(1), 1–18.

Crainiceanu, C., Ruppert, D., Claeskens, G., and Wand, M. P. 2005. Exact likelihood ratio tests for penalised splines. Biometrika, **92**, 91–103.

Currie, I. D., and Durban, M. 2002. Flexible smoothing with P-splines: a unified approach. Statistical Modelling, **4**, 333–349.

De Bastiani, F., Stasinopoulos, D. M., Rigby, R. A., Heller, G. Z., and Silva, L. A. 2022. Bucket plot: A visual tool for skewness and kurtosis comparisons. Brazilian Journal of Probability and Statistics, **36**(3), 421–440.

Dean, C., Lawless, J. F., and Willmot, G. E. 1989. A mixed Poisson–inverse Gaussian regression model. Canadian Journal of Statistics, **17**(2), 171–181.

Diggle, P. J., Tawn, J. A., and Moyeed, R. A. 1998. Model-based geostatistics. Applied Statistics, **47**(3), 229–350.

Dobson, A. J., and Barnett, A. G. 2018. An Introduction to Generalized Linear Models. 4th edn. Chapman and Hall/CRC.

Efron, B., and Tibshirani, R. J. 1994. An Introduction to the Bootstrap. CRC Press.

Efron, B., Hastie, T., Johnstone, I., and Tibshirani, R. 2004. Least angle regression. The Annals of Statistics, **32**, 407–451.

Eilers, P. H. C., and Marx, B. D. 1996. Flexible smoothing with B-splines and penalties. Statistical Science, 89–102.

Eilers, P. H. C., and Marx, B. D. 2021. Practical Smoothing: The Joys of P-splines. Cambridge University Press.

Ellenbach, N., Boulesteix, A.-L., Bischl, B., Unger, K., and Hornung, R. 2021. Improved outcome prediction across data sources through robust parameter tuning. Journal of Classification, **38**(2), 212–231.

Engle, R. 2001. GARCH 101: The use of ARCH/GARCH models in applied econometrics. Journal of Economic Perspectives, **15**(4), 157–168.

Engle, R. F. 1982. Autoregressive conditional heteroscedasticity with estimates of the variance of United Kingdom inflation. Econometrica: Journal of the Econometric Society, 987–1007.

Fahrmeir, L., and Kaufmann, H. 1985. Consistency and asymptotic normality of the maximum likelihood estimator in generalized linear models. The Annals of Statistics, **13**, 342–368.

Fahrmeir, L., and Kneib, T. 2009. Propriety of posteriors in structured additive regression models: Theory and empirical evidence. Journal of Statistical Planning and Inference, **139**, 843–859.

Fahrmeir, L., and Kneib, T. 2011. Bayesian Smoothing and Regression for Longitudinal, Spatial and Event History Data. Oxford University Press.

Fahrmeir, L., and Lang, S. 2001. Bayesian inference for generalized additive mixed models based on Markov random field priors. Applied Statistics, **50**, 201–220.

Fahrmeir, L., Kneib, T., and Lang, S. 2004. Penalized structured additive regression for space-time data: a Bayesian perspective. Statistica Sinica, **14**(3), 731–762.

Fahrmeir, L., Kneib, T., Lang, S., and Marx, B. 2021. Regression – Models, Methods and Applications. Berlin: Springer.

Faschingbauer, F., Dammer, U., Raabe, E., Schneider, M., Faschingbauer, C., Schmid, M., Mayr, A., Schild, R. L., Beckmann, M. W., and Kehl, S. 2015. Sonographic weight estimation in fetal macrosomia: influence of the time interval between estimation and delivery. Archives of Gynecology and Obstetrics, **292**, 59–67.

Fasiolo, M., Wood, S. N., Zaffran, M., Nedellec, R., and Goude, Y. 2021. Fast calibrated additive quantile regression. Journal of the American Statistical Association, **116**(535), 1402–1412.

Fenske, N., Kneib, T., and Hothorn, T. 2011. Identifying risk factors for severe childhood malnutrition by boosting additive quantile regression. Journal of the American Statistical Association, **106**(494), 494–510.

Fox, J. 2003. Effect displays in R for generalised linear models. Journal of Statistical Software, **8**(15), 1–27.

Francis, B. J., Green, M., and Payne, C. 1993. GLIM 4: The Statistical System for Generalized Linear Interactive Modelling. Oxford: Clarendon Press.

Fredriks, A. M., van Buuren, S., Wit, J. M., and Verloove-Vanhorick, S. P. 2000a. Body index measurements in 1996–7 compared with 1980. Archives of Childhood Diseases, **82**, 107–112.

Fredriks, A. M., van Buuren, S., Burgmeijer, R. J. F., Meulmeester, J. F., Beuker, R. J., Brugman, E., Roede, M. J., Verloove-Vanhorick, S. P., and Wit, J. M. 2000b. Continuing positive secular change in The Netherlands, 1955–1997. Pediatric Research, **47**, 316–323.

Freund, Y. 1995. Boosting a weak learning algorithm by majority. Information and Computation, **121**(2), 256–285.

Frühwirth-Schnatter, S., Frühwirth, R., Held, L., and Rue, H. 2009. Improved auxiliary mixture sampling for hierarchical models of non-Gaussian data. Statistics and Computing, **19**, 479–492.

Gamerman, D. 1997. Efficient sampling from the posterior distribution in generalized linear mixed models. Statistics and Computing, **7**, 57–68.

Gelman, A. 2006. Prior distributions for variance parameters in hierarchichal models. Bayesian Analysis, **1**, 515–533.

Gelman, A., Carlin, J. B., Stern, H. S., Dunson, D. B., Vehtari, A., and Rubin, D. B. 2013. Bayesian Data Analysis. Boca Raton: Chapman & Hall/CRC.

Gholami, A. M., Hahne, H., Wu, Z., Auer, F. J., Meng, C., Wilhelm, M., and Kuster, B. 2013. Global proteome analysis of the NCI-60 cell line panel. Cell Reports, **4**(3), 609–620.

Gilks, W. R., and Wild, P. 1992. Adaptive rejection sampling for Gibbs sampling. Journal of the Royal Statistical Society, Series C (Applied Statistics), **41**(2), 337–348.

Gneiting, T., and Raftery, A. E. 2007. Strictly proper scoring rules, prediction, and estimation. Journal of the American Statistical Association, **102**, 359–378.

Good, I. J. 1976. Contribution to the discussion of Savage, L.J. (1976). On rereading R. A. Fisher. The Annals of Statistics, **4**(3), 441–500.

Gosset, W. S. 1908. The probable error of a mean. Biometrika, **6**(1), 1–25.

Green, P. J. 1987. Penalized likelihood for general semi-parametric regression models. International Statistical Review, **55**, 245–259.

Greven, S., and Kneib, T. 2010. On the behaviour of marginal and conditional Akaike information criteria in linear mixed models. Biometrika, **97**, 773–789.

Gu, C. 2002. Smoothing Spline ANOVA Models. Springer Verlag.

Gupta, R. D., and Richards, D. St. P. 2001. The history of the Dirichlet and Liouville distributions. International Statistical Review, **69**(3), 433–446.

Hardin, J. W., and Hilbe, J. M. 2002. Generalized Estimating Equations. Chapman and Hall/CRC.

Harrell, F. E. Jr. 2015. Regression Modeling Strategies: With Applications to Linear Models, Logistic and Ordinal Regression, and Survival Analysis. 2nd edn. Springer.

Hastie, T., and Tibshirani, R. 1986. Generalized additive models. Statistical Science, 297–310.

Hastie, T., Tibshirani, R., and Tibshirani, R. 2020. Best subset, forward stepwise or lasso? Analysis and recommendations based on extensive comparisons. Statistical Science, **35**(4), 579–592.

Hastie, T. J., and Tibshirani, R. J. 1990. Generalized Additive Models. London: Chapman & Hall.

Hastie, T. J., Tibshirani, R. J., and Friedman, J. 2009. The Elements of Statistical Learning: Data Mining, Inference and Prediction. 2nd edn. New York: Springer.

Hastings, W. K. 1970. Monte Carlo sampling methods using Markov chains and their applications. Biometrika, **57**, 97–109.

Held, L., and Sabanés Bové, D. 2012. Applied Statistical Inference. Springer Verlag.

Heller, G. Z., Couturier, D.-L., and Heritier, S. R. 2019. Beyond mean modelling: Bias due to misspecification of dispersion in Poisson-inverse Gaussian regression. Biometrical Journal, **61**(2), 333–342.

Hepp, T., Schmid, M., Gefeller, O., Waldmann, E., and Mayr, A. 2016. Approaches to regularized regression – A comparison between gradient boosting and the lasso. Methods of Information in Medicine, **55**(5), 422–430.

Hepp, T., Schmid, M., and Mayr, A. 2019. Significance Tests for Boosted Location and Scale Models with Linear Base-Learners. The International Journal of Biostatistics, **15**(1).

Hoerl, A., and Kennard, R. 1988. Ridge regression. Pages 129–136 of: Encyclopedia of Statistical Sciences, vol. 8. New York: Wiley.

Hofner, B., Hothorn, T., Kneib, T., and Schmid, M. 2011. A framework for unbiased model selection based on boosting. Journal of Computational and Graphical Statistics, **20**, 956–971.

Hofner, B., Mayr, A., Robinzonov, N., and Schmid, M. 2014. Model-based boosting in R: a hands-on tutorial using the R package mboost. Computational Statistics, **29**(1), 3–35.

Hofner, B., Boccuto, L., and Göker, M. 2015. Controlling false discoveries in high-dimensional situations: boosting with stability selection. BMC Bioinformatics, **16**(1), 1–17.

Hofner, B., Mayr, A., and Schmid, M. 2016a. gamboostLSS: An R package for model building and variable selection in the GAMLSS framework. Journal of Statistical Software, **74**(1), 1—-31.

Hofner, B., Kneib, T., and Hothorn, T. 2016b. A unified framework of constrained regression. Statistics and Computing, **26**, 1–14.

Hohberg, M., Donat, F., Marra, G., and Kneib, T. 2021. Beyond unidimensional poverty analysis using distributional copula models for mixed ordered-continuous outcomes. Journal of the Royal Statistical Society, Series C (Applied Statistics), **70**, 1365—-1390.

Holmes, C. C., and Held, L. 2006. Bayesian auxiliary variable models for binary and multinomial regression. Bayesian Analysis, **1**, 145–168.

Homan, M. D., and Gelman, A. 2014. The No-U-Turn sampler: Adaptively setting path lengths in Hamiltonian Monte Carlo. Journal of Machine Learning Research, **15**(1), 1593–1623.

Hothorn, T. 2020. Transformation boosting machines. Statistics and Computing, **30**, 141–152.

Hothorn, T., Kneib, T., and Bühlmann, P. 2014. Conditional transformation models. Journal of the Royal Statistical Society, Series B (Statistical Methodology), **76**(1), 3–27.

Hothorn, T., Möst, L., and Bühlmann, P. 2018. Most likely transformations. Scandinavian Journal of Statistics, **45**(1), 110–134.

Hu, W., Swanson, B. A., and Heller, G. Z. 2015. A statistical method for the analysis of speech intelligibility tests. PLoS ONE, **10**(7).

Hüllermeier, E., and Waegeman, W. 2021. Aleatoric and epistemic uncertainty in machine learning: an introduction to concepts and methods. Machine Learning, **110**, 457—-506.

Huzurbazar, V. S. 1950. Probability distributions and orthogonal parameters. Pages 281–284 of: Mathematical Proceedings of the Cambridge Philosophical Society, vol. 46. Cambridge University Press.

Ichimura, H. 1993. Semiparametric least squares (SLS) and weighted SLS estimation of single-index models. Journal of Econometrics, **58**(1), 71–120.

James, G., Witten, D., Hastie, T., and Tibshirani, R. 2013. An Introduction to Statistical Learning: with Applications in R. Springer.

Jarque, C. M., and Bera, A. K. 1987. A test for normality of observations and regression residuals. International Statistical Review, **55**(2), 163–172.

Joe, H. 1997. Multivariate Models and Dependence Concepts. London: Chapman and Hall.

Kammann, E. E., and Wand, M. P. 2003. Geoadditive models. Journal of the Royal Statistical Society, Series C (Applied Statistics), **52**, 1–18.

Kauermann, G., Krivobokova, T., and Fahrmeir, L. 2009. Some asymptotic results on generalized penalized spline smoothing. Journal of the Royal Statistical Society, Series B (Statistical Methodology), **71**, 487–503.

Klein, N., and Kneib, T. 2016a. Scale-dependent priors for variance parameters in structured additive distributional regression. Bayesian Analysis, **11**, 1107–1106. 10.1214/15-BA983.

Klein, N., and Kneib, T. 2016b. Simultaneous inference in structured additive conditional copula regression models: A unifying Bayesian approach. Statistics and Computing, **26**, 841–860.

Klein, N., Kneib, T., and Lang, S. 2015a. Bayesian generalized additive models for location, scale and shape for zero-inflated and overdispersed count data. Journal of the American Statistical Association, **110**, 405–419. 10.1080/01621459.2014.912955.

Klein, N., Kneib, T., Klasen, S., and Lang, S. 2015b. Bayesian structured additive distributional regression for multivariate responses. Journal of the Royal Statistical Society, Series C (Applied Statistics), **64**, 569–591. 10.1111/rssc.12090.

Klein, N., Kneib, T., Lang, S., and Sohn, A. 2015c. Bayesian structured additive distributional regression with an application to regional income inequality in Germany. The Annals of Applied Statistics, **9**(2), 1024–1052.

Klein, N., Carlan, M., Kneib, T., Lang, S., and Wagner, H. 2021. Bayesian effect selection in structured additive distributional regression models. Bayesian Analysis, **16**(2), 545–573. 10.1214/20-BA1214.

Klein, N., Hothorn, T., Barbanti, L., and Kneib, T. 2022. Multivariate conditional transformation models. Scandinavian Journal of Statistics, **49**(1), 116–142.

Kneib, T. 2013. Beyond mean regression. Statistical Modelling, **13**(4), 275–303.

Kneib, T., and Fahrmeir, L. 2007. A mixed model approach for geoadditive hazard regression. Scandinavian Journal of Statistics, **34**, 207–228.

Kneib, T., Hothorn, T., and Tutz, G. 2009. Variable selection and model choice in geoadditive regression models. Biometrics, **65**(2), 626–634.

Kneib, T., Klein, N., Lang, S., and Umlauf, N. 2019. Modular regression – a Lego system for building structured additive distributional regression models with tensor product interactions. TEST, **28**(1), 1–39.

Kneib, T., Silbersdorff, A., and Säfken, B. 2023. Rage against the mean – A review of distributional regression approaches. Econometrics and Statistics, **26**, 99–123. 10.1016/j.ecosta.2021.07.006.

Koenker, R. 2005. Quantile Regression. Econometric Society Monographs. Cambridge University Press. 10.1017/CBO9780511754098.

Koenker, R., Chernozhukov, V., He, X., and Peng, L. (eds). 2020. Handbook of Quantile Regression. CRC Press.

Kozumi, H., and Kobayashi, G. 2011. Gibbs sampling methods for Bayesian quantile regression. Journal of Statistical Computation and Simulation, **81**, 1565–1578.

Krivobokova, T., Crainiceanu, C. M., and Kauermann, G. 2008. Fast adaptive penalized splines. Journal of Computational and Graphical Statistics, **17**, 1–20.

Krivobokova, T., Kneib, T., and Claeskens, G. 2010. Simultaneous confidence bands for penalized spline estimators. Journal of the American Statistical Association, **105**(490), 852–863. 10.1198/jasa.2010.tm09165.

Kullback, S., and Leibler, R. A. 1951. On Information and Sufficiency. The Annals of Mathematical Statistics, **22**, 79–86.

Laird, N. M., and Ware, J. H. 1982. Random-effect models for longitudinal data. Biometrics, **38**, 963–974.

Lang, S., Umlauf, N., Wechselberger, P., Harttgen, K., and Kneib, T. 2014. Multilevel structured additive regression. Statistics and Computing, **24**, 223–238.

Lorah, J., and Womack, A. 2019. Value of sample size for computation of the Bayesian information criterion (BIC) in multilevel modeling. Behavior Research Methods, **51**(1), 440–450.

Marra, G., and Radice, R. 2020. GJRM: Generalised Joint Regression Modelling. R package version 0.2-2.

Marra, G., and Wood, S. N. 2012. Coverage properties of confidence intervals for generalized additive model components. Scandinavian Journal of Statistics, **39**, 53–74.

Marra, G., Radice, R., Bärnighausen, T., Wood, S. N., and McGovern, M. E. 2017. A simultaneous equation approach to estimating HIV prevalence with nonignorable missing responses. Journal of the American Statistical Association, **112**(518), 484–496.

Mayr, A., and Hofner, B. 2018. Boosting for statistical modelling – A non-technical introduction. Statistical Modelling, **18**(3-4), 365–384.

Mayr, A., Fenske, N., Hofner, B., Kneib, T., and Schmid, M. 2012a. Generalized additive models for location, scale and shape for high-dimensional data – A flexible approach based on boosting. Journal of the Royal Statistical Society, Series C (Applied Statistics), **61**(3), 403–427.

Mayr, A., Hofner, B., and Schmid, M. 2012b. The importance of knowing when to stop. Methods of Information in Medicine, **51**(02), 178–186.

Mayr, A., Binder, H., Gefeller, O., and Schmid, M. 2014a. The evolution of boosting algorithms. Methods of Information in Medicine, **53**(6), 419–427.

Mayr, A, Binder, H, Gefeller, O, and Schmid, M. 2014b. Extending statistical boosting. Methods of Information in Medicine, **53**(6), 428–435.

Mayr, A., Schmid, M., Pfahlberg, A., Uter, W., and Gefeller, O. 2015. A permutation test to analyse systematic bias and random measurement errors of medical devices via boosting location and scale models. Statistical Methods in Medical Research. 10.1177/0962280215581855.

Mayr, A., Hofner, B., and Schmid, M. 2016. Boosting the discriminatory power of sparse survival models via optimization of the concordance index and stability selection. BMC Bioinformatics, **17**(1), 1–12.

Mayr, A., Hofner, B., Waldmann, E., Hepp, T., Meyer, S., and Gefeller, O. 2017. An update on statistical boosting in biomedicine. Computational and Mathematical Methods in Medicine, **2017**. 10.1155/2017/6083072.

McCullagh, P., and Nelder, J. A. 1989. Generalized Linear Models. 2nd edn. London: Chapman & Hall.

Meinshausen, N., and Bühlmann, P. 2010. Stability selection (with discussion). Journal of the Royal Statistical Society, Series B (Statistical Methodology), **72**(4), 417–473.

Min, Y., and Agresti, A. 2002. Modeling nonnegative data with clumping at zero: a survey. Journal of the Iranian Statistical Society, **1**(1), 7–33.

Moro, S., Rita, P., and Vala, B. 2016. Predicting social media performance metrics and evaluation of the impact on brand building: A data mining approach. Journal of Business Research, **69**(9), 3341–3351.

Muggeo, V. M. R., and Ferrara, G. 2008. Fitting generalized linear models with unspecified link function: A P-spline approach. Computational Statistics & Data Analysis, **52**(5), 2529–2537.

Mullahy, J. 1986. Specification and testing of some modified count data models. Journal of Econometrics, **33**(3), 341–365.

Müller, S., Scealy, J. L., and Welsh, A. H. 2013. Model selection in linear mixed models. Statistical Science, **28**(2), 135–167.

Murdoch, W. J., Singh, C., Kumbier, K., Abbasi-Asl, R., and Yu, B. 2019. Definitions, methods, and applications in interpretable machine learning. Proceedings of the National Academy of Sciences, **116**(44), 22071–22080.

Nagelkerke, N. J. D. 1991. A note on a general definition of the coefficient of determination. Biometrika, **78**(3), 691–692.

Neal, R. M. 2003. Slice sampling. The Annals of Statistics, **31**, 705–767. 10.1214/aos/1056562461.

Nelder, J. A., and Pregibon, D. 1987. An extended quasi-likelihood function. Biometrika, **74**, 221–232.

Nelder, J. A., and Wedderburn, R. W. M. 1972. Generalized linear models. Journal of the Royal Statistical Society, Series A, **135**, 370–384.

Nelsen, R. B. 2007. An Introduction to Copulas. Springer Science & Business Media.

Newey, W. K., and Powell, J. L. 1987. Asymmetric least squares estimation and testing. Econometrica: Journal of the Econometric Society, **55**(4), 819–847.

Oelker, M., and Tutz, G. 2017. A uniform framework for the combination of penalties in generalized structured models. Advances in Data Analysis and Classification, **11**, 97—-120.

Ospina, R., and Ferrari, S. L. P. 2010. Inflated beta distributions. Statistical Papers, **51**, 111–126.

Pinheiro, J. C., and Bates, D. M. 2000. Mixed-Effects Models in S and S-Plus. Springer.

Plummer, M., Best, N., Cowles, K., and Vines, K. 2006. CODA: Convergence Diagnosis and Output Analysis for MCMC. R News, **6**(1), 7–11.

Puka, L. 2011. Kendall's tau. Pages 713–715 of: Lovric, M. (ed), International Encyclopedia of Statistical Science. Berlin, Heidelberg: Springer.

Pya, N., and Wood, S. N. 2015. Shape constrained additive models. Statistics and Computing, **25**, 543–559.

Ramires, T. G., Nakamura, L. R., Righetto, A. J., Pescim, R. R., Mazuchelli, J., Rigby, R. A., and Stasinopoulos, D. M. 2021. Validation of stepwise-based procedure in GAMLSS. Journal of Data Science, **19**(1), 96–110.

Ramsay, J. O., and Silverman, B. W. 2005. Functional Data Analysis. 2nd edn. Springer.

Rigby, R. A., and Stasinopoulos, D. M. 1996. A semi-parametric additive model for variance heterogeneity. Statistics and Computing, **6**, 57–65.

Rigby, R. A., and Stasinopoulos, D. M. 2005. Generalized additive models for location, scale and shape. Journal of the Royal Statistical Society, Series C (Applied Statistics), **54**(3), 507–554.

Rigby, R. A., and Stasinopoulos, D. M. 2006. Using the Box–Cox t distribution in GAMLSS to model skewness and kurtosis. Statistical Modelling, **6**(3), 209.

Rigby, R. A., and Stasinopoulos, D. M. 2013. Automatic smoothing parameter selection in GAMLSS with an application to centile estimation. Statistical Methods in Medical Research, **23**(4), 318–332.

Rigby, R. A., Stasinopoulos, M. D., Heller, G. Z., and De Bastiani, F. 2019. Distributions for Modeling Location, Scale, and Shape: Using GAMLSS in R. Boca Raton: Chapman & Hall/CRC.

Robert, C. P., and Casella, G. 2010. Introducing Monte Carlo Methods with R. Springer.

Rodrigues, T., and Fan, Y. 2017. Regression adjustment for noncrossing Bayesian quantile regression. Journal of Computational and Graphical Statistics, **26**(2), 275–284.

Royston, P. 2007. Profile likelihood for estimation and confidence intervals. Stata Journal, **7**(3), 376–387.

Rubin, D. B. 1981. The Bayesian bootstrap. The Annals of Statistics, 130–134.

Rue, H., and Held, L. 2005. Gaussian Markov Random Fields: Theory and Applications. CRC Press.

Rue, H., Martino, S., and Chopin, N. 2009. Approximate Bayesian inference for latent Gaussian models by using integrated nested Laplace approximations. Journal of the Royal Statistical Society, Series B (Statistical Methodology), **71**(2), 319–392.

Rügamer, D., and Greven, S. 2020. Inference for L_2-boosting. Statistics and Computing, **30**(2), 279–289.

Ruppert, D., Wand, M. P., and Carroll, R. J. 2003. Semiparametric Regression. Cambridge University Press.

Säfken, B., Rügamer, D., Kneib, T., and Greven, S. 2021. Conditional Model Selection in Mixed-Effects Models with cAIC4. Journal of Statistical Software, **99**(8), 1–30.

Schall, R. 1991. Estimation in generalized linear models with random effects. Biometrika, **78**, 719–727.

Schapire, R. E. 1990. The strength of weak learnability. Machine Learning, **5**(2), 197–227.

Schapire, R. E., and Freund, Y. 2012. Boosting: Foundations and Algorithms. MIT Press.

Scheipl, F., and Kneib, T. 2009. Locally adaptive Bayesian P-splines with a normal-exponential-gamma prior. Computational Statistics & Data Analysis, **53**(10), 3533–3552.

Scheipl, F., Fahrmeir, L., and Kneib, T. 2012. Spike-and-slab priors for function selection in structured additive regression models. Journal of the American Statistical Association, **107**, 1518–1532.

Schnabel, S. K., and Eilers, P. 2009. Optimal expectile smoothing. Computational Statistics & Data Analysis, **53**, 4168–4177.

Schnabel, S. K., and Eilers, P. H. C. 2013. Simultaneous estimation of quantile curves using quantile sheets. AStA Advances in Statistical Analysis, **97**(1), 77–87.

Schwarz, G. E. 1978. Estimating the dimension of a model. The Annals of Statistics, **6**(2), 461–464.

Shah, R. D., and Samworth, R. J. 2013. Variable selection with error control: another look at stability selection. Journal of the Royal Statistical Society, Series B (Statistical Methodology), **75**(1), 55–80.

Siegfried, S., and Hothorn, T. 2020. Count transformation models. Methods in Ecology and Evolution, **11**(7), 818–827.

Simpson, D., Rue, H., Martins, T. G., Riebler, A., and Sørbye, S. H. 2017. Penalising model component complexity: A principled, practical approach to constructing priors. Statistical Science, **32**(1), 1–28.

Smith, G. 2018. Step away from stepwise. Journal of Big Data, **5**(1), 32.

Smithson, M., and Verkuilen, J. 2006. A better lemon squeezer? Maximum-likelihood regression with beta-distributed dependent variables. Psychological Methods, **11**(1), 54.

Smyth, G. K. 1989. Generalized linear models with varying dispersion. Journal of the Royal Statistical Society, Series B (Statistical Methodology), **51**, 47–60.

Sobotka, F., and Kneib, T. 2012. Geoadditive expectile regression. Computational Statistics & Data Analysis, **56**, 755–767.

Sørbye, S. H., and Rue, H. 2014. Scaling intrinsic Gaussian Markov random field priors in spatial modelling. Spatial Statistics, **8**, 39–51.

Sottile, G., and Frumento, P. 2021. Parametric estimation of non-crossing quantile functions. Statistical Modelling, 1471082X211036517.

Speed, T. 1991. Comment on Robertson (1991): "That BLUP is a good thing: The estimation of random effects". Statistical Science, **6**, 42–44.

Speller, J., Staerk, C., and Mayr, A. 2022. Robust statistical boosting with quantile-based adaptive loss functions. The International Journal of Biostatistics.

Spiegel, E., Kneib, T., and Otto-Sobotka, F. 2019. Generalized additive models with flexible response functions. Statistics and Computing, **29**(1), 123–138.

Spiegelhalter, D. J., Best, N. G., Carlin, B. P., and van der Linde, A. 2002. Bayesian measures of model complexity and fit. Journal of the Royal Statistical Society, Series B (Statistical Methodology), **64**(4), 583–639.

Spinnato, J. A., Allen, R. D., and Mendenhall, H. W. 1988. Birth weight prediction from remote ultrasound examination. Obstetrics & Gynecology, **71**(6), 893–898.

Stadlmann, S., and Kneib, T. 2021. Interactively visualizing distributional regression models with distreg.vis. Statistical Modelling.

Stadlmann, S., Heller, G. Z., Kneib, T., and Koens, L. 2023. Parameter orthogonality transformations in distributional regression models. PREPRINT (Version 1) available at Research Square. 10.21203/rs.3.rs-3197667/v1.

Stasinopoulos, D. M., Rigby, R. A., Heller, G. Z., Voudouris, V., and De Bastiani, F. 2017. Flexible Regression and Smoothing: Using GAMLSS in R. Boca Raton: Chapman & Hall/CRC.

Stasinopoulos, D. M., Rigby, R. A., Giorgikopoulos, N., and De Bastiani, F. 2022. Principal component regression in GAMLSS applied to Greek–German government bond yield spreads. Statistical Modelling, **22**(1-2), 127–145.

Strömer, A., Staerk, C., Klein, N., Weinhold, L., Titze, S., and Mayr, A. 2022. Deselection of base-learners for statistical boosting — with an application to distributional regression. Statistical Methods in Medical Research, **31**(2), 207–224.

Sun, Q., Zhou, W.-X., and Fan, J. 2020. Adaptive Huber regression. Journal of the American Statistical Association, **115**(529), 254–265.

Thomas, J., Hepp, T., Mayr, A., and Bischl, B. 2017. Probing for sparse and fast variable selection with model-based boosting. Computational and Mathematical Methods in Medicine, Article ID 1421409.

Thomas, J., Mayr, A., Bischl, B., Schmid, M., Smith, A., and Hofner, B. 2018. Gradient boosting for distributional regression: Faster tuning and improved variable selection via noncyclical updates. Statistics and Computing, **28**(3), 673–687.

Thompson, R., and Baker, R. J. 1981. Composite link functions in generalized linear models. Journal of the Royal Statistical Society, Series C (Applied Statistics), **30**(2), 125–131.

Tibshirani, R. 1996. Regression shrinkage and selection via the Lasso. Journal of the Royal Statistical Society, Series B (Statistical Methodology), **58**(1), 267–288.

Tutz, G., and Petry, S. 2016. Generalized additive models with unknown link function including variable selection. Journal of Applied Statistics, **43**(15), 2866–2885.

Umlauf, N., Klein, N., and Zeileis, A. 2018. BAMLSS: Bayesian additive models for location, scale and shape (and beyond). Journal of Computational and Graphical Statistics, **27**, 612–627. 10.1080/10618600.2017.1407325.

van Buuren, S. 2018. Flexible Imputation of Missing Data. Chapman and Hall/CRC.

van Buuren, S., and Fredriks, M. 2001. Worm plot: A simple diagnostic device for modelling growth reference curves. Statistics in Medicine, **20**, 1259–1277.

Venzon, D. J., and Moolgavkar, S. H. 1988. A method for computing profile-likelihood-based confidence intervals. Journal of the Royal Statistical Society, Series C (Applied Statistics), **37**(1), 87–94.

Wahba, G. 1978. Improper priors, spline smoothing and the problem of guarding against model errors in regression. Journal of the Royal Statistical Society, Series B (Statistical Methodology), **40**, 364–372.

Waldmann, E., Kneib, T., Yue, Y. R., Lang, S., and Flexeder, C. 2013. Bayesian semiparametric additive quantile regression. Statistical Modelling, **13**(3), 223–252.

Watanabe, S. 2010. Asymptotic equivalence of Bayes cross validation and widely applicable information criterion in singular learning theory. The Journal of Machine Learning Research, **11**, 3571–3594.

Wood, S. N. 2017. Generalized Additive Models. An Introduction with R. 2nd edn. Boca Raton: Chapman & Hall/CRC.

Wood, S. N., Pya, N., and Säfken, B. 2016. Smoothing parameter and model selection for general smooth models. Journal of the American Statistical Association, **111**(516), 1548–1563.

Yang, Z. 2014. Predicting birth weight from ultrasound examination and maternal characteristics. Unpublished Master of Biostatistics thesis, Macquarie University, Sydney.

Yee, T. W. 2019. VGAM: Vector Generalized Linear and Additive Models. R package version 1.1-2.

Yee, T. W., and Wild, C. J. 1996. Vector generalized additive models. Journal of Royal Statistical Society, Series B (Statistical Methodology), **58**, 481–493.

Yu, K., and Moyeed, R. A. 2001. Bayesian quantile regression. Statistics & Probability Letters, **54**, 437–447.

Yu, Y., and Ruppert, D. 2002. Penalized spline estimation for partially linear single-index models. Journal of the American Statistical Association, **97**(460), 1042–1054.

Yu, Y., Wu, C., and Zhang, Y. 2017. Penalised spline estimation for generalised partially linear single-index models. Statistics and Computing, **27**(2), 571–582.

Yue, Y., and Rue, H. 2011. Bayesian inference for additive mixed quantile regression models. Computational Statistics & Data Analysis, **55**, 84–96.

Zhu, J., Wen, C., Zhu, J., Zhang, H., and Wang, X. 2020. A polynomial algorithm for best-subset selection problem. Proceedings of the National Academy of Sciences, **117**(52), 33117–33123.

Ziel, F., Muniain, P., and Stasinopoulos, M. 2021. gamlss.lasso: Extra Lasso-Type Additive Terms for GAMLSS. R package version 1.0-1.

Zou, H., and Hastie, T. 2005. Regularization and variable selection via the elastic net. Journal of the Royal Statistical Society, Series B (Statistical Methodology), **67**(2), 301–320.

Index

Printed in the United States
by Baker & Taylor Publisher Services